2011 Asia Pacific Conference on Postgraduate Research in Microelectronics and Electronics

(PrimeAsia 2011)

**Macao, China
6 – 7 October 2011**

IEEE Catalog Number:	CFP1144H-PRT
ISBN:	978-1-4577-1608-9

**Copyright © 2011 by the Institute of Electrical and Electronic Engineers, Inc
All Rights Reserved**

Copyright and Reprint Permissions: Abstracting is permitted with credit to the source. Libraries are permitted to photocopy beyond the limit of U.S. copyright law for private use of patrons those articles in this volume that carry a code at the bottom of the first page, provided the per-copy fee indicated in the code is paid through Copyright Clearance Center, 222 Rosewood Drive, Danvers, MA 01923.

For other copying, reprint or republication permission, write to IEEE Copyrights Manager, IEEE Service Center, 445 Hoes Lane, Piscataway, NJ 08854. All rights reserved.

***This publication is a representation of what appears in the IEEE Digital Libraries. Some format issues inherent in the e-media version may also appear in this print version.*

IEEE Catalog Number: CFP1144H-PRT
ISBN 13:   978-1-4577-1608-9
ISSN: 2159-2144

Additional Copies of This Publication Are Available From:

Curran Associates, Inc
57 Morehouse Lane
Red Hook, NY 12571 USA
Phone: (845) 758-0400
Fax: (845) 758-2633
E-mail: curran@proceedings.com
Web: www.proceedings.com

2011 Asia Pacific Conference on Postgraduate Research in Microelectronics and Electronics (PrimeAsia 2011)

Macao, China
6-7 October 2011

IEEE Catalog Number: CFP1144H-POD
ISBN: 978-1-45771-608-9

TABLE OF CONTENTS

SESSION A: DATA CONVERTER TECHNIQUES

A Nonlinearity Digital Background Calibration Algorithm for 2.5bit/stage Pipelined ADCs With Opamp Sharing Architecture 1
  Yuan Fei, Sai-Weng Sin, Seng-Pan U, Rui P. Martins

A Calibration Technique for Mismatch of Capacitor Arrays in A/D and D/A Converters 5
  Libing Zhou, Liyuan Liu, Dongmei Li

A Time-Efficient Dither-Injection Scheme for Pipelined SAR ADC 9
  Rui Wang, U-Fat Chio, Chi-Hang Chan, Li Ding, Sai-Weng Sin, Seng-Pan U, Rui P. Martins, Zhihua Wang

A Hardware-effective Digital Decimation Filter Implementation for 24-bit ΔΣ ADC 13
  Yafei Ye, Ting Li, Zhihua Wang, Liyuan Liu, Dongmei Li

SESSION B – POWER

A 3.3V to 3.19V Low-Dropout Regulator with Frequency Compensation Strategy without Trimming 17
  Yiwei Zhang, Liyuan Liu, Dongmei Li

Investigation of LC-Hybrid Active Power Filters in Resonances Prevention and Compensation Capabilities 21
  Chi-Seng Lam, Man-Chung Wong, Ying-Duo Han

A FPGA-Based Power Electronics Controller for Hybrid Active Power Filters 25
  Bo Sun, U-Fat Chio, Chi-Seng Lam, Ning-Yi Dai, Man-Chung Wong, Chi-Kong Wong, Sai-Weng Sin, Seng-Pan U, R. P. Martins

SESSION C – DIGITAL SIGNAL PROCESSING

An Algorithm for the Design of Low Power Linear Phase FIR Filters 29
  Wenbin Ye, Ya Jun Yu

A New Motion Compensation Algorithm and New Test Algorithm to Compare the Result of Motion Compensation 33
  Chunchun Chen, Xuewei Chen, Eryan Yang

On-Chip Process and Temperature Compensation and Self-Adjusting Slew Rate Control for Output Buffer 37
  Ron-Chi Kuo, Hsin-Yuan Tseng, Jen-Wei Liu, Chua-Chin Wang

SESSION D - CAD AND MEMS

Theorems on the Global Convergence of the Nonlinear Homotopy Method for MOS Circuits 41
  Dan Niu, Guangming Hu, Yasuaki Inoue

Stability Analysis of MEMS Gyroscope Drive Loop Based on CPPLL 45
  Huan-Ming Wu, Hai-Gang Yang, Tao Yin, Hui Zhang

Electromechanical Closed-Loop with High-Q Capacitive Micro-Accelerometers and Pulse Width Modulation Force Feedback 49
  Zhen-Hua Ye, Hai-Gang Yang, Fei Liu, Tao Yin, Qi-Song Wu

SESSION E – VISUAL SIGNAL PROCESSING

CUDA-Based Acceleration of Post Deblocking Filter 53
  Ting Liu, Chunchun Chen, Eryan Yang

High Performance JPEG Decoder Based on FPGA 57
  Junming Shan, Duyao Wang, Eryan Yang

A Fast Noise Variance Estimation Algorithm 61
  Wenjiang Liu, Tao Liu, Mengtian Rong, Ruolin Wang, Hao Zhang

SESSION F – BIOMEDICAL CIRCUITS AND SYSTEMS

A Real-Time Heart Beat Detector and Quantitative Investigation based on FPGA 65
Cheng Dong, Chio In Ieong, Mang I. Vai, Peng Un Mak, Pui In Mak, Feng Wan

The Design of a Universal and Configurable ASIC for Biological Stimulation 70
Weifeng Zhang, Yinan Dong, Jo-Yu Wu, Kea-Tiong Tang, Guoxing Wang

A New ECG Signal Processing Scheme for Low Power Wearable ECG Devices 74
Yibin Hong, Iniyal Rajendran, Yong Lian

SESSION G - ANALOG SIGNAL PROCESSING

A SiGe BiCMOS Class A Power Amplifier Targeting 5.5GHz Application 78
Qiong Yan, Lin Hua, Lei Chen, Ying Ruan, Jie Su, Shulin Zhang, Wei Zhang, Shengfu Liu, Zongsheng Lai

NTF Zero Compensation Technique for Passive Sigma-delta Modulator 82
Arshad Hussain, Sai-Weng Sin, Seng-Pan U, Rui P. Martins

A 0.7 V DTMOS-Based Class AB Current Mirror 86
Arnon Kanjanop, Varakorn Kasemsuwan

A High-Gain Fully-Differential Thermal Noise-Canceling CMOS Front-End Amplifier 90
Puttachai Chimpleekul, Varakorn Kasemsuwan

SESSION H - VLSI SYSTEMS AND APPLICATIONS

Compact Distributed RLCG Interconnect Model -- Signal Transient Response for Dispersionless Interconnect 94
Jing Xia, Chi Liu, Xinnan Lin, Wei Zhao, Yu Han, Aixi Zhang, Jin He

Multiple Continuous Error Correct Code for High Performance Network-on-chip 98
Bin Wang, Jing Xie, Zhigang Mao, Qin Wang

Non-Linear Partitioning for Decimal Logarithm Approximation 102
Chetan Kumar Vudadha, Sreehari Veeramachaneni, M. B. Srinivas

Reconfigurable Adders for Binary/BCD Addition/Subtraction 106
Syed Ershad Ahmed, Sreehari Veeramachaneni, Moorthy Muthukrishnan, M. B. Srinivas

Conditional Sum Block for High Sparse Adders 110
Sai Phaneendra Parlapalli, Sreehari Veeramachaneni, Moorthy Muthukrishnan, M. B. Srinivas

SESSION I - CIRCUITS AND SYSTEMS FOR COMMUNICATIONS – 1

A 3.9 pJ/Pulse Differential IR-UWB Pulse Generator in 90 nm CMOS 115
Kin Keung Lee, Øivind Næss, Tor Sverre Lande

A Novel Digital Predistortion Technique for Class-E PA with Delay Mismatch Estimation 119
U-Wai Lok, Pui-In Mak, Wei-Han Yu, Rui P. Martins

A 0.13μm CMOS 0.8-10.6GHz Low Noise Amplifier with Active Balun for Multi-Standard Applications 123
Kaichen Zhang, Wei Li, Fan Ye, Ning Li, Junyan Ren

SESSION J - CIRCUITS AND SYSTEMS FOR COMMUNICATIONS – 2

Memory Efficient LDPC Decoder Design 127
Yuan Yao, Wei Liang, Fan Ye, Junyan Ren

A Double Active-Decoupling Technique for Reducing Package Effects in a Cognitive-Radio Balun-LNA 131
Miao Liu, Pui In Mak, Yao Hua Zhao, Rui P. Martins

An Effective Buffer Space Management in Serial RapidIO Endpoint 135
Fengfeng Wu, Song Jia, Yuan Wang

Author Index

MESSAGE FROM THE GENERAL CHAIR

Welcome to the 3rd IEEE Asia Pacific Conference on Postgraduate Research in Microelectronics & Electronics - *PrimeAsia*, jointly organized by the IEEE Circuits And Systems (CAS) Society, the IEEE Macau Section, the IEEE Macau Joint-Chapter on CAS / COM, the IEEE Macau Solid-State Circuits Chapter and the University of Macau. Besides these institutions the conference had also the financial support of Macau Foundation.

PrimeAsia is a new initiative of the IEEE Circuits and Systems Society for engaging students in Asia, the first two editions took place in Shanghai, China, and the fourth will be organized next year in Hyderabad, India. It aims to provide opportunities for postgraduate students to present their research works and to interact with people in the research community and industry. PRIMEASIA 2011, sponsored by the IEEE CAS Society, will be held from 6 to 7 October 2011, here in Macao, China, a convenient location close to the prestigious UNESCO World Heritage sites. Papers in PRIMEASIA 2011 will be presented by postgraduate students, and experienced people from academia and industry are also participating, thus creating a stimulating environment for favoring exchange of knowledge and mentoring of young researchers.

As the General Chairman of PRIMEASIA 2011 I would like to present my sincere thanks to everybody that contributed for its success. In particular, I would like to thank Prof. Yong Lian for his original idea of organizing this conference in Macao, as well as for his enthusiastic support since the beginning. Besides them, a special word also to Prof. Seng-Pan U (TPC Chair), Prof. Pui-In Mak (Publications Chair), Dr. Sai-Weng Sin (Finance Chair), Prof. Mang-I Vai (Publicity Chair), Prof. Man-Chung Wong (Student Liaison Chair), for their efficient work, as well as to the *indefatigable* Fan Ng, Leo (Local Arrangement Chair). This year's event is only possible with the great support of this excellent local team from University of Macau, leading Higher Education Institution of Macao SAR, China, commemorating in 2011 its 30th Anniversary, and being one of the oldest universities in the Pearl River Delta Region, heiress of the former St. Paul's University College, founded in 1594. I hope you find PRIMEASIA 2011 an interesting forum for exchange of ideas and also enjoy your stay in Macau.

Rui Martins

PrimeAsia 2011, General Chairman

TECHNICAL PROGRAM CO-CHAIRS' MESSAGE

On behalf of the Technical Program Committee (TPC) of The 2011 Asia Pacific Conference on Postgraduate Research in Microelectronics & Electronics (PrimeAsia), it is our great honor and pleasure to welcome you to this UNESCO world heritage Macau city.

With the endeavor from members of Technical Program Committee, the final technical program is consisted by 34 papers spread over 10 lecture sessions during Oct. 6-7, 2011 with 2 parallel sessions in the first day. The final program was reviewed and selected from totally 47 paper submissions from 8 countries and covers a wide range of topics including analog circuits, analog, digital and visual signal processing, CAD & MEMS, Biomedical and Communications Circuits and Systems. PrimeAsia follows the tradition of Prime conference and recruits postgraduate authors from Asia and around the world to present their most recent research results in the field of microelectronics and electronics. To honor and encourage students, the top 10% papers will receive Gold Leaf Certificates, the next 10% for Silver Leaf Certificates and the following 10% for Bronze Leaf Certificates.

We are very honor to organize 2 outstanding keynote speeches: firstly from Prof. José Franca from Institute Superior Tecnico, Portugal and also the Co-Founder and ex-President & CEO of Chipidea Microelectronics, S.A., for an inspiring talk of "From the Lab to the Market - Changing the algorithm to drive semiconductor entrepreneurship", and then by Prof. Enrico Macii from Politecnico di Torino, Italy for an technical talk of "Power-Gating for Leakage Control and Beyond in Nanometer CMOS Circuits".

The success of this technical program won't be feasible without the tremendous effort from every technical committee members and session chairs who have fulfilled their strong commitments and responsibilities. Special thanks to Fan Ng (Leo) for his effort and support on the technical program paper review process.

Last but not least, we thank all the authors and attendees for your contribution and coming, we do wish you enjoy the technical and social programs to exchange your ideas and also meet friends as well as build up your professional friendship, and also, have a great time for both the historic sights of this UNESCO world heritage sites with full of perfect mixing of East and West cultures.

Seng-Pan U (Ben), Junyan REN, Franco MALOBERTI and Yong LIAN
PrimeAsia 2011 Technical program Co-Chairs

ORGANIZING COMMITTEE

General Chair
Rui Martins
University of Macau

Technical Program Co-Chairs
Seng-Pan U
University of Macau

Junyan Ren
Fudan University

Franco Maloberti
University of Pavia

Yong Lian
National University of Singapore

Finance Chair
Sai-Weng Sin
University of Macau

Publication Chair
Pui-In Mak
University of Macau

Publicity Chair
Mang-I Vai
University of Macau

Student Liaison Co-Chairs
Man-Chung Wong
University of Macau

Local Arrangement Chair
Fan Ng
University of Macau

Asian Liaison Chair
M.B. Srinivas
BITS-Pilani, Hyderabad Campus

Industrial Liaison Chair
Wenwu Zhu
Microsoft Research(China)

International Advisor

David Allstot
University of Washington, USA

Nobuo Fujii
Tokyo Institute of Technology, Japan

Chen He
Shanghai Jiao Tong University , China

Tor S. Lande
University of Oslo, Norway

Yong Ching Lim
Nan Yang Technological University, Singapore

Bin-Da Liu
National Cheng Kung University, Taiwan

Mohamad Sawan
Polytechnique, University of Montreal, Canada

Wouter A. Serdjin
Delft University of Technology, Netherlands

Gianluca Setti
University of Ferrara, Italy

Mani Soma
University of Washington, USA

Myung H. Sunwoo
Ajou University, Korea

Ljiljana Trajkovic
Simon Fraser University, Canada

TECHNICAL PROGRAM COMMITTEE

Baoyong Chi	*Tsinghua University*
Tara J. Hamilton	*University of New South Wales*
Weifeng He	*Shanghai Jiao Tong University*
Yong Hei	*Institute of Micro-Electronics, Chinese Academy of Sciences*
Pui-In Mak	*University of Macau*
Rui Martins	*University of Macau*
Shanthi Pavan	*Indian Institute of Technology, Madras*
Sai-Weng Sin	*University of Macau*
Chien-Cheng Tseng	*National Kaoshiung First University of Science and Technology*
Mang-I Vai	*University of Macau*
Man-Chung Wong	*University of Macau*
Haigang Yang	*Institute of Electronics, Chinese Academy of Sciences*
Zhiyi Yu	*Fudan University*
Yuzhe Liu	*University of Notre Dame*
Peng-Un Mak	*University of Macau*
Man-Kay Law	*Hong Kong University of Science and Technology*
Feng Wan	*University of Macau*

KEYNOTE SPEECHES

Keynote Session -- Thursday, October 6, 2011, 09:10 – 11:10
University of Macau Library (UM Library), Auditorium STDM

Professor José Franca
Department of Electrical and Computer Engineering
IST / TU of Lisbon, Portugal

KEYNOTE I: FROM THE LAB TO THE MARKET – CHANGING THE ALGORITHM TO DRIVE SEMICONDUCTOR ENTREPRENEUSHIP

Abstract:

The Semiconductor Industry is, and will remain for the foreseeable future, one of the key strategic industries of the century. Over the past 25 years it has undergone a dramatic transformation of its value chain, giving rise to increased specialized layers of industry segments and business models, from simple design shops to fabless chip companies, steered by a wave of semiconductor-driven entrepreneurship on a global scale.

Alongside the transformational dynamics of the semiconductor industry, there has been worldwide a remarkable dissemination of research & development activities in integrated circuit (IC) design based on solid educational foundations of human capital. While solid technology foundations and, above all, internationally recognized track records of achievements, safeguard the future of continued academic excellence, they can, and should also be, a platform for local economic development and wealth creation. For this to materialize, however, it is needed more than just world-class technology. Above all, that relies on a strong entrepreneurship culture especially based on key strategic aspects of business and management in semiconductor enterprises.

This lecture addresses some of those key aspects of a changing algorithm, from academic publications to addressing world market needs, to foster semiconductor entrepreneurship in an increasingly complex and sophisticated value chain. It examines different business models and how they can create value, and discusses key strategic organizational and operational issues, implementation and tactics that are needed to leverage assets and resources, both internal and external to the start-up organization.

Biography:

Dr. José Franca co-founded "CHIPIDEA Microelectrónica, S.A." in February 1997, and became Chairman of the Board and CEO. From a small analog design team he led Chipidea to eventually become a world leader in analog/mixed-signal Semiconductor Intellectual Property with over 300 engineers in design centers in Portugal, China, Poland, Belgium, France and Norway. He eventually left the company in September 2008 after its acquisition by MIPS Technologies Inc. in August 2007.

Before that, in 1987, he founded the Integrated Circuits and Systems Group at IST and, in 1994, the IST Centre of Microsystems. He was a member of IST's Executive Council between 1987 and 1991, Vice-President of ITEC (Technology Institute for the European Community), between 1988 and 1991, and a member of the Senate of the Technical University of Lisbon between 1990 and 1992. In June 2000 he became a Full Professor of the Department of Electrical and Computer engineering of IST.

Among the numerous international and national academic and industry honors Dr. José Franca has received was being named **Fellow of the IEEE** in 1997, for "contributions to Multirate Analog Signal Processing and Engineering Education", and being the recipient of the **Golden Jubilee Medal** of the IEEE Circuits and Systems Society in 1999. He was made **Doctor Honoris Causa in Science** by the University of Macao in 2006 and honored with the IEEE Circuits and Systems Society **Industrial Pioneer Award in 2010** for "pioneering contributions to the development of Analog/Mixed-Signal Integrated Circuits and to their widespread use in the semiconductor industry by starting up an Analog IP design company". In 2006, the President of the Republic of Portugal bestowed upon him the high distinction of "**Grande Oficial da Ordem do Mérito**". He received the **2006 Entrepreneur of the Year Award** established by the Alumni Association of the prestigious INSEAD Business School and the **2007 Ernst & Young International Entrepreneur of the Year Award**, and in 2008 was the recipient of the prestigious **Universidade de Coimbra Award**.

Dr. Franca was a member of the **Board of Governors** of the IEEE Circuits and Systems Society between 1997 and 1999 and of the **Executive Council** of the European Circuits and Systems Society between 1998 and 2001. He served in various consulting assignments to the European Commission Scientific and Technology Programs, namely the **Information Society Technologies Advisory Group** (ISTAG) of the European Commission in 2008 and 2009, and was member of the Steering and Technical Program Committees of several International Conferences. He was General Chair of the International Electronics Conference of the IEEE Circuits and Systems Society (Lisbon, 1998), General Chair of the Design, Automation and Test in Europe (DATE) Conference (Paris, 2002), and General Chair of the ESSCIRC/ESSDERC Conferences (Lisbon, 2003).

Dr. Franca holds a degree in Electrical Engineering from Instituto Superior Técnico (IST) of the Technical University of Lisbon (UTL) (1978), a Ph.D. degree from Imperial College of Science and Technology in London (1985) and the degree of "Agregado" from UTL (1992). In 1992, he also completed an Executive Education program at MIT on "Management of Research and Technology based Innovation".

Dr. Franca was State Secretary for Education of the Government of Portugal in 1991/92. He is affiliated with Universities in Portugal and abroad, namely CUHK (Hong Kong), FEUP (Porto, Portugal) and IST (Lisbon, Portugal), and is a member of the Advisory Board of the Department of Engineering and Industrial Management of IST, the Strategic Board of the Department of Electrical and Computer Engineering of IST and the General Council of the Business and Management School of Universidade Lusófona. He is also a non-executive Director of Banco Espírito Santo in Portugal.

Professor Enrico Macii
Computer Engineering
Politecnico di Torino, Italy

KEYNOTE II: POWER-GATING FOR LEAKAGE CONTROL AND BEYOND IN NANOMETER CMOS CIRCUITS

Abstract:

CMOS devices are approaching the size of atoms, which is a fundamental barrier for the scaling of the bulk technology. The research community is actively pursuing alternative fabrication processes and device materials to be used as substitute in the electronics market. However, while the debate on which of emerging technology will prevail is still open, CMOS is still the technology of choice for the large majority of electronic systems manufactured today.

The 2009 ITRS Roadmap reported that static power consumption and variability are the most serious concerns for the design of nanometer ICs below the 40nm feature size. Static power due to leakage mechanisms inside active devices represents the main source of consumption when the circuit is not switching. This makes electronic devices power-hungry even when they are idle. Variability, instead, refers to the marked tendency of a manufactured circuit to show a deviation from its nominal behavior, thus resulting in lower fabrication yield and lower reliability. Main sources of variability include random and systematic process variations, environmental variations (e.g., temperature and Vdd fluctuations), and aging effects (e.g., Negative Bias Temperature Instability – NBTI – and Hot Carrier Injection – HCI).

Static power and variability are not new in the EDA community, and various options for their management are already available. However, while in the past they could be considered separately, advanced technology nodes require a strong synergy between the two in order to provide concurrent static power optimization and variability compensation. Unfortunately, this challenge is complicated by the fact that most of the design solutions for compensating variability are intrinsically power inefficient: Fault-tolerant approaches, such as "fail and correct" strategies, and adaptive techniques, such as Dynamic Voltage Scaling and Adaptive Body Biasing are based on the concept of redundancy, which does contrast with low-power.

The main goal of this talk is to show how power-gating, well known to be the most effective technique to reduce static power consumption in CMOS circuits, represents also a valuable solution to mitigate the effects of variability, thus maximizing the fabrication yield and the reliability of digital systems. Power-gating, if properly designed, enables the implementation of circuits with better aging profiles, hence longer lifetimes, and the capability to be dynamically adjusted in order to meet the timing specifications. The conventional way of implementing power-gating, in fact, is based on the insertion of dedicated MOS switches, called sleep transistors, between the gated block and the ground rail. This provides the circuit with two power modes: A low-power mode, during which the sleep transistor is turned-off, thus guaranteeing minimum leakage power consumption, and an active mode, during which the sleep transistor is turned-on and it is transparent from a functional point of view. In terms of variability, the sleep transistor shows important electrical properties: During the low-power mode, its effect is to make the gated circuit immune by the aging mechanisms (NBTI and HCI); during the active periods, it can act as a supply-voltage regulator which enables a natural, yet low-cost way of implementing adaptive control strategies for process variation mitigation. Based on these important properties, the talk introduces some new power-gating strategies and circuit optimizations techniques that enhance standard design flows with effective solutions for the implementation of low-power, reliable, and aging-free CMOS circuits.

Biography:

Enrico Macii is a Full Professor of Computer Engineering at Politecnico di Torino, Torino, Italy. Prior to that, he was an Associate Professor (from 1998 to 2001) and an Assistant Professor (from 1993 to 1998) at the same institution. From 1991 to 1995 he was also an Adjunct Faculty at the University of Colorado at Boulder. He holds a Dr. Eng. degree in Electrical Engineering from Politecnico di Torino, a Dr. Sc. degree in Computer Science from Università di Torino and a PhD degree in Computer Engineering from Politecnico di Torino. Since year 2007, he is the Vice Rector for Research, Technology

Transfer and EU Affairs at Politecnico di Torino, and a Member of the Rector's Advisory Board.

His research interests are in the design automation of digital circuits and systems, with particular emphasis on low-power design aspects. In the fields above, he has authored over 350 scientific publications, including the book: "Ultra Low-Power Electronics and Design" by Kluwer. He received the Best Paper Award for articles presented at IEEE EURODAC-96 and at ACM/IEEE GLS-VLSI-08.

He was the Editor-in-Chief of the IEEE Transactions on CAD/ICAS for the term 2006-2009. Prior to that, he was an Associate Editor for the same journal (1997-2005) and an Associate Editor for the ACM Transactions on Design Automation of Electronic Systems (2000-2005). Currently, he is an Associate Editor for the IEEE Transactions on Computers.

He was the Technical Program Co-Chair (in 1999) of the IEEE Alessandro Volta Memorial Workshop on Low Power Design, the Technical Program Co-Chair (in 2000) and the General Chair (in 2001) of the ACM/IEEE International Symposium on Low Power Electronics and Design (ISLPED), the General Chair (in 2003) and the Technical Program Chair (in 2004) of the IEEE PATMOS Workshop, the General Co-Chair (in 2007) and the Technical Program Co-Chair (in 2008) of the ACM/IEEE Great Lakes Symposium on VLSI (GLS-VLSI), the Vice Technical Program Chair (in 2010) and the Technical Program Chair (in 2011) of the IEEE Design Automation and Test (DATE) Conference, and the Technical Program Co-Chair (in 2010) of the IEEE International Conference of Electronics Circuits and Systems (ICECS). Currently, he is the Vice General Chair of DATE 2012.

Enrico Macii is a Fellow of the IEEE. He was a Member of the Board of Governors of the IEEE Circuits and Systems Society for two consecutive terms of duty (2002-2004 and 2005-2007), and a Member of the Board of Governors of the IEEE Council on EDA for the period 2006-2009. Currently, he is the Vice President for Publications of the IEEE Circuits and Systems Society for the term 2010-2011. He has already served in this position during the previous term (2009-2010).

A Nonlinearity Digital Background Calibration Algorithm for 2.5bit/stage Pipelined ADCs With Opamp Sharing Architecture

Yuan Fei, Sai-Weng Sin, Seng-Pan U, Rui Paulo Martins[1]

Analog and Mixed Signal VLSI Laboratory (http://www.fst.umac.mo/en/lab/ans_vlsi/index.html)
Faculty of Science and Technology, University of Macau, Macao, China
Tel:+853 83978796, Fax: +853 83978797, email: ma66581@umac.mo
1 – On leave from Instituto Superior Técnico/TU of Lisbon, Portugal

Abstract— This paper presents a new digital background calibration algorithm for 2.5bits/stage pipelined analog-to-digital converters (ADCs) with opamp sharing architecture. Background calibration can extract calibration data without interrupting ADCs normal conversion operation. Digital calibration can relax the design difficulty of analog circuits of ADCs, and gains the improvement of technology scaled down. This algorithm provides a method to effectively estimate the nonlinearity of opamp, and calibrates it in digital domain. For a 10bit 2.5bit/stage pipelined ADCs with opamp sharing architecture, only one opamp need to be calibrated to achieve 10bit resolution. Simulation results show that the ENOB can be improved from 5.54b to 8.80b by the proposed algorithm.

I. INTRODUCTION

Digital background calibration techniques haves been applied to pipelined analog-to-digital converters (ADCs) to improve resolution and/or reduce power dissipation. A pipelined ADC comprises some cascaded stages. For example, a 10bits pipelined ADC with 2.5bits per-stage. It comprises 4 2.5bits stages and a 2bits flash ADC. The 4 stages have 0.5 bit overlap to obtain an 8bits digital output with digital error correction. And for each 2.5bits stage, it comprises a sub-ADC which quantizes the stage's analog input, and a sub DAC which generates a corresponding analog signal with digital output of sub-ADC. Analog input subtract corresponding analog signal of sub-DAC to obtain the residue signal. This residue signal multiply a gain factor ("4" for 2.5 bit per stage), and output signal pass to next stage to do the same process until the final 2 bits flash ADC. Digital calibration corrects the digital output of pipelined ADC, and yielding a linear analog-to-digital (A/D) conversion characteristic. Opamp sharing architecture utilizes the opamp that is unnecessary for sampling phase of pipelined ADC to achieve one opamp sharing with two stages, which can obtain more power efficiency and less area cost.

There are some different background calibration schemes. In some schemes, an extra signal is injected into the signal path, and an additional signal range is required [3]. It needs a long time to converge. In other schemes, an additional ADC channel is required. This channel is used to be a reference to calibrate the main ADC channel [4]. But all of them only extract one gain error factor for each stage. For the high finite gain of opamp, gain error is dominated. But for low finite gain of opamp, nonlinearity becomes more dominated. The conventional linear calibration cannot calibrate gain error efficiently.

In [1], it's an algorithm for 1.5bit pipelined ADCs. The first five stages should be calibrated to achieve the resolution. Proposed algorithm apply to a 2.5bit pipelined ADCs with opamp sharing architecture, this architecture can further improve the calibration speed. To calibrate the nonlinear issue of opamp and further relax the design tradeoff of speed and accuracy of opamp, it needs at least two gain error factors for each stage. With these two gain error factors, the nonlinearity of gain error factor array of opamp can be estimated with a second order polynomial. Using this nonlinear model, the nonlinear gain error of each stage can be calibrated with corresponding output levels.

II. NONLINEARITY OF PIPELINED ADCs

For general pipeline stage, the jth stage analog input V_j, is quantized by a sub-ADC. Its digital output D_j, drives a sub-DAC to obtain an analog signal $V_j^{da}(D_j)$. Analog input V_j subtract $V_j^{da}(D_j)$ to obtain a residue signal, and this residue signal multiply gain factor G_j to generate the output signal of jth stage V_{j+1}. This is an ideal operation of a pipeline stage. It can be expressed as (1). Generally, switched capacitor circuit is applied to achieve the function of analog input subtract DAC output and multiply gain factor. So the ideal gain factor G_j can be expressed as (2)

$$V_{j+1} = G_j \times [V_j - V_j^{da}(D_j)] \quad (1)$$

$$G_j = \frac{C_s + C_f}{C_f} \quad (2)$$

978-1-4577-1608-9/11 $26.00 © 2011 IEEE

C_s, C_f are sampling capacitors of sample and hold circuit. $D_j \in \{-3, -2, -1, 0, +1, +2, +3\}$ is determined with comparison of V_j with the $-5/8V_r$, $-3/8V_r$, $-1/8V_r$ and $+1/8V_r$, $+3/8V_r$, $+5/8V_r$, But in reality, the nonlinear finite DC gain A_{0j} and parasitic capacitance of negative input node of opamp C_p also should be considered in gain factor calculation. So the gain factor also can be expressed as (3). And output of DAC $V_j^{da}(D_j)$ can be written as (4).

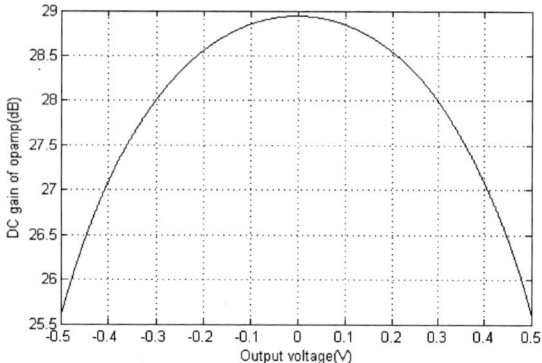

Figure 1.   Example of nonlinear DC gain of opamp v.s. output voltage

$$G_j' = \frac{C_s + C_f}{C_f} \times \frac{1}{1 + \frac{1}{A_{0j}} \frac{C_s + C_f + C_p}{C_f}} \quad (3)$$

$$V_j^{da}(D_j) = V_r \cdot \frac{C_s}{C_s + C_f} \times D_j \quad (4)$$

The offset effect due to the input-referred offset voltage of the opamp, the charge injection from the analog switches, and the offset of sub-DAC can be summarized as an offset voltage. This offset voltage can be corrected by digital error correction algorithm. So it is not considered in this transfer function. The output of jth stage can be expressed as (5). For 2.5 bit stage, assume that $C_s = 3C_f$. (5) can be rewritten as (6). In (6), the analog input of jth stage just multiply a constant factor G_j. If the gain error G_e can be measured, and the transfer function of jth stage can be calibrated to ideal transfer function (1). This is the concept of calibration.

$$V_{j+1} = G_j' \times [V_j - V_j^{da}(D_j)] \quad (5)$$

$$V_{j+1} = G_e \times [G_j \cdot V_j - V_r \cdot D_j] \quad (6)$$

with

$$G_e = \frac{1}{1 + \frac{1}{A_{0j}} \frac{C_s + C_f + C_p}{C_f}} \quad (7)$$

In (7), G_e include the information of nonlinear gain of opamp and input parasitic capacitor of opamp. Assume A_{0j} is a constant. So the gain error G_e also can be written as a constant. But in real opamp design A_{0j} is even function of jth stage output V_{j+1}, which can be expressed as,

$$A_{0j} = A_{dc} + \sum_{i=1}^{\infty} a_i \cdot V_{j+1}^{2i} \quad (8)$$

A_{dc} represents DC gain of opamp. a_i are gain coefficients of polynomial of A_{0j}(Figure 1). If this output dependent gain is considered as a constant for linear calibration algorithm. After calibration, the output signal which is close to $\pm V_r$ has more Inaccurate than the signal around common-mode level, so nonlinear calibration is applied to achieve higher accuracy of pipelined ADCs.

III.   PROPOSED GAIN ERROR EXTRACTION METHOD

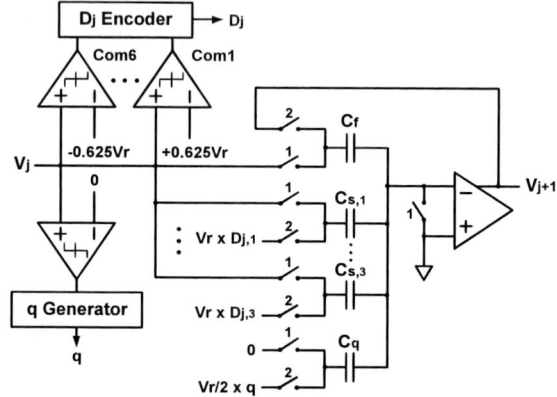

Figure 2.   SC pipeline stage with single stage operation

In Figure 2, there is structure applied in 2.5 bit pipeline stage. Six comparators are quantized input signal as typical 2.5bit pipeline stage, and one more comparator and capacitor Cq are added. This comparator controls pseudorandom number generator for injecting pseudorandom number into signal path. The structure can solve the over-range issue when pseudorandom number is injected to signal path. As shown in Figure 3, the comparator at common-mode divide the full rang into two parts, when input signal less than common-mode level, pseudorandom number q can be +1 or 0. When input signals greater than common-mode level, pseudorandom number q can be -1 or 0. If all output signals are collected and be calculated, then one gain error factor can be extracted after calibration. In the proposed work, one more extracted gain factor is needed. The output signals in region A (form $-5/8V_r$ to $+5/8V_r$) are collected to calculate another gain factor.

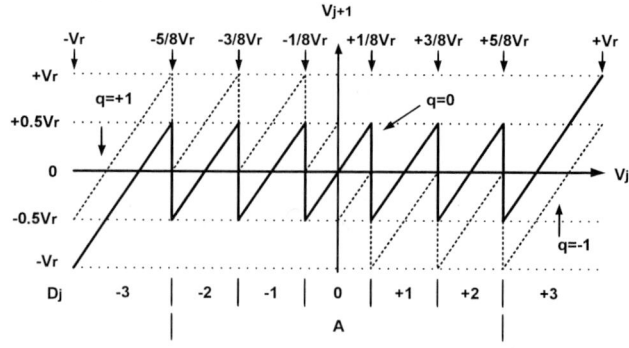

Figure 3.   Transfer characteristcis of Figure 2's pipeline stage

978-1-4577-1608-9/11 $26.00 © 2011 IEEE

And then, there are two gain error factors can be extracted. One is gain error factor of mid-region A of opamp's outputs, another is gain error factor of full-range of opamp's output. This range can be set to different values for each stage. With these two gain error factors, the nonlinearity of opamp can be estimated and calibrated.

IV. NONLINEARITY CALIBRATION ALGRITHM

As eq. (7) shown, all capacitance values are constant for pipeline stage. The only variable in eq. (7) is A_{0j}. It's not difficult to proof if A_{0j} is a even function, that gain error factor G_e has the same characteristic. So the same model can be used to estimate gain error factor G_e.

$$G_{ej} = G_{e0} + \sum_{i=1}^{\infty} b_i \cdot V_{j+1}^{2i} \qquad (9)$$

In [2], the digital background calibration extracts gain error factors with correlation-based. This algorithm needs a large number of input samples. For N-bit ADC, the order of samples is on the order of 2^{2N}. So the final convergence gain error factor can be considered as the average of these samples. The two gain factors is extracted in Section III also can be considered as the gain average of different output range. Ordinary, the gain error factors of region A is larger than that of full range. (Figure 4)

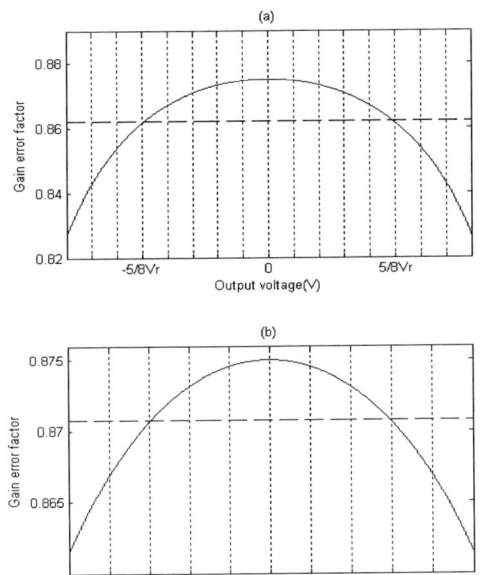

Figure 4. (a) Extraced gain error factor (dash line) and gain error factor(solid line) at full range (b) Extraced gain error factor(dash line) and gain error factor(solid line) at region A .

Two extracted gain error factors can be considered the average of these nonlinear gain error factors at different output range. So, in Figure 4, the area below the dash line (extracted gain error factor) should be equal to the area below the solid line (nonlinear gain error factor curve). With (9), if second order polynomial estimation is taken to approximate the gain error factor curve, the following equations can be written,

$$\begin{cases} \int_{-Vr5/8}^{Vr5/8} G_{e0} + b_1 \cdot V_{j+1}^2 \, dV_{j+1} = G_{e1} \cdot 2 \cdot (Vr5/8) \\ \int_{-Vr}^{Vr} G_{e0} + b_1 \cdot V_{j+1}^2 \, dV_{j+1} = G_{e2} \cdot 2Vr \end{cases} \qquad (10)$$

In (10), G_{e1}, G_{e2} are extracted gain error factors for region A and full range, respectively. There are two unknown values G_{e0} and b_1 and two equations, so the values of this two coefficients can be obtained by solving equations. And put these two coefficients back to second polynomial estimation of G_e, the rebuild approximation curve can be drawn as Figure 5. With this second order polynomial gain error factor model, each stage is calibrated with an array of 2^{bit} gain error factor values for corresponding output levels, which instead of a single gain error factor value in linear gain calibration. With digital binary output of pipelined ADCs, the corresponding gain error factors of each stage can be identified. With these nonlinear gain error factors and output signals, the rest part of digital calibration is identical with conventional linear digital background calibration works.

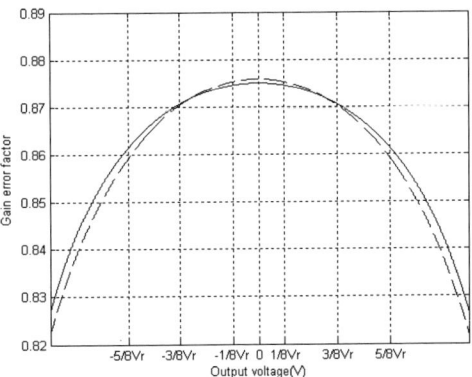

Figure 5. Second order polynomial rebuild G_e,(dash line), and higher order model of G_e for pipeline stage simulation (solid line)

V. SIMULATION RESULTS

The Matlab model of 10bit 200MHz pipelined ADC is applied to test the performance of proposed digital background nonlinearity calibration with first 4 stages of pipelined ADCs. Because there are two gain error factors should be extracted in propose algorithm, the convergence time is longer than conventional linear calibration. So this 2.5bit/stage pipelined ADC with opamp sharing architecture is applied to improve the calibration speed. The first two stages need to be calibrated and only one opamp need to be calibrated with this opamp sharing architecture to achieve 10 bit precision. In opamp model, the higher order polynomial is applied to build nonlinearity of opamp' open-loop gain. The DC gain in opamp model is only 30 dB. Input frequency is 99.8MHz. In table I, it shows that the original opamp with 30dB open loop gain and nonlinear effect, SNDR is only 35.14dB. With linear calibration, finite open loop gain issue can be improved so much and achieve a higher SNDR 47.32dB. But harmonic distortion is still a dominate issue of output. With proposed

978-1-4577-1608-9/11 $26.00 © 2011 IEEE

nonlinear gain calibration, the harmonic distortion issue also can be suppressed and achieve 54.71dB SNDR.

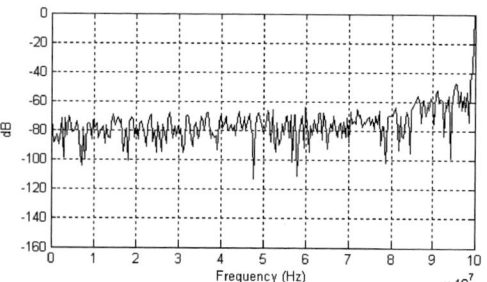

Figure 6.   The FFT of output without calibration

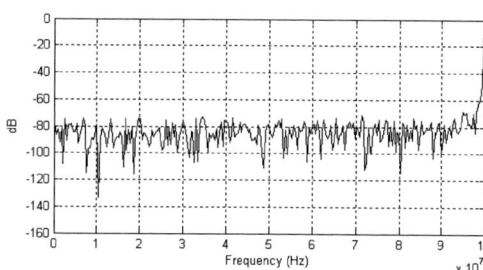

Figure 7.   The FFt of output with conventional digital background linear calibration

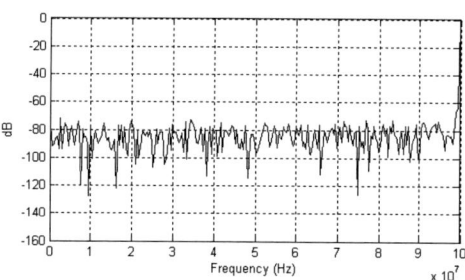

Figure 8.   The FFT of output with proposed digital background nonlinear calibration

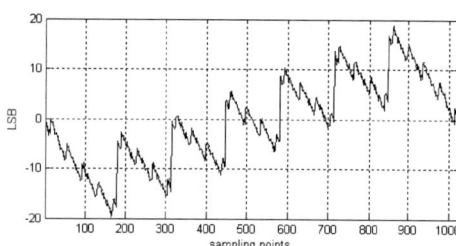

Figure 9.   INL of output without calibration

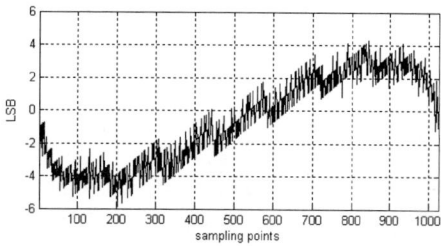

Figure 10. INL of output with conventional digital background linear calibration

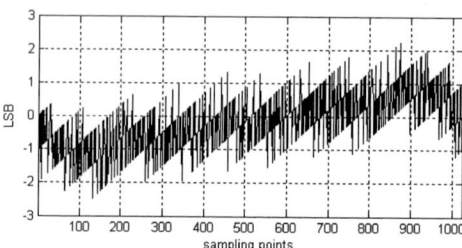

Figure 11. INL of output with proposed digital background nonlinear calibration

TABLE I.   SIMULATION RESULTS OF DIFFERENT ALGORITHM

	Comparison of different Algorithm		
	Original output without calibration	*Digital background linear calibration*	*Digital background nonlinear calibration*
SNDR(dB)	35.14	47.32	54.71.
SFDR(dB)	37.35	48.64	63.53
ENOB(bit)	5.54	7.57	8.80
THD(dB)	-36.81	-48.14	-61.53
INL(LSB)	-19.7~+18.8	-5.9~+4.3	-2.5~+2.2

VI.   CONCLUSIONS

This paper describes an algorithm of nonlinear digital background calibration for 10bit 2.5bits/stage pipelined ADC with opamp sharing architecture. Comparing with linear digital background calibration, this algorithm can further relax the design difficulty of opamp in pipeline stage or achieve a lower common-mode DC gain and/or higher speed.

VII.   ACKNOWLEDGMENT

This work was financially supported by Research Grants of University of Macau and Macau Science & Technology Development Fund (FDCT).

VIII.   REFERENCES

[1] Yuan Fei, Sai-Weng Sin, Seng-Pan U; Martins, R.P., "A digital background nonlinearity calibration algorithm for pipelined ADCs," *IEEE Conferences, Primeasia*, pp. 115–118, Shanghai, China, Sep. 2010.

[2] Jen-Lin Fan,Chung-Yin Wang, Jieh-Tsorng Wu, "A Robust and Fast Digital Background Calibration Technique for Pipelined ADCs," *IEEE Transactions on circuit and systems*, vol.54, No.6, pp. 1213–1223, June 2007.

[3] H.-C. Liu, Z.-M. Lee, and J.-T.Wu, "A 15-b 40-MS/s CMOS pipelined analog-to-digital converter with digital background calibration," *IEEE J. Solid-State Circuits*, vol. 40, no. 5, pp. 1047–1056, May 2005.

[4] X. Wang, P. J. Hurst, and S. H. Lewis, "A 12-bit 20-MSample/s pipelined analog-to-digital converter with nested digital background calibration," *IEEE J. Solid-State Circuits*, vol. 39, no. 11, pp. 1799–1807, Nov. 2004.

A Calibration Technique for Mismatch of Capacitor Arrays in A/D and D/A Converters

Libing Zhou, Liyuan Liu and Dongmei Li
Department of Electronic Engineering
Tsinghua University, Beijing, 100084, China
zhoulb87@gmail.com

Abstract — **This paper proposes a new calibration technique for capacitor array SAR (Successive Approximation Register) ADC, here we focus on the true value and the true weight of each capacitor. By measuring the capacitor, we calculate the true weight (W_i) of each weighted-capacitor, and map this weight into m-bit code C_i ($m \geq n$). With the normal conversion code D_i (n-bit), we define the first n-bit of $\sum D_i * C_i$ as the ultimate output code. The simulation indicates that this method can greatly suppress harmonics, and the ENOB gets improved greatly.**

I. INTRODUCTION

Thanks to the rapid expansion of portable battery-powered electronic devices, the ADC and DAC, which connect the analog world and the digital, get more and more widely used. Among all kinds of ADCs, the SAR ADC based on charge redistribution gets extremely popular to satisfy high energy efficiency. However, because of mismatch between capacitor, SAR ADC with conversion resolution beyond 12 bits is difficult to achieve, and what's worse is that Laser-trim which can revise capacitor is expensive. Therefore a lot of calibration techniques are proposed [1],[2],[4].

Fig. 1 shows a classical self-calibrating block diagram [1]. First, it converts the capacitor mismatch to residual voltage, then measures the residual voltage, and computes the error voltage of each capacitor, stores all those error voltage in digital memory. Within subsequent normal conversion cycles, the corresponding error voltage is subtracted from the DAC output according to the digital input code of DAC, so the effects of mismatch can be removed. There are two deficiencies for this method: first, it needs an ADC with higher resolution to quantify the error voltage, which is relatively weak; on the other hand, within normal conversion cycles, the error voltage should be subtracted from the output of DAC in real time, as may slow down the conversion rate for settling time.

This paper proposes a new technique of calibrating the effects of Capacitor Array mismatch, as shown in Fig. 2. Differing from traditional calibration method, here this method doesn't attempt to measure the mismatch between capacitor, but focuses on the true weight of each capacitor.

With the true weight, we correct the ADC output code to achieve the goal of calibration. Since we don't need a higher resolution ADC, don't interfere with normal conversion as well, we could enjoy the advantage in both hardware and conversion rate.

The rest of this paper is organized as follows. Section II describes the calibration technique. Section III gives the circuit for measuring capacitor and describes its working process. Section IV gives the results of Matlab simulation. Section V draws conclusions.

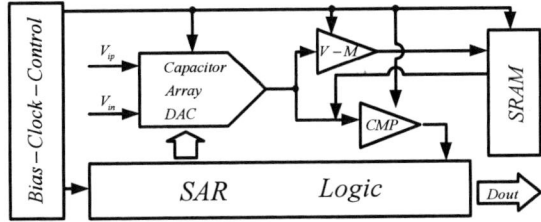

Figure 1. Conventional SAR-ADC with calibration

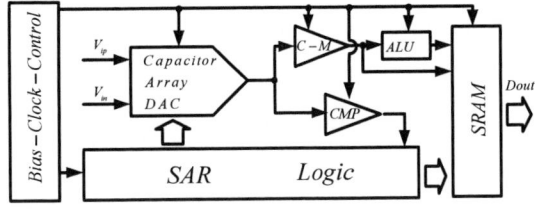

Figure 2. Proposed SAR-ADC with calibration

II. CALIBRATION TECHNIQUE

A. Weight of Binary-Weighted Capacitor Array

Fig. 3 shows Binary-Weighted Capacitor Array, which consists of $2*n$ weighted-capacitor ($C_{N(P)}^1 \sim C_{N(P)}^n$) and two redundancy capacitors ($C_{N(P)}^1$).

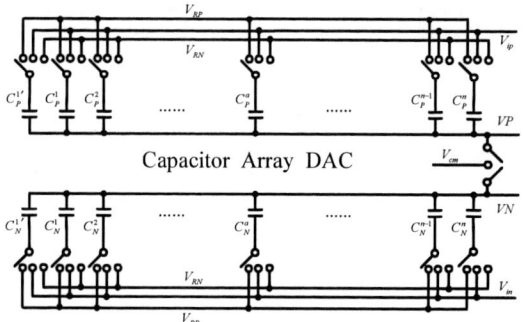

Figure 3. Binary-Weighted capacitor array

Taking the process variation into account [3], the actual value of each pair of capacitor is as follows:

$$\begin{cases} {C_{P(N)}^1}' = c_0(1+{\varepsilon_{P(N)}^1}') \\ C_{P(N)}^i = 2^{i-1}c_0(1+\varepsilon_{P(N)}^i) \end{cases} \quad (1)$$

Here we define c_0 as the unit capacitor when ε_N^i and ε_P^i perform as the deviation factors of capacitor due to process. So each weight of capacitor can be expressed as follows:

$$\begin{cases} W_{P(N)}^i = \dfrac{C_{P(N)}^i}{C_{totP(N)}} \\ C_{totP(N)} = {C_{P(N)}^1}' + \displaystyle\sum_{i=1}^n C_{P(N)}^i \end{cases} \quad (2)$$

B. Weight of Split Binary-Weighted Capacitor Array

Fig. 4 shows Split Binary-Weighted Capacitor Array [4], which consists of 2*n weighted-capacitor ($C_{P(N)L}^1 \sim C_{P(N)L}^a$, $C_{P(N)M}^1 \sim C_{P(N)M}^{n-a}$, and n=2*a) and two redundancy capacitors ($C_{P(N)M}^0$) and two bridge capacitors ($C_{P(N)S}$). In a similar way, here c_0 is the unit capacitor, ε performs as the deviation factor due to process.

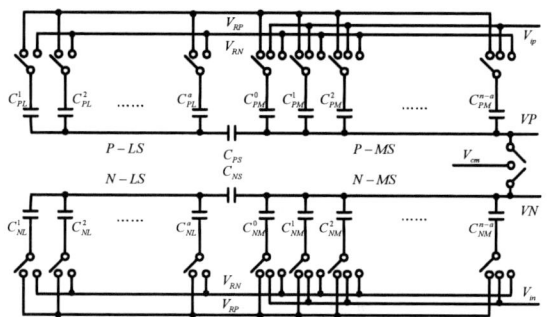

Figure 4. Split Binary-Weighted capacitor array

The actual value of each pair of capacitor is as follows:

$$\begin{cases} C_{P(N)M}^0 = c_0(1+\varepsilon_{P(N)M}^0) \\ C_{P(N)M}^i = 2^{i-1}c_0(1+\varepsilon_{P(N)M}^i) \\ C_{P(N)S} = c_0(1+\varepsilon_{P(N)S}) \\ C_{P(N)L}^i = 2^{i-1}c_0(1+\varepsilon_{P(N)L}^i) \end{cases} \quad (3)$$

And the weight of each weighted-capacitor is as follows:

$$\begin{cases} W_{P(N)}^i = \dfrac{C_{P(N)M}^{i-a}}{\displaystyle\sum_{j=0}^{n-a} C_{P(N)M}^j}, \quad a+1 \le i \le n \\ W_{P(N)}^i = \dfrac{C_{P(N)S}*C_{P(N)L}^i}{(C_{P(N)S}+\displaystyle\sum_{i=1}^a C_{P(N)L}^i)*\displaystyle\sum_{j=0}^{n-a} C_{P(N)M}^j}, \quad 1 \le i \le a \end{cases} \quad (4)$$

C. The ultimate weight for calibration

For a SAR ADC with n bits resolution, after n times comparison and approximation, the voltage of N port and P port can be expressed as follows:

$$\begin{cases} VN = \alpha*\left[-V_{in}+V_{RN}*\displaystyle\sum_{i=1}^n D_i*W_i^N+V_{RP}*(1-\displaystyle\sum_{i=1}^n D_i*W_i^N)\right] \\ VP = \alpha*\left[-V_{ip}+V_{RN}*(1-\displaystyle\sum_{i=1}^n D_i*W_i^P)+V_{RP}*\displaystyle\sum_{i=1}^n D_i*W_i^P\right] \end{cases} \quad (5)$$

Here α is a constant determined by capacitor array, we assume $VN \approx VP$, so (5) becomes

$$V_{ip}-V_{in}+V_{RP}-V_{RN} = (V_{RP}-V_{RN})*\sum_{i=1}^{12} D_i*(W_i^N+W_i^P) \quad (6)$$

And we get

$$\begin{cases} V_{in} = V_{cm}-V_{pp}Sin(\omega t) \\ V_{ip} = V_{cm}+V_{pp}Sin(\omega t) \end{cases} \quad (7)$$

Then, (6) becomes

$$\frac{V_{RP}-V_{RN}}{2}+V_{pp}Sin(\omega t) = (V_{RP}-V_{RN})*\sum_{i=1}^{12} \frac{D_i*(W_i^N+W_i^P)}{2} \quad (8)$$

There is only a DC shift between (8) and the signal V_{ip}, so defining $W_i = \dfrac{W_i^N+W_i^P}{2}$ as the ultimate weight, the initial signal V_{ip} can be recovered.

Considering that $C_{totP} \approx C_{totN}$, then for Binary-Weighted Capacitor Array, the ultimate weight can be expressed:

$$W_i = \frac{W_N^i+W_P^i}{2} = \frac{C_N^i+C_P^i}{C_{totN}+C_{totP}}, 1 \le i \le n \quad (9)$$

For Split Binary-Weighted Capacitor Array, following are two reasonable assumptions:

$$\begin{cases} C_{PS}+\displaystyle\sum_{i=1}^a C_{PL}^i = C_{NS}+\displaystyle\sum_{i=1}^a C_{NL}^i \\ \displaystyle\sum_{j=0}^{n-a} C_{PM}^j = \displaystyle\sum_{j=0}^{n-a} C_{NM}^j \end{cases} \quad (10)$$

The ultimate weight can be expressed:

$$\begin{cases} W_i = \dfrac{C_{PM}^{i-a}+C_{NM}^{i-a}}{\displaystyle\sum_{j=0}^{n-a} C_{PM}^j+\displaystyle\sum_{j=0}^{n-a} C_{NM}^j}, \quad a+1 \le i \le n \\ W_i = \dfrac{(C_{PS}+C_{NS})}{(C_{PS}+C_{NS}+\displaystyle\sum_{i=1}^a C_{PL}^i+\displaystyle\sum_{i=1}^a C_{NL}^i)}*\dfrac{(C_{PL}^i+C_{NL}^i)}{(\displaystyle\sum_{j=0}^{n-a} C_{PM}^j+\displaystyle\sum_{j=0}^{n-a} C_{NM}^j)}, \quad 1 \le i \le a \end{cases} \quad (11)$$

Ideally, all weighted-capacitor match perfectly, that is $\varepsilon = 0$, so (9) and (11) become

$$W_i = \frac{1}{2^{n+1-i}}, 1 \leq i \leq n \tag{12}$$

While taking the mismatch into account, the ultimate weight defined by (9) or (11) can eliminate the nonlinearity of capacitors.

III. MEASURING CAPACITOR

Fig. 5 shows the circuit for measuring capacitor [3],[5]. C_{Tn} and C_{Tp} are the capacitors to be measured, C_{Rn} and C_{Rp} are the reference capacitors, C_{In} and C_{Ip} are the integration capacitors.

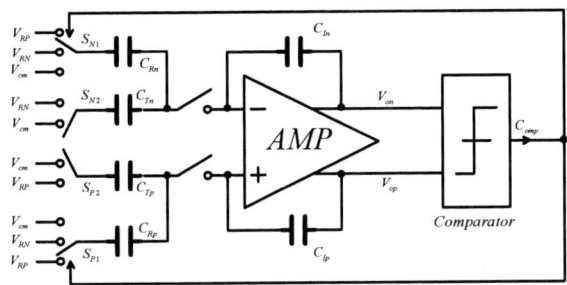

Figure 5. Circuit for measuring capacitor

Following lists the steps for measurement:

1.1) S_{N1} switches to the voltage of V_{cm}, S_{N2} switches to V_{RN}; S_{P1} switches to the voltage of V_{cm}, S_{P2} switches to V_{RP}; the output voltage of amplifier are V_{op} and V_{on}.

1.2) After the charging is stable, S_{N2} and S_{P2} switch to V_{cm}; according to the results of comparator output, S_{N1} and S_{P1} switch to V_{RP} or V_{RN}. For example, if Comp=1, then S_{N1} switches to V_{RN}, while S_{P1} switches to V_{RP}.

1.3) Then the output voltage of amplifier change to $V_{op}{}'$ and $V_{on}{}'$.

1.4) Repeat steps 1.1) ~ 1.3) for X times (X>2^n, n represent the resolution of ADC).

For one measurement, according to charge conservation law, we get

$$\begin{cases} (V_{RP} - V_{cm})C_{Tp} + (V_{cm} - C_{omp}V_{RP} - \overline{C_{omp}}V_{RN})C_{Rp} = (V_{op}' - V_{op})C_{Ip} \\ (V_{RN} - V_{cm})C_{Tn} + (V_{cm} - C_{omp}V_{RN} - \overline{C_{omp}}V_{RP})C_{Rn} = (V_{on}' - V_{on})C_{In} \end{cases} \tag{13}$$

Repeat X times, we get

$$\begin{cases} [V_{cm} - V_{RP}\frac{\sum Comp}{X} - V_{RN}(1 - \frac{\sum Comp}{X})]C_{Rp} + (V_{RP} - V_{cm})C_{Tp} = \frac{(V_{op}^{X} - V_{op})C_{Ip}}{X} \\ [V_{cm} - V_{RN}\frac{\sum Comp}{X} - V_{RP}(1 - \frac{\sum Comp}{X})]C_{Rn} + (V_{RN} - V_{cm})C_{Tn} = \frac{(V_{on}^{X} - V_{on})C_{In}}{X} \end{cases} \tag{14}$$

And considering $V_{cm} = \frac{V_{RP} + V_{RN}}{2}$, for X is large enough, it's reasonable that assuming the two formula of (14) are equal, finally we get

$$C_{Tp} + C_{Tn} = \frac{2 * \sum_{i=1}^{X} Comp - X}{X}(C_{Rp} + C_{Rn}) \tag{15}$$

Though we can't get the true value of $C_{Rp} + C_{Rn}$, but all pairs of capacitors in Fig. 3 or Fig. 4 can be expressed by the ratio $\frac{2 * \sum_{i=1}^{X} Comp - X}{X}$, and the accuracy of measurement is determined by the number of measuring.

Here, as long as X is large enough, the effect of finite gain of the amplifier and mismatch of C_{In} and C_{Ip} can be eliminated. What's more, change the way of sampling and feedback, we can measure the value of $C_{Tp} - C_{Tn}$. In other words, we can calculate C_{Tp} and C_{Tn} separately.

Changing step 1.1), S_{N1} switches to the voltage of V_{cm}, S_{N2} switches to V_{RP}; S_{P1} switches to the voltage of V_{cm}, S_{P2} switches to V_{RP}; the output voltage of amplifier are V_{op} and V_{on}. And the rest steps keep the same, repeat X times, too. Then we get

$$\begin{cases} [V_{cm} - V_{RP}\frac{\sum Comp'}{X} - V_{RN}(1 - \frac{\sum Comp'}{X})]C_{Rp} + (V_{RP} - V_{cm})C_{Tp} = \frac{(V_{op}^{X'} - V_{op})C_{Ip}}{X} \\ [V_{cm} - V_{RN}\frac{\sum Comp'}{X} - V_{RP}(1 - \frac{\sum Comp'}{X})]C_{Rn} + (V_{RP} - V_{cm})C_{Tn} = \frac{(V_{on}^{X'} - V_{on})C_{In}}{X} \end{cases} \tag{16}$$

Finally, we get

$$C_{Tp} - C_{Tn} = \frac{2 * \sum_{i=1}^{X} Comp' - X}{X}(C_{Rp} + C_{Rn}) \tag{17}$$

Combine (15) with (17), we can even calculate the capacitor C_{Tn} and C_{Tp} separately. So it is possible to calculate the weight of P-port and N-port separately.

Even if $V_{cm} \neq \frac{V_{RP} + V_{RN}}{2}$, here we change step 1.1), repeat the measurement for another $2*X$ times, we can still get $C_{Tn} + C_{Tp}$ and $C_{Tn} - C_{Tp}$, followings list the method.

Changing step 1.1), S_{N1} switches to the voltage of V_{cm}, S_{N2} switches to V_{RP}; S_{P1} switches to the voltage of V_{cm}, S_{P2} switches to V_{RN}. And the rest steps keep the same, repeat X times, too. Then we get

$$\begin{cases} [V_{cm} - V_{RP}\frac{\sum Comp''}{X} - V_{RN}(1 - \frac{\sum Comp''}{X})]C_{Rp} + (V_{RP} - V_{cm})C_{Tp} = \frac{(V_{op}^{X''} - V_{op})C_{Ip}}{X} \\ [V_{cm} - V_{RN}\frac{\sum Comp''}{X} - V_{RP}(1 - \frac{\sum Comp''}{X})]C_{Rn} + (V_{RP} - V_{cm})C_{Tn} = \frac{(V_{on}^{X''} - V_{on})C_{In}}{X} \end{cases} \tag{18}$$

Combine (14) with (18), it's easy to get

$$C_{Tp} + C_{Tn} = \frac{\sum_{i=1}^{X}(Comp - Comp'')}{X}(C_{Rp} + C_{Rn}) \tag{19}$$

Changing step 1.1), S_{N1} switches to the voltage of V_{cm}, S_{N2} switches to V_{RN}; S_{P1} switches to the voltage of V_{cm}, S_{P2} switches to V_{RN}. And the rest steps keep the same, repeat X times, too. Then we get

978-1-4577-1608-9/11 $26.00 © 2011 IEEE

$$\begin{cases} [V_{cm}-V_{RP}\dfrac{\sum Comp'''}{X}-V_{RN}(1-\dfrac{\sum Comp'''}{X})]C_{Rp}+(V_{RN}-V_{cm})C_{Tp}=\dfrac{(V_{op}^{X'''}-V_{op})C_{Ip}}{X} \\[2mm] [V_{cm}-V_{RN}\dfrac{\sum Comp'''}{X}-V_{RP}(1-\dfrac{\sum Comp'''}{X})]C_{Rn}+(V_{RN}-V_{cm})C_{Tn}=\dfrac{(V_{on}^{X'''}-V_{on})C_{In}}{X} \end{cases} \quad (20)$$

Combine (16) with (20), it's easy to get

$$C_{Tp}-C_{Tn}=\frac{\sum\limits_{i=1}^{X}(Comp'-Comp''')}{X}(C_{Rp}+C_{Rn}) \quad (21)$$

With (19) and (21), we can calculate C_{Tn} and C_{Tp} more precisely, the only drawback is that the number of measurement increases to 4*X for each pairs of weighted-capacitor.

IV. MATLAB SIMULATION

In this simulation model, we adopt the structure shown in Fig. 4. Assuming that the unit capacitor c_0 obey the law of normal distribution, Table I lists the simulation parameters. Fig. 6 shows the ADC output spectrum without calibration, and Fig. 7 shows the spectrum after calibration.

TABLE I.   SIMULATION PARAMETERS

c_0	μ	35
	σ^2	0.3
n		12
a		6
C_I		$128*c_0$
C_R		$40*c_0$
X		2^{14}
$Finite-Gain$		10000

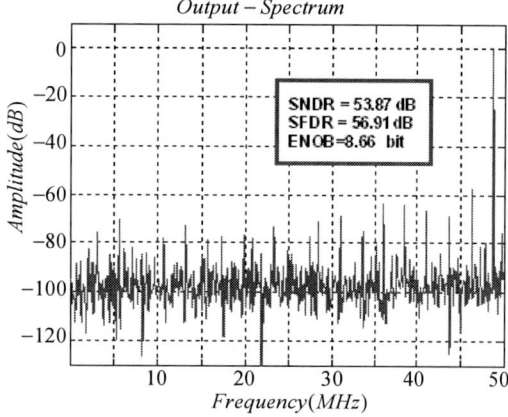

Figure 6.   Output spectrum without calibration

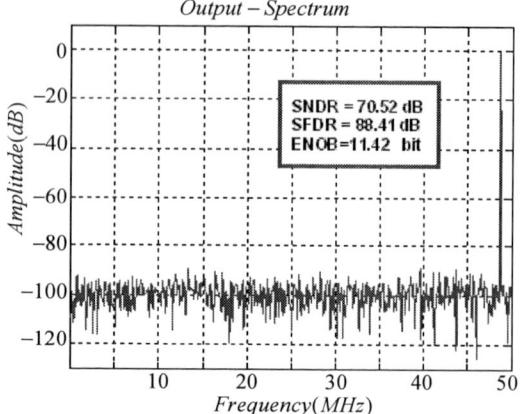

Figure 7.   Output spectrum after calibration

Compare Fig. 7 with Fig. 6, it is obvious that after calibration, the harmonics are greatly inhibited. According to the true value of each pairs of weighted-capacitor, changing the weight can remove the effect of nonlinearity due to mismatch between capacitor. The only drawback of this technique is that before normal successive approximation conversion, it is necessary to measure capacitor, calculate and store the weight.

V. CONCLUSIONS

A new calibration technique for capacitor array SAR ADC is proposed in this paper, which focuses on the true value and the true weight of capacitor. This calibration technique can greatly suppress harmonics and improve ENOB. Comparing with Lee and Hodges's design, we don't have to accumulate the error voltage, and subtract from the actual output voltage of DAC, thus the conversion rate is not restricted by the calibration. And higher resolution ADC isn't needed, too.

REFERENCES

[1] Hae-seung Lee, David A. Hodges, Self-Calibration Technique for A/D Converters, Circuits and Systems Letters, 1983

[2] Gerhard, Dieter Herbst, Error Cancellation Technique for Capacitor Arrays in A/D and D/A Converters, IEEE TRANSACTIONS ON CIRCUITS AND SYSTEMS, VOL. 35, NO. 6, JUNE 1988

[3] Boby George, V. Jagadeesh Kummar, Analysis of the Switched-Capacitor Dual-Slope Capacitance-to-Digital Converter, IEEE TRANSACTIONS ON INSTRUMENTATION AND MEASURE-MENT, VOL. 59, NO. 5, MAY 2010

[4] Yanfei Chen, Xiaolei Zhu, Hirotaka Tamura etc, Split Capacitor DAC Mismatch Calibration in Successive Approximation ADC, IEEE Custom Integrated Circuits Conference, 2009

[5] Jipeng Li, Un-Ku Moon, Background Calibration Techniques for Multistage Pipelined ADCs With Digital Redundancy, IEEE TRANSACTIONS ON CIRCUITS AND SYSTEMS- II :ANALOG AND DIGITAL SINGAL PROCESSING, VOL. 50, NO. 9, SEPTEMBER 2003.

A Time-Efficient Dither-Injection Scheme for Pipelined SAR ADC

Rui Wang[1,2], U-Fat Chio[2], Chi-Hang Chan[2], Li Ding[2], Sai-Weng Sin[2], Seng-Pan U[2], Zhihua Wang[1],
Rui Paulo Martins[2,3]

1. Institute of Microelectronic, Tsinghua University, Beijing, China

[1]raywang0923@gmail.com

2. *State-Key Laboratory of Analog and Mixed Signal VLSI (http://www.fst.umac.mo/en/lab/ans_vlsi/website/index.html)*

Faculty of Science and Technology, University of Macau, Macao, China

Tel:+853 83978796, Fax: +853 83978797, Email: terryssw@umac.mo

3. On leave from Instituto Superior Técnico/TU of Lisbon, Portugal

Abstract—**This paper presents a time-efficient dither-injection scheme in digital domain for pipelined successive approximation register analog-to-digital converter (SAR ADC). Compared with the conventional dither injection method, the proposed method can achieve faster injection speed and reduce the disturbance during the quantization of the ADC. Only 1 LSB dither injection is discussed in this method. Simulation results show more than 8 times speed improvement comparing to the conventional configuration.**

Keywords- SAR ADC; pipelined; digital calibration; dither injection.

I. INTRODUCTION

Pipeline ADCs have been the most prevalent topology, which utilize in the aspect of high speed and high resolutions converters designs [1-3]. In order to realize pipelined structure, N-bit resolution ADC usually consists of N stages of flash ADC and N-1 operational amplifiers, which is noted as single bit structure. In the single bit structure, the burden of the precision requirement has been transferred from flash ADCs to the first few stages of operation amplifiers. For relaxing the requirements of operational amplifiers, multi-bit pipeline ADC structure is adopted in [4] where every single stage can contain two types of ADC such as flash ADCs and SAR ADCs. By utilizing flash ADC at each stage, the fastest conversion can be achieved at the cost of lager areas and more comparators, whose existence introduces another nonlinear error in need of the help from dynamic element matching (DEM) technique. As utilizing SAR ADC at each stage, smaller die areas with less power consumptions can be achieved in the tradeoff of conversion speed, whose limit will become less critical due to the trend of technology scaling.

With the target of high resolution ADC implementation in Nano-meter technology, digital calibration has to be applied into the traditional ADC designs. Among the various types of the digital calibration methods, dither injection is one of the popular schemes [5-7]. Dither injection, generally, is used to extract both the information of gain error and nonlinear error of operational amplifiers, which function with digital output codes in certain algorithm to calibrate errors. On the purpose of injecting the calibration signal into the input of the operational

amplifiers, dither codes need to be transformed into residues. There are two ways of dither transformation: digital domain and analog domain. Compared with analog domain, digital domain method only requires the modification of the digital circuits with the cost of disturbance to the normal conversion of ADC.

This paper presents a digital-domain dither-injection scheme, which is applied in pipelined SAR ADC. With the proposed scheme, the speed of the dither injection can be increased and the disturbance, which causes by dither injection, to both the regular SAR ADC and operational amplifier operations, is reduced.

II. DITHER INJECTION FOR PIPELINED SAR ADC

A. Pipelined SAR ADC architecture

As shown in Fig.1, a pipelined SAR ADC consists of N sub-ADC stages, operational amplifiers and a digital encoder block [8]. Although some modifications are made for adapting to extra phase of amplification for pipelined structure, every sub-ADC is a traditional SAR ADC [9], which contains the SAR logic, the capacitive DAC array and the comparator. Because every sub-stage is one SAR ADC, pipelined SAR ADC can be categorized into multi-bit structure [10]. V_{ref} is the reference voltage and V_{in} is input signal of each stage. The operational amplifier in each stage is used to transfer the residue from the

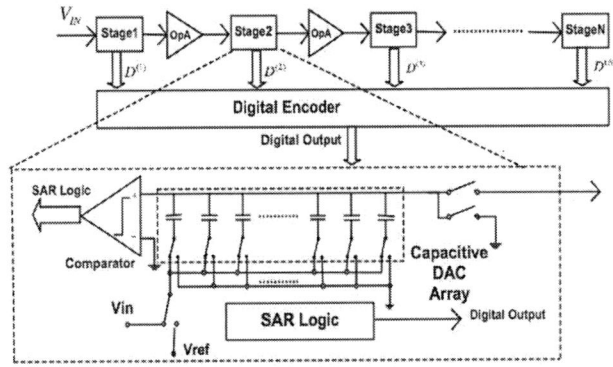

Fig.1. Pipelined SAR ADC without dither injection

This research work was financially supported by Research Grants of University of Macau and Macao Science & Technology Development Fund (FDCT).

last stage with gain to the next stage in pipeline fashion, which is achieved by feeding part of the capacitors in DAC array to the outputs of the operational amplifier depending on the close loop gain.

B. SAR Logic Operation with Conventional Dither Injection

In order to inject dither signal in digital domain, digital circuits of each pipelined SAR stage have to be modified. Therefore, the modification of each stage is focused on the SAR logic part.

Fig.2 shows a self-timing SAR logic with a conventional dither injection method, which is directly derived from traditional pipeline flash ADC [6]. It includes a pulse generator, shift registers and bit registers. The pulse generator produces the self-timing strobe phase Φ_{SAR} and activates the shift registers to generate multiple shifted clocks CLK_1 to CLK_n. The bit registers are turned on/off by the shifted clocks CLK_1 to CLK_n, which record the conversion result of each bit from SAR ADC and transfer it to switch the DAC through buffers. For the dither injection part, the adders are directly inserted between bit registers and buffers. Besides, the dither signal is required to be ready before the regular SAR operation. Although this structure of dither injection in the traditional pipeline flash ADCs can be effective, it causes significant performance degradation in the pipelined SAR ADC architecture.

Because of the dither addition operation is directly inserted into SAR operation loop, each bit of SAR operation will be affected, which causes extra charging and discharging power consumption. Besides, for the critical path of SAR logic operation, the final bit signal need to pass through a series of n bit adders of each stage in total. Therefore, either the resolution of each stage after dither injection is limited or the whole speed of the pipelined SAR ADC is decreased.

C. Delay-reduced Post-dither Code Selection Network

To avoid the problems in the SAR logic with conventional dither injection, a new technique named Delay-reduced Post-dither Code Selection Network (DPCSN) is proposed, which can directly replace the conventional bit adders group without adjusting other circuits in the SAR logic as shown in Fig.3.

DPCSN provides path one for the bit registers to pass the digital bits from bit registers to buffers quickly, and it also provides path two to perform the dither injection by DPCSN core. Two paths have to be chosen by multiplexers controlled

by signal S1. During the first n-1 bits quantization, the path one has been chosen, and the first n-1 bits are not affected by dither injection. Although the DPCSN core is not mingled into the path one, it is still working with the input of both dither and input codes without interrupting the regular SAR operation. After the n-th bit has been generated from comparator, the path two is chosen. Unlike the conventional dither injection method affecting the whole SAR operation, the proposed DPCSN only affects the n-th bit SAR operation. Nevertheless, the following discussion shows that disturbance to the n-th bit caused by dither injection will be greatly reduced through the DPCSN core.

Assuming the sub-ADC is n-bit and the dither of 1 LSB has been injected before SAR operation, where the number of n can be expressed as

$$n = mK + 1 \qquad (1)$$

Formula (1) illustrates that the whole n bits can be divided into m of K bits each with one extra bit and the block diagram of the DPSCN core derived from it is shown in Fig.4:

As shown in fig.4, there are m sub-blocks to synthesize the Code_New<1:n-1>, which stands for Code<1:n-1> through dither injection. Since the final bit of 1 LSB dither signal always equals to one, the polarity of Code_New<n> is always opposite to the polarity of Code<n>. Since there are no related signals between each block, all of the m+1 sub-blocks are independent between each other.

Fig.5 shows the detailed description of i-th sub block of the DPCSN core, which contains two K-bit full adders, one mux group and one XOR tree Traditionally, after Code<iK-K+1:iK> has been generated from i-th sub-stage SAR operation, digital series derived from both dither signal and carry signal of unknown Code<iK+1:n> will affect Code<iK-K+1:iK>, producing complex addition operations. However, the digital series mentioned above can only be three combinations, which are $\underset{K-1}{\underbrace{0\cdots0}}1$, $\underset{K}{\underbrace{1\cdots1}}$ and $\underset{K}{\underbrace{0\cdots0}}$. The first combination happens when Code<iK+1:n> are all ones and dither signal is positive. The second combination happens when Code<iK+1:n> are all zeros and dither signal is negative. In other scenarios, the third combination is satisfied.

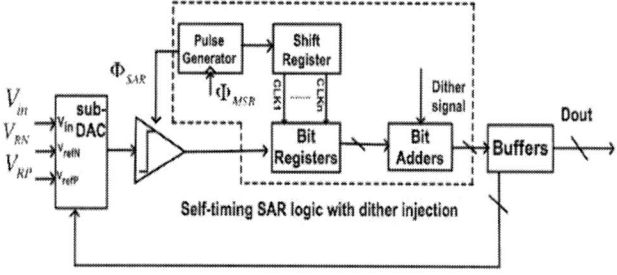

Fig.2. SAR logic with conventional dither injection.

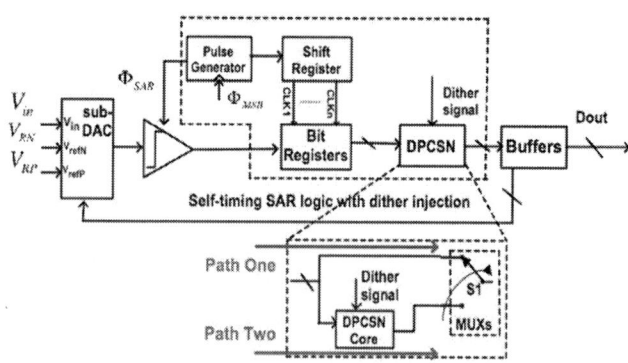

Fig.3. Proposed SAR logic with dither injection.

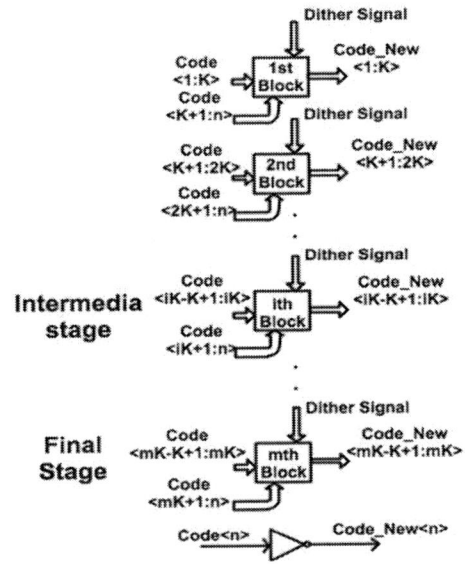

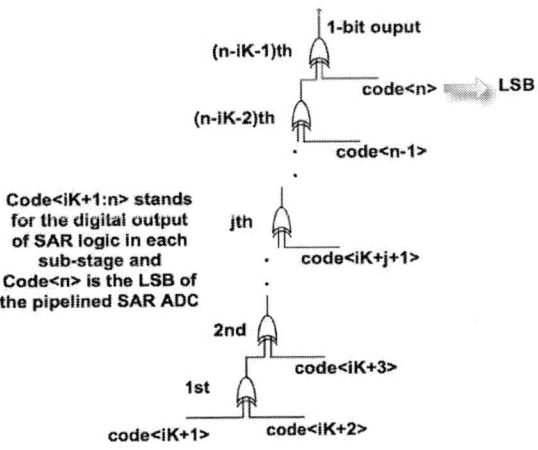

Fig.4. Proposed block diagram of DPCSN core

Fig.6. i-th XOR tree

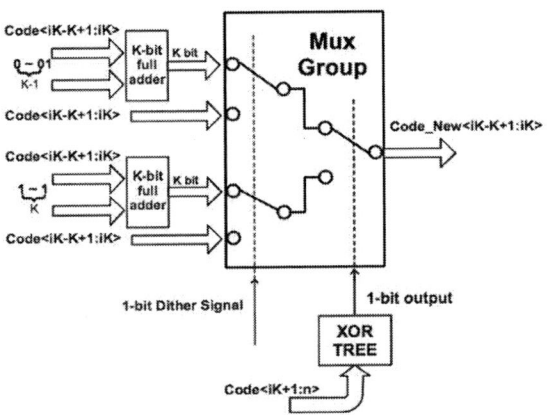

Fig.5. i-th sub-block of the DPCSN core

As shown in Fig.5, the K-bit output codes of the two K-bit full adders and Code< iK-K+1:iK >, which reflect the three possible digital series, will appear at the input of the multiplexers group for the latter digital codes to select from in order to generate Code_New<iK-K+1:iK> directly. Therefore, the addition operation of dither injection in the i-th sub block can be finished before the generation of Code<iK-K+1:n>. Then, only multiplexers operation of selecting the three possible series is needed and thus reduces the penalty of addition operations. However, another selecting signal besides dither signal has to be generated by XOR tree as shown in Fig.6.

From the three scenarios mentioned above, it is noteworthy that the three cases can be distinguished only by the XOR operation of the Code<iK+1:n>. Besides, after realizing that each of the Code<iK+1:n> is generated subsequently, the number of n-iK+1 input XOR operation can be redistributed as a XOR tree in order to minimize the critical path illustrated in red.

III. DESIGN STRATEGY OF DPCSN

As the formula (1) illustrated in section II, K needs to be chosen according to both the delay of one-bit adders and the interim between consecutive comparator trigger pulses, whose value is related to the silicon process. To minimize the influence of dither injection, K has to satisfy the condition as below

$$K < \frac{T_\Delta - (T_{comparator} + T_{lock})}{T_{adder}} \qquad (2)$$

where T_Δ, $T_{comparator}$, T_{adder}, T_{lock} denote the period between two consecutive trigger pulses of the comparator, the comparison time of the comparator, the delay time of one bit adder and the time for bit registers to lock the comparator output respectively. Actually, formula (2) demonstrates a design strategy, that addition operation can be distributed into the interim of SAR operation of subsequence bit in order to prevent inserting it into the final SAR operation, which is related to the critical path.

Once the formula (2) is satisfied, the addition operation of the first m sub-blocks had been finished before the Code<N> is generated. Besides, for each sub-block, before the generation of the Code<N>, the inputs of the final XOR gate in XOR tree had been ready. Therefore, the critical path that Code<N> signal passes one multiplexer and one XOR gate is the same for every sub-block. Combining the whole SAR logic and transmission gates are implemented as multiplexers, we can conclude that the delay time of dither injection has been reduced to a couple of inverters' delay plus one multiplexer's, which is much less than the conventional dither injection method.

If the following condition is also satisfied, m>2 can be degraded to m=2 and less sub-block can be used without changing critical path mentioned above.

$$(N - K - 1)T_{adder} < K[T_\Delta - (T_{comparator} + T_{lock})] \qquad (3)$$

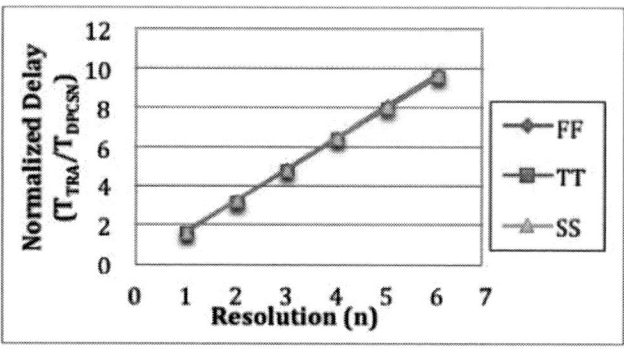

Fig.7 Normalized delay versus the pipelined stage resolution

Formula (3) illustrates that after grouping, addition operation can also be distributed into the interim of the SAR operation of succeeding groups.

IV. ANALYSIS AND COMPARISON

Assume the resolution of each stage is n-bit in Pipelined SAR ADC. As mentioned in Section III, the critical path of traditional dither injection method is that the n-th bit signal will pass through the whole n-bit full adder and thus affects the additions of previous n-1 bits. Therefore, the delay time after the generation of the n-th bit, T_{TRA} should be:

$$T_{TRA} = n \times T_{adder} \qquad (4)$$

With the implementation of the DPCSN, as long as the formula (2) is satisfied, the critical path of proposed dither injection method is that the final bit signal just passes through one XOR gate, one 2-to-1 multiplexer and one inverter. T_{inv}, T_{mux}, and T_{XOR} denote the delay time of the inverter, the 2-to-1 multiplexer and of XOR gate respectively. Eventually, the delay time T_{DPCSN} after n-th bit quantization can be expressed as:

$$T_{DPCSN} = T_{mux} + T_{XOR} + T_{inv} \qquad (5)$$

Using the 65nm CMOS technology, the delay of T_{inv}, T_{XOR} and $Tmux$ can be obtained by the post layout simulation. T_{inv} equals to 14ps, 20ps and 25ps with respect to FF, TT and SS corner respectively. T_{mux} equals to 17ps, 24ps and 30ps with respect to FF, TT and SS corner respectively. T_{XOR} equals to 40ps, 50ps and 60ps with respect to FF, TT and SS corner respectively. T_{adder} equals to 115ps, 150ps and 185ps with respect to FF, TT and SS corner respectively. And Then, the delay of T_{DPCSN} is a factor of T_{TRA} by the calculation of (4) and (5). Fig.7 shows the normalize curve of T_{TRA}/T_{DPCSN} versus the pipelined stage resolution n. As the increasing of n, the factor will become bigger, which manifests the benefit of DPCSN furthermore.

V. CONCLUSION

This paper proposes a time-efficient dither-injection scheme used by pipelined SAR ADC. This method not only prevents the disturbance of dither injection to the first n-1 bits but also reduces the delay time induced by dither injection to the level of a few inverters, thus minimizes the disturbance to the final bit. As a result, the speed of SAR ADC and settling time of operational amplifiers will not be affected by the dither injection, which speeds up the whole pipelined SAR ADC operation.

ACKNOWLEDGMENT

We gratefully appreciate Mr. Fan Ng (Leo) for computer-related support and Guohe Yin for document consultancy.

REFERENCES

[1] B.-S.Song,M.Tompsett,andK.Lakshmikumar,"A12-bit1-Msample/s capacitor error-averaging pipelined A/D converter," *IEEE J. Solid-State Circuits*, vol. 23, pp. 1324–1333, Dec. 1988.

[2] Y. Chiu, P. Gray, and B. Nikolic, "A 1.8 V 14 b 10 MS/s pipelined ADC in 0.18m CMOS with 99 dB SFDR," in *ISSCC Dig. Tech. Papers*, Feb. 2004, pp. 458–459.

[3] P. C. Yu and H.-S. Lee, "A 2.5 V 12 b 5 Msample/s pipelined CMOS ADC," *IEEE J. Solid-State Circuits*, vol. 31, pp. 1854–1861, Dec. 1996.

[4] A.Panigada and I.Galton,"Digital background correction of harmonic distortion in pipelined ADCs," *IEEE Trans. Circuits Syst. I: Reg. Papers*, vol. 53, no. 9, pp. 1885–1895, Sep. 2006.

[5] H. S. Fetterman et al., "CMOS pipelined ADC employing dither to improve linearity," in *CICC* 1999, pp. 109-112.

[6] E. 1. Siragusa and 1. Galton, "A digitally enhanced 1.8V 15b 40MS/s CMOS pipelined ADC," in *ISSCC* 2004, pp. 452-453.

[7] Y.-D. Jeon, S.-C. Lee, K.-D. Kim, J.-K. Kwon, and J. Kim, "A 5-mW 0.26- mm 10-bit 20-MS/s pipelined CMOS ADC with multi-stage amplifier sharing technique," in *Proc. Eur. Solid-State Circuits Conf.*, Montreux, Switzerland, 2006, pp. 544–547

[8] Young-Hwa Kim, Jaewon Lee and SeongHwan Cho, "A 10-bit 300MSample/s Pipelined ADC using Time-Interleaved SAR ADC for Front-End Stages," in ISCAS 2010, pp.4041-4044.

[9] J. Craninckx and G. V. Plas, "A 65 fJ/conversion-step 0-to-50 MS/s 0-to-0.7 mW 9 b charge-sharing SAR ADC in 90 nm digital CMOS," in *IEEE ISSCC Dig. Tech. Papers*, Feb. 2007, pp. 246–247.

[10] J. Li and U.-K. Moon, "Background calibration techniques for multi-stage pipelined ADC's with digital redundancy," *IEEE Trans. Circuits Sys. II*, vol. 50, pp. 531–538, Sept. 2003.

A Hardware-effective Digital Decimation Filter Implementation for 24-bit $\Delta\Sigma$ ADC

Yafei Ye, Ting Li, Zhihua Wang
Institution of Microelectronics
Tsinghua University
Beijing, China
ye-yafei@163.com

Liyuan Liu, Dongmei Li
Department of Electronic Engineering
Tsinghua University
Beijing, China
lidmei@tsinghua.edu.cn

Abstract—**A hardware-effective digital decimation filter implementation used in the 24-bit $\Delta\Sigma$ ADC for audio application is described in this paper. Composing of four comb filters and two half-band Finite Impulse Response (FIR) filters, the digital decimation filter uses multistage structure to relax the filter design. Since the multipliers are the most hardware consuming components in the digital filters, the coefficients of the FIR filters are coded by Canonical Signed Digit (CSD) which can make the filter multiplier-free. Meanwhile, time-multiplexing method is adopted in the filter to further reduce the hardware consumption. The proposed design is synthesized in 180nm CMOS process and occupies a die area of 1.44 mm². This implementation is well suited for VLSI and can be applied to many other high resolution $\Delta\Sigma$ ADC.**

I. INTRODUCTION

With the increasing demand for high precision audio A/D and D/A in consumer electronics, the $\Delta\Sigma$ ADC which can be easily integrated within the digital system is becoming increasingly popular. The digital decimation filter is an important part of the $\Delta\Sigma$ ADC, which can effectively filter the high-frequency quantization noise and does not cause signal distortion. Because the decimation filter usually occupies more than half the chip area, improving and optimizing the design of decimation filter becomes a key factor in reducing area consumption. A high hardware-efficiency implementation of a digital decimation filter for 24-bit $\Delta\Sigma$ ADC is described in this paper.

The system block diagram of this digital decimation filter is illustrated in Figure 1. The decimation filter uses multi-stage structure to relax the filter design. The front of the decimation filter is the comb ones including four cascade comb sub-filters. The following FIR filters consist of two half-band FIR sub-filters. Each of the sub-filters decimates the signal by a factor of 2. Thus the decimation filter realizes a down-sampling of 64.

The two half-band FIR filters determine the overall performance of the decimation filter, thus they should be carefully designed. In order to achieve the requirement of 24-bit $\Delta\Sigma$ ADC, the noise should be attenuated under -140 dB level. Hence two half-band filters need a cut-off frequency of 0.2fs, a stopband rejection of at least -100 dB and a narrow transition bandwidth. Those requirements will make the order of the filter especially high. In the traditional ways, the high order half-band filter will take up a large area. In this design both half-band filters employ the multiplier-free structure. The function of the half-band FIR filter is based on the procedure described by Saramaki[1].

In session II, the design of the decimation filter will be described. Session III presents the simulation results while the IV shows the layout of the decimation filter. The conclusion will be drawn in session V.

II. DIGITAL DECIMATION FILTER

A. Architecture

Figure 1. shows the detail diagram of the decimation filter. The comb filters use the traditional structure to simplify the filter design. The half-band structure is chosen because about half of the total coefficients are zeros. The two half-band filters and the back stages of the comb filters work at low frequency compared to the master clock, thus it allows the operation units to be multiplexed.

B. Comb Filters

The structure of Cascaded Integrator Comb (CIC) is adopted in the design of comb filters [2]. The first two stages are 4th-order comb filters, and the rest two stages are 5th-order comb filters. The 4th-order and 5th-order comb filters'

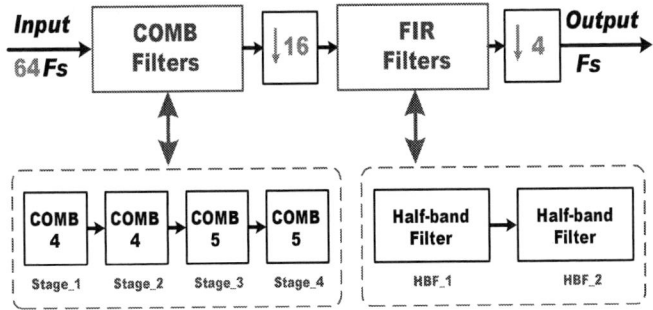

Figure 1. Block diagram of the multi-stage decimation filter

978-1-4577-1608-9/11 $26.00 © 2011 IEEE

frequency response are shown in the Figure 2. (a) and Figure 2. (b) respectively.

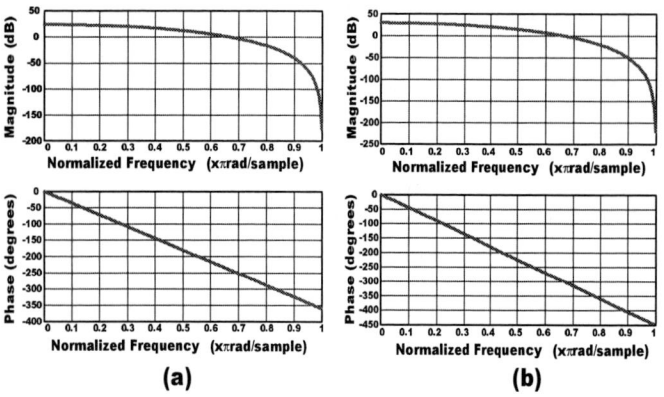

Figure 2. Frequency response of the comb filters

The transfer function of the 4ᵗʰ-order comb filter is:

$$H_4 = (1+z^{-1})^4 = (1+z^{-4}) + 4(z^{-1}+z^{-3}) + 6z^{-2} \qquad (1)$$

The transfer function of the 5ᵗʰ-order comb filter is:

$$H_5 = (1+z^{-1})^5 = (1+z^{-5}) + 5(z^{-1}+z^{-4}) + 10(z^{-2}+z^{-3}) \qquad (2)$$

Because all the coefficients of comb filters are integers, it's easy to realize the filter by simple shifters and adders. The implementation of these stages are similar, thus the fourth stage comb filter is chosen to illustrate the scheduling of the comb filters[3][4].

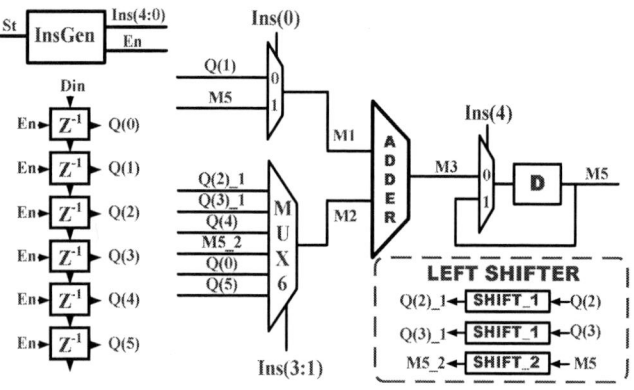

Figure 3. Proposed structure of the fourth stage comb filter

Figure 3. shows the structure of the fourth stage comb filter. This stage filter works under 1/16 of the master clock, so only one adder is needed by using time-multiplexing method. In this stage all the actions of the shift are toward the left. St is a counter signal. When the rising edge of the clock occurs, St is updated by St+1. Ins (4:0) is the global control signal, which is generated by the module of InsGen. TABLE I. shows the Ins (4:0) changes as the St(3:0) signal.

TABLE I.   THE INS (4:0) CHANGES AS THE ST(3:0) SIGNAL

St(3:0)	Ins(4:0)
0000	00000
0001	00011
0010	00101
0011	00111
0100	01001
0101	01011
Others	10000

C. Half-band Filters

The FIR filters perform the final decimation by 4 and determine the overall performance of the decimation filter. Because the high order FIR filter is difficult to design by traditional methods, this work is based on the cascading the identical sub-filters. The two half-band filters use the same structure which is shown in Figure 4. (a). The F2 is the sub-filter of the half-band filter, which is shown in Figure 4. (b)[5].

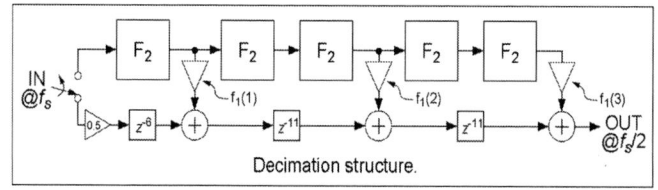

(a)

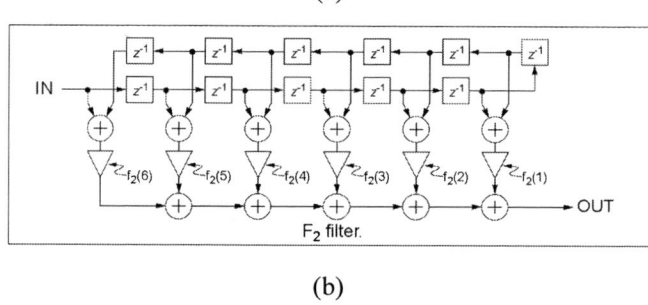

(b)

Figure 4. Block diagram of the half-band filter

The half-band filter's frequency response is shown in the Figure 5. The filter achieves 110dB of attenuation in the stop-band which satisfies the requirement of the decimation filter.

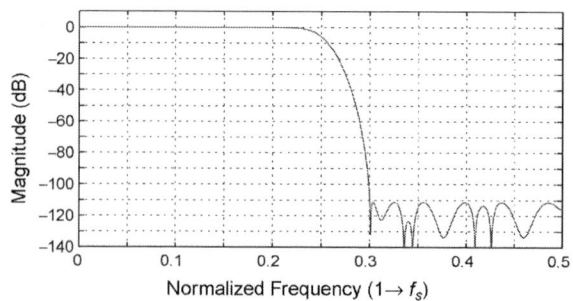

Figure 5. Frequency response of the half-band filters

Since multipliers are the most power consuming components in the digital filters, the coefficients of this filter are coded by CSD which can make the filter multiplier-free, so the hardware consumption is significantly reduced [6]. The coefficients and their signed-digit decompositions are summarized in the TABLE II. All the coefficients of the filter use only 3 signed digits, so any of the multiplication can be replaced by 2 binary additions. Thus, the proposed half-band filter only needs $3 \times 3 + 5(3 \times 6 + 6 - 1) = 124$ additions.

TABLE II.   COEFFICIENTS AND THEIR SIGNED-DIGIT DECOMPOSITIONS

Coefficients	Value	Decomposition
f11	0.9453	$2^0 - 2^{-4} + 2^{-7}$
f12	-0.6406	$-2^{-1} - 2^{-3} - 2^{-6}$
f13	0.1953	$2^{-2} - 2^{-4} + 2^{-7}$
f21	0.6211	$2^{-1} + 2^{-3} - 2^{-8}$
f22	-0.1895	$-2^{-2} + 2^{-4} - 2^{-9}$
f23	0.0957	$2^{-3} - 2^{-5} + 2^{-9}$
f24	-0.0508	$-2^{-4} + 2^{-7} + 2^{-8}$
f25	0.0269	$2^{-5} - 2^{-8} - 2^{-11}$
f26	-0.0142	$-2^{-6} + 2^{-9} - 2^{-11}$

The first stage half-band filter and the second stage half-band filter works at 1/32 and 1/64 of the master clock frequency respectively. Take the second stage as an example, there are 64 clock cycles to complete the operation. By the operation units multiplexing, each sub-filter needs 1 adder and the half-band filter only needs $5 \times 1 + 2 = 7$ adders. Through the use of this structure and the hardware multiplexing method, the hardware consumption is greatly reduced.

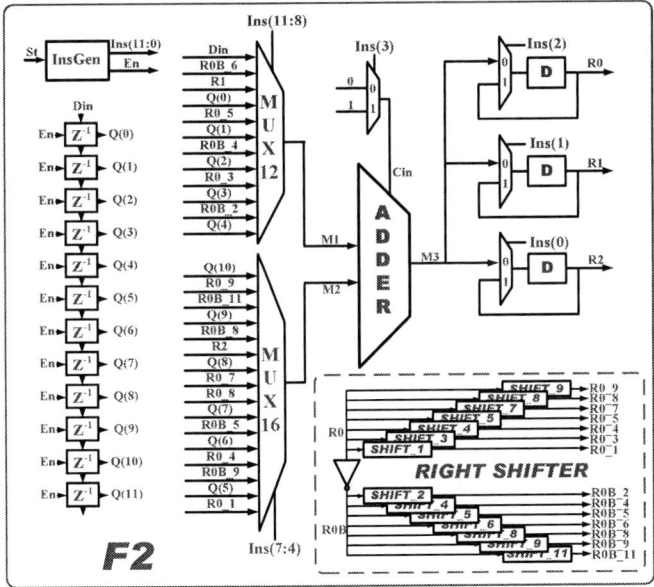

Figure 6.   Proposed structure of the sub-filter F2

Figure 6. shows the proposed structure of the sub-filter F2. In this sub-filter, all the actions of the shift are toward the right. Ins (11:0) also is the global control signal. TABLE III. summarizes the Ins (11:0) changes as the St (4:0) signal.

TABLE III.   THE INS (11:0) CHANGES AS THE ST(4:0) SIGNAL

St(4:0)	Ins(11:0)	St(4:0)	Ins(11:0)
00000	000000000011	01100	100010101101
00001	000100011101	01101	000100011101
00010	001000101110	01110	001001010110
00011	001100110011	01111	100110110011
00100	010001001101	10000	101011001101
00101	001000101101	10001	001011011101
00110	001001010110	10010	001001010110
00111	010101100011	10011	101111100011
01000	011001111101	10100	100011110101
01001	001010000101	10101	001001001101
01010	001001010110	10110	001001010110
01011	011110010011	Others	000000000111

The proposed structure of the half-band filter is shown in Figure 7. In this stage, all the actions of the shift are also toward the right. There are only 2 adders to complete all the operations. TABLE IV. summarizes the Ins (11:0) changes as the St (5:0) signal.

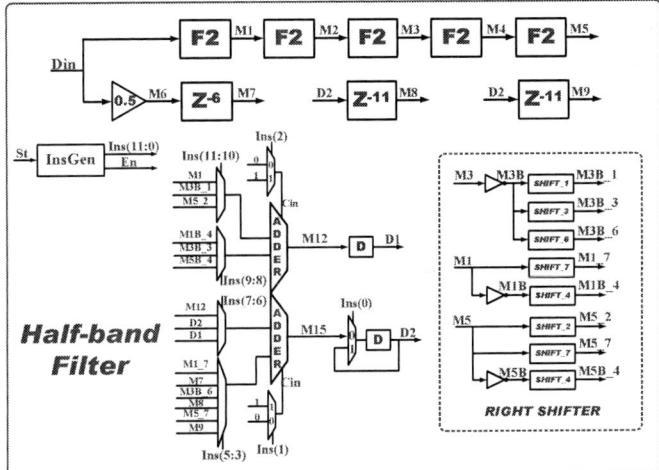

Figure 7.   Proposed structure of the half-band filter

TABLE IV.   THE INS (11:0) CHANGES AS THE ST(5:0) SIGNAL

St(5:0)	Ins(11:0)
100001	000000000100
100010	000001001100
100011	010100010110
100100	101001011110
100101	101010100000
100110	101001101000
Others	000000000001

D. *Parallel-to-Serial Conversion*

Because 24-bit parallel output needs to occupy 24 PADs，the decimation filter introduces a module which converts the 24-bit parallel data into serial data to reduce the hardware consumption. The working principle of this module of parallel-to-serial conversion is described in Figure 8. The port of serial output is connected with the least significant bit of the 24-bit parallel data. When the rising edge of the clock occurs, 24-bit parallel data are shifted to the right one bit and the most significant bit is padded with zero. After repeating this process 24 times, the procedure of converting 24-bit parallel data into serial data is completed.

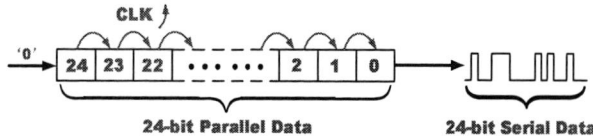

Figure 8.   Working principle of the parallel to serial

III.   SIMULATION

The decimation filter is simulated in the ModelSim SE 6.2e. The input data is generated from a 3rd-order single loop $\Delta\Sigma$ modulator.

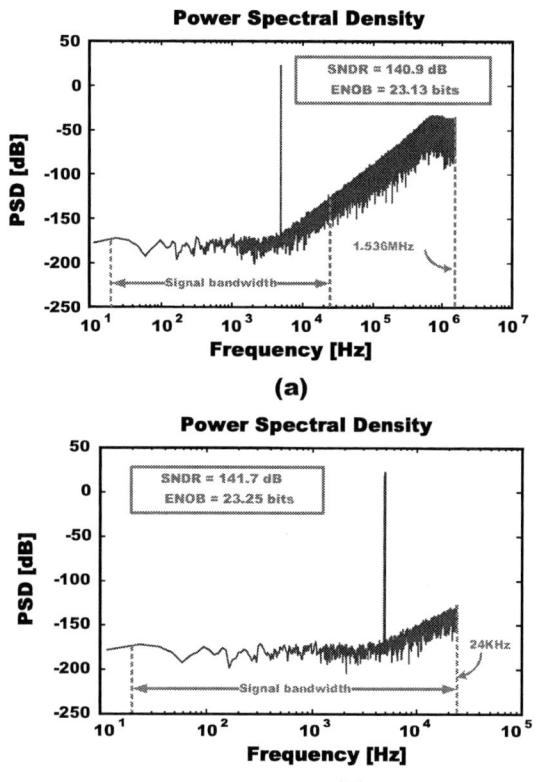

Figure 9.   Power spectral density

The power spectral density of the input and output are shown in the Figure 9. (a) and Figure 9. (b) respectively. The comparison of spectrum indicates the noise beyond signal bandwidth is filtered and the SNDR in the signal band has no attenuation. Thus, the proposed decimation filter can be well qualified for a 24-bit $\Delta\Sigma$ ADC.

IV.   LAYOUT

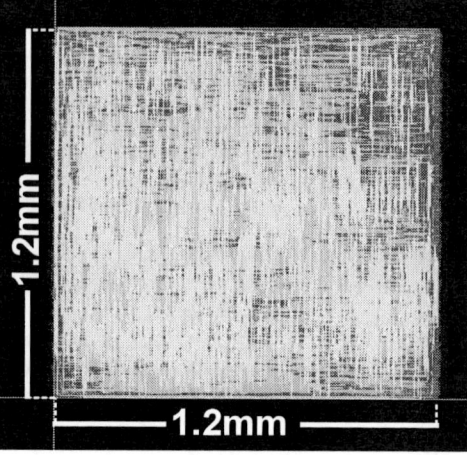

Figure 10.   Layout of the digital decimation filter

Figure 10. shows the layout of the digital decimation filter. This design is implemented in 180nm CMOS process. The digital decimation filter is described by VHDL, which is synthesized by using Synopsys Design Complier. The layout of the filter is obtained through using Cadence SOC Encounter. The total filter area is 1.44 mm^2.

V.   CONCLUSION

Low hardware consumption has been achieved in design and implementation of a digital decimation filter for a 24-bit $\Delta\Sigma$ ADC. The proposed architecture of the digital decimation filter reduces the chip area significantly. The CSD multiplier eliminates the use of high hardware consuming multiplies. The time-multiplexing method is used to distribute the computations in time and hence greatly reduce the number of operation units.

[1]   T. Saramaki, "Design of FIR filters as a tapped cascaded interconnection of indentical subfilters," IEEE Transactions on Circuits and Systems, vol. 34, pp. 1011-1029, 1987.

[2]   Hogenauer E "An Economical Class of Digital Filters for Decimation and Interpolation," Acoustics, Speech and Signal Processing, IEEE Transactions on, 1981.

[3]   Liyuan Liu, Run Chen, Dongmei Li, "A 20-Bit Sigma-Delta D/A for Audio Applications in 0.13um CMOS," IEEE International Symposium on Circuits and Systems, ISCAS 2007.

[4]   Liyuan Liu, Run Chen, Dongmei Li, "A cost-effective digital front-end realization for 20-bit ΣΔ DAC in 0.13 μm CMOS," IEEE Custom Integrated Circuits Conference,CICC 2007 .

[5]   R.Schreier and G. C. Temes, "Understanding Delta-Sigma Data Converters," John Wiley & Sons, New York, 2004 .

[6]   Henry Samueli, "An Improved Search Algorithm for the Design of Multiplierless FIR Filters with Powers-of-Two Coefficients," in IEEE Trans.on Circuits and Systems, vol.36, no.7, Jul, 1989.

A 3.3V to 3.19V Low-Dropout Regulator with frequency compensation strategy without trimming

Yiwei Zhang[1], Liyuan Liu[2], Dongmei Li[2]
1 Institution of Microelectronics
Tsinghua University
Beijing, China
2 Electronic Engineering Department
Tsinghua University
Beijing, China

Abstract—**This paper introduces a 3.3V to 3.19V low-dropout regulator (LDO) with frequency compensation strategy implemented in 0.18μm 1P6M COMS process. The LDO circuit has the frequency compensation strategy to keep stable. There is a bandgap designed in the LDO to generate the reference voltage and the bias current for the LDO. The maximum output current of the LDO is 70mA when the output voltage could reach 3.18V. And the output voltage of the LDO could still reach 3.19V, where the supply voltage is 3.20V and the output current is 0.8mA. The quiescent current is 0.04mA when there is no output current and the VDD voltage is 3.3V. Because the LDO of this paper could change the power supply from 3.3V to 3.19V, the dropout voltage is only 0.11V and the power efficiency could be increased significantly. The active core area of the LDO is 0.46mm×0.7mm and the whole area is 0.86mm×1.1mm.**

I. INTRODUCTION

Low-dropout regulators are widely used in the field of power management as a result of the high efficiency and stability.

Because of the circuit structure of the LDO there is a problem with the stability. Usually the distribution of poles of the LDO will lead to the poor phase margin. So it is needed to design the compensation strategy to enhance the phase margin. The compensation strategy presented in this can give the LDO enough phase margin within the design range of the output current. And the compensation strategy dose not depends on the ESR (Equivalent Series Resistance) of the off-chip capacitors. Within the 10Ω of the ESR, there is enough phase margin above 60° to keep the stability.

The power efficiency of the LDO is an important index about the utilization ratio of the power supply. In application of the mobile electronic equipments it is needed that the LDO should provide sufficient and constant power supply. Meanwhile, the LDO should consume the power supply as little as possible. So the dropout voltage should be as low as

possible. The LDO proposed in this paper could convert the power supply voltage from 3.3V to 3.19V, where the dropout voltage is only 0.11V.

II. COMPENSATION STRATEGY

The compensation strategy presented in this paper is shown in Fig.1. M1~M11 in Fig.1 form the differential amplifier comparing the V_{REF} and $V_{feedback}$, where $V_{feedback}$ is the voltage of the feedback network and V_{REF} is the reference voltage of the bandgap. M13 is the power transistor, which receive the modulating voltage and control the output current. The voltage of the R_2 is $V_{feedback}$. And the voltage of R_1 and R_2 is V_{out}. M12 and M14 compose the buffer to enhance the stability of the LDO [1].

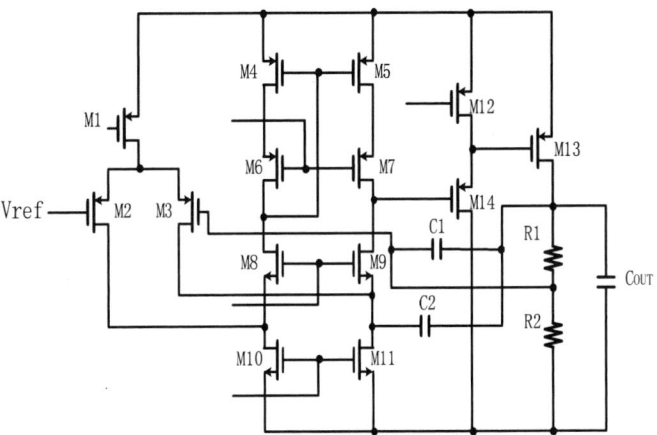

Fig.1 Compensation strategy

In this LDO of this paper there are two capacitors in the frequency compensation strategy to keep the stability [2]. The simulation of the frequency compensation strategy without ESR is shown in Fig.2. The simulation in Fig.2 shows that the phase margin of the LDO during the output current range is

978-1-4577-1608-9/11 $26.00 © 2011 IEEE

more than 60° within 60mA. And when the output current is 70mA the phase margin is still above 50°.

There is an off-chip capacitor at the output terminal of the LDO. Usually an off-chip capacitor has a certain ESR, so it is needed to analyze the effect on the stability of the value of the ESR. Considering the ESR of the off-chip capacitor, it is needed to make sure that the LDO can get enough phase margin within a range of values of the ESR. Fig.3 and Fig.4 have shown the simulations of the phase margin of the LDO with different values of the ESR, and it is shown that this frequency compensation strategy can keep the LDO get enough phase margin in these conditions referred in Fig.3 and Fig.4.

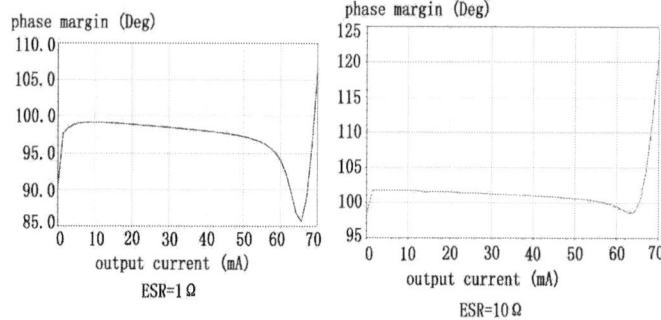

Fig.4 Simulation results of the phase margin with ESR=1Ω and ESR=10 Ω

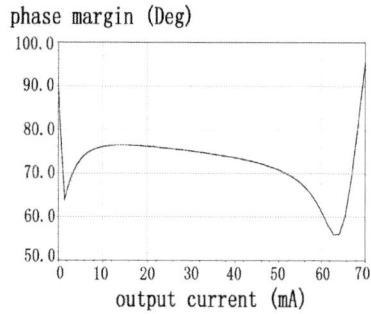

Fig.2 Simulation results of the phase margin without ESR

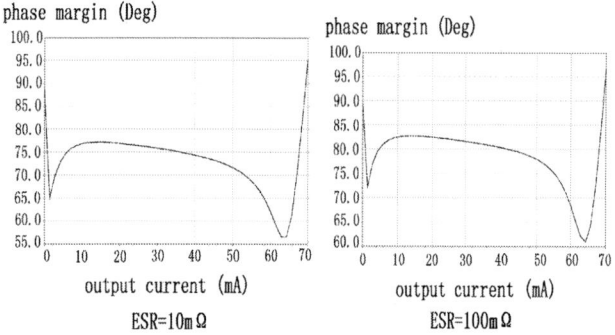

Fig.3 Simulation results of the phase margin with ESR=10mΩ and ESR=100mΩ

III.   BANDGAP

Because of the low temperature coefficient of the voltage generated by the bandgap, the bandgap circuit is widely used to provide the reference voltage. Usually the reference voltage of the bandgap is about 1.25V [3]. However, considering the matching of the resisters of the LDO, the same resister has been chosen to constitute the R1 and R2 of the feedback network. Because the design value of the output voltage of the LDO is 3.2V, the R1 and R2 can be composed of different numbers of the same resisters, and the ratio is (3.2-1.2):1.2=5:3, if the reference voltage of the bandgap is 1.2V.

This bandgap introduced in [4] could provide a 1.2V reference voltage for the LDO. In this design the bandgap circuit topology is shown Fig.5, and this circuit structure could provide the 1.2V reference voltage by modulating the ratio of R_5 and R_6. At the same time the bandgap could provide the bias current for the operational amplifier and the buffer of the LDO shown in Fig.1.

The operational amplifier topology of the bandgap is shown in Fig.6. In the operational amplifier there is also a Miller capacitor C_4 to keep the bandgap stable. And the M15, M16, M17, M18, M19 and M20 compose the starting circuit for the bandgap.

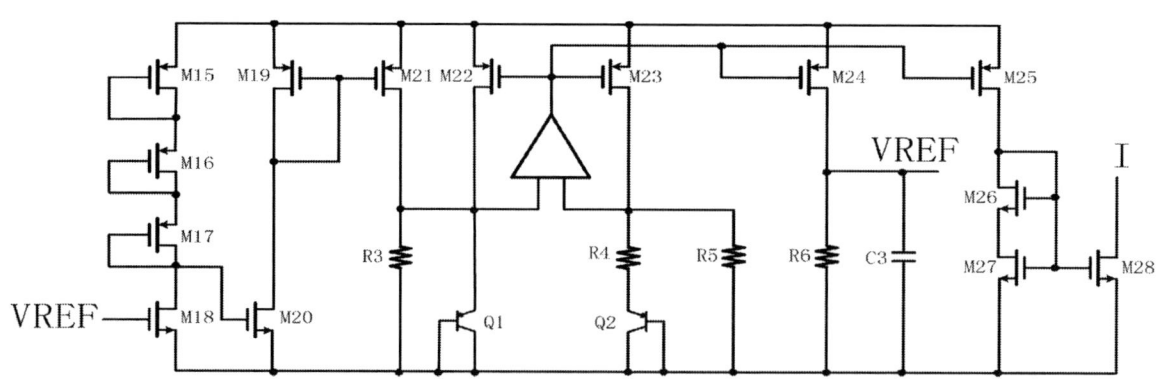

Fig.5 Bandgap topology

The M25, M26, M27 and M28 are used to generate the bias current. The M28 is used to reproduce the I_{ds} of the M25 according to a certain proportion.

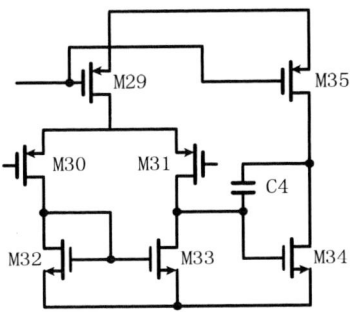

Fig.6 Operational amplifier topology of the bandgap

IV. EFFICIENCY

The efficiency is an important index that indicates the ability of converting the power supply. This LDO could convert 3.3V power supply to 3.19 so the drop-out voltage is only 0.11V. The power efficient is defined as

$$\eta = \frac{V_{out} \bullet I_{out}}{V_{in} \bullet I_{in}} = \frac{V_{out} \bullet I_{out}}{V_{in} \bullet (I_{quiescent} + I_{out})} \quad (1)$$

$$V_{drop-out} = V_{in} - V_{out} \quad (2)$$

So the low drop-out voltage could increase the power efficiency of the LDO. The $I_{quiescent}$ is the current that the LDO consumes except the output current. The simulation of the $I_{quiescent}$ is shown in Fig.7.

The table I shows the $I_{quiescent}$ of different output current of the simulation. The simulation show that the power efficient is 96.5% when the output current is 30mA and the power efficient is 96.3% when the output current is 70mA.

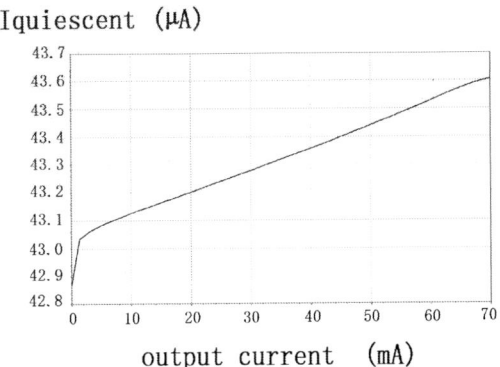

Fig.7 Simulation result of the $I_{quiescent}$ with different output current

TABLE I. $I_{QUIESCENT}$ WITH DIFFERENT OUTPUT CURRENT

output current (mA)	$I_{quiescent}$ (μA)
10	43.12
20	43.2
30	43.28
40	43.36
50	43.33
60	43.53
70	43.61

V. MEASUREMENT RESULTS

The Micrograph of the LDO circuit in this paper is shown in Fig.8. And this LDO circuit is implemented in 0.18μm 1P6M COMS process. The active core area of this LDO is 0.46mm×0.7mm and the whole area is 0.86mm×1.1mm. To reduce the influence of the parasitic resistance there are five pads for the VDD terminal, five pads for the V_{OUT} terminal and three pads for the GND terminal of the LDO layout.

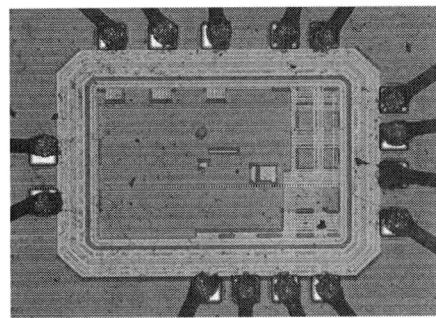

Fig.8 Micrograph of the LDO

The test circuit is shown in Fig.9. The capacitor C_5 is 2.2μF and the load capacitor C_{OUT} is 2.64μF. The R_L changes from high resistance value to low resistance value. In this way, the output current can rise from low current value to high current value. During the changes, the output voltages have been measured. The quiescent current is 0.04mA when there is no output current and the VDD voltage is 3.30V.

The table II show the output voltage values with different output current when the VDD voltage is 3.30V. The measurement results of the LDO circuit are tested without trimming. When there is no output current and the VDD voltage is 3.30V, the test value of the output voltage of the LDO is 3.19V.

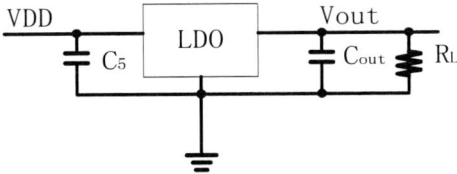

Fig.9 Test circuit

The Fig.10, Fig.11, Fig.12 and Fig.13 show the output voltage values with the different VDD voltage values. The measurement results show that the dropout voltage of this LDO is quite low. When the VDD voltage is 3.20 and the R_L

978-1-4577-1608-9/11 $26.00 © 2011 IEEE   19

is 3.94kΩ (output current=0.8mA) the output voltage could still reach 3.19V.

TABLE II.   OUTPUT VOLTAGE WITH DIFFERENT OUTPUT CURRENT

VDD=3.30V	
output current (mA)	output voltage (V)
no output current	3.19
0.1	3.19
0.8	3.19
7	3.19
70	3.18

output voltage (V)

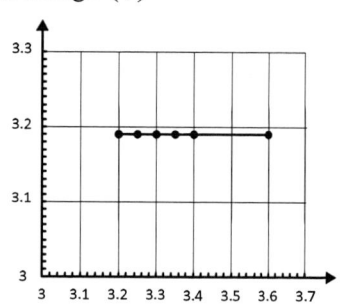

VDD voltage (V)
Fig.10 R_L=26.3kΩ

output voltage (V)

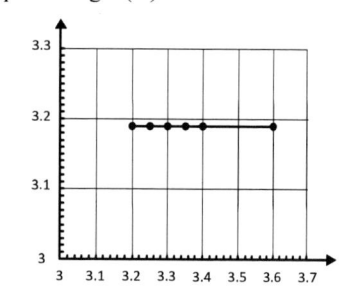

VDD voltage (V)
Fig.11 R_L=3.94kΩ

output voltage (V)

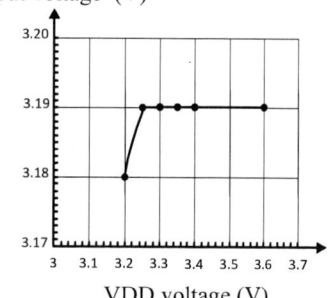

VDD voltage (V)
Fig.12 R_L=0.45kΩ

output voltage (V)

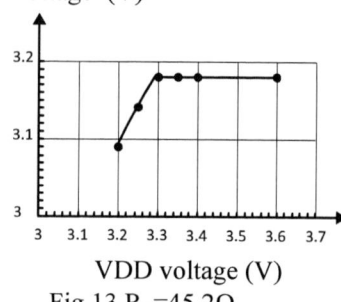

VDD voltage (V)
Fig.13 R_L=45.2Ω

VI.   CONCLUSION

The compensation strategy presented in this paper could provide enough phase margin for the LDO within the design range of the output current and a large range of ESR of the off-chip capacitor. The LDO could convert the power supply from 3.3V to 3.19V, so the dropout voltage is only 0.11V. The low dropout voltage could increase the power efficiency significantly. And the LDO could export 70mA current with 3.18V output voltage, when the VDD voltage is 3.30V. The quiescent current of the circuit is 0.04mA when there is no output current and the VDD voltage is 3.30V.

REFERENCES

[1] G.palmsano, G.palumbo, "An optimized compensation strategy for two-stage CMOS opamplifiers," IEEE Transactions on Circuits and Systems I :vol.42, pp.178–182. March 1995.

[2] Ka Nang Leung and Philip K.T.Mok, "A Capacitor-Free CMOS Low-Dropout Regulator With Damping-Factor-Control Frequency Compensation," IEEE Journal of Solid-State Circuits, vol.38, NO.10, pp. 1691–1702, October 2003.

[3] Behzad Razavi "Design of Analog CMOS Integrated Circuits," McGraw-Hill Companies 2001.

[4] Piero Malcovati, Franco Maloberti, Carlo Fiocchi and Marcello Pruzzi, "Curvature-compensated BiCMOS Bandgap with 1-V Supply Voltage," IEEE Journal of Solid-State Circuits, vol. 36, NO.7, pp. 1076–1081, July 2001.

[5] Hironoti Banba, Hitoshi Shiga, Akira Umezawa, Takeshi Miyaba, Toru Tanzawa, Shigeru Atsumi and Koji Sakui, "A CMOS Bandgap Reference Circuit with Sub-1-V Operation," IEEE Journal of Solid-State Circuits, vol.34, NO.5, pp.670–674 May 1999.

[6] Gabriel A. Rincon-Mora, "Active Capacitor Multiplier in Miller-Compensated Circuits," IEEE Transactions on Solid-State Circuits, vol.35, NO.1, pp. 26–32 January 2000.

[7] Gabriel A. Rincon-Mora and Phillip E. Allen, "A Low-Voltage, Low Quiescent Curren, Low Drop-Out Regulator," IEEE Journal of Solid-State Circuits, vol.33, NO.1, pp.36–44 Janualy 1998.

[8] Gabriel A. Rincon-Mora, "Active Capacitor Multiplier in Miller-Compensated Circuits," IEEE Transactions on Solid-State Circuits, vol.35, NO.1, pp. 26–32 January 2000.

[9] Ka Nang Leung and Philip K.T.Mok, "A CMOS Voltage Reference Based on Weighted ΔV_{GS} for CMOS Low-Dropout Linear Regulators," IEEE Journal of Solid-State Circuits, vol.38, NO.1, pp. 146–150, January 2003.

[10] Sai Kit Lau, Philip K.T.Mok and Ka Nang Leung, "A Low-Dropout Regulator for SoC With Q-Reduction," IEEE Journal of Solid-State Circuits, vol.42, NO.3, pp. 658–664, March 2007.

978-1-4577-1608-9/11 $26.00 © 2011 IEEE

Investigation of LC-Hybrid Active Power Filters in Resonances Prevention and Compensation Capabilities

Chi-Seng Lam[1], Man-Chung Wong[1], Ying-Duo Han[1,2]

1 - Department of Electrical and Computer Engineering, University of Macau, Macau, SAR, P. R. China
2 - Department of Electrical Engineering, Tsinghua University, Beijing, P. R. China
E-mail: cslam@umac.mo

Abstract—This paper presents the harmonic resonances prevention and compensation capabilities of three-phase four-wire center-spilt LC coupling hybrid active power filter (LC-HAPF). Firstly, a single-phase harmonic equivalent circuit model of the LC-HAPF is deduced and built. Based on the circuit model, the LC-HAPF compensation characteristics are studied and analyzed in details, which shows a superior compensation characteristic compared with its pure passive power filter (PPF) part. Finally, simulation results for the pure PPF part and LC-HAPF are given to verify all the analyses.

I. INTRODUCTION

Since the first installation of passive power filters (PPFs) in the mid 1940's, PPFs have been widely used to compensate current quality problems in distribution power systems [1] due to their low cost, simplicity and high efficiency. However, they have disadvantages such as low dynamic performance, resonance problems, etc. [2] – [5]. Since the concept "Active ac Power Filter" was first developed by L. Gyugyi in 1976 [1], [3], the research studies of the active power filters (APFs) are prospering since then. APFs can overcome the disadvantages inherent in PPFs, but their initial costs are relatively high [2] – [4] because the dc-link operating voltage should be higher than the system voltage. In order to lower the cost of APFs, different hybrid active power filter (HAPF) topologies have been proposed. The HAPF topologies in [2] – [5] consist of many passive components, thus increasing the whole system cost. A LC coupling HAPF (LC-HAPF) has been recently proposed for current quality compensation and harmonic damping [6] – [9], because it has less passive components and the dc-link operating voltage can be much lower than the APF.

In this paper, the compensating performances for a LC-HAPF and its PPF part will be studied, analyzed and compared with four evaluation indexes: capabilities to prevent parallel resonance, series resonance, improve the filtering performances and enhance the system robustness. Firstly, a single-phase harmonics equivalent circuit model of a three-phase four-wire center-spilt LC-HAPF is deduced and built. Based on the model, the compensation performances under either LC-HAPF or its pure PPF part operation will be studied. At last, their simulated current quality compensation results will be given to verify all the deduced and analyzed results.

II. ANALYSIS OF LC-HAPF COMPENSATION PERFORMANCES

A. LC-HAPF Single-phase Harmonic Circuit Model

Fig. 1 shows a three-phase four-wire center-spilt LC-HAPF, where the subscript 'x' denotes phase $x = a, b, c, n$. v_{sx} and v_x are the system and load voltage, L_s is the system inductance. i_{sx}, i_{Lx} and i_{cx} are the system, load and inverter current for each phase. C_{c1}, L_{c1} and R_{c1} are the coupling part capacitance, inductance and internal resistance. C_{dc1}, V_{dc1_U} and V_{dc1_L} are the dc capacitance, upper and lower dc capacitor voltages with $V_{dc1_U} = V_{dc1_L} = 0.5 V_{dc1}$. From Fig. 1, the inverter line-to-ground voltages $v_{inv1x-g}$ will be equal to the inverter line-to-neutral voltages $v_{inv1x-n}$ because the neutral point n is connected to the dc-link midpoint g.

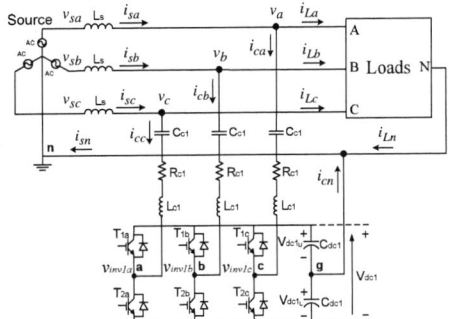

Figure 1. Configuration of a three-phase four-wire center-spilt LC-HAPF

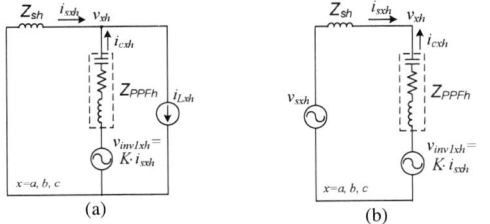

Figure 2. Simplified LC-HAPF single-phase harmonic circuit model: (a) when only i_{Lxh} is considered, (b) when only v_{sxh} is considered

978-1-4577-1608-9/11 $26.00 © 2011 IEEE

Fig. 2 show the simplified LC-HAPF single-phase harmonic circuit model due to loading harmonic current i_{Lxh} and system harmonic voltage v_{sxh}, where the subscript "h" represents harmonic components. The loading and inverter are modeled as current and voltage sources. Z_{sh} and Z_{PPFh} are the harmonic impedance of the system and PPF part. Provided that the inverter is controlled by hysteresis PWM with hysteresis error band $H = 0$, the inverter can be modeled as a current control voltage source.

When $(i_{cxh}* - i_{cxh}) \geq H$, i.e. $(i_{cxh}* - i_{cxh}) \geq 0$,

$$v_{invx1h} = 0.5V_{dc1} = K_1 \cdot (i_{cxh}* - i_{cxh}), \; K_1 > 0 \qquad (1)$$

When $(i_{cxh}* - i_{cxh}) < H$, i.e. $(i_{cxh}* - i_{cxh}) < 0$,

$$v_{invx1h} = -0.5V_{dc1} = K_2 \cdot (i_{cxh}* - i_{cxh}), \; K_2 > 0 \qquad (2)$$

Where v_{invx1h} represents the inverter harmonic output voltage, $i_{cxh}*$ and i_{cxh} represent the reference and actual harmonic compensating currents. Since K_1 and K_2 are both in positive, v_{invx1h} can be expressed into a general form as:

$$v_{invx1h} = K \cdot (i_{cxh}* - i_{cxh}), \; K > 0 \qquad (3)$$

From Fig. 2, $i_{sxh} + i_{cxh} = i_{Lxh}$. In ideal compensation case, $i_{cxh}* = i_{Lxh}$. Thus, v_{invx1h} can also be expressed as:

$$v_{inv1xh} = K \cdot i_{sxh}, \; K > 0 \qquad (4)$$

B. LC-HAPF Harmonic Circuit Model Due to i_{Lxh} only

For the system voltage v_{sx} does not contain harmonic components ($v_{sxh} = 0$), the LC-HAPF single-phase harmonic circuit model due to i_{Lxh} is shown in Fig. 2(a). From Fig. 2(a), the i_{sxh} and i_{cxh} due to i_{Lxh} only can be expressed as:

$$K_{sxh\_i} = \frac{i_{sxh}}{i_{Lxh}} = \frac{Z_{PPFh}}{K + Z_{sh} + Z_{PPFh}} \qquad (5)$$

$$K_{cxh\_i} = \frac{i_{cxh}}{i_{Lxh}} = \frac{K + Z_{sh}}{K + Z_{sh} + Z_{PPFh}} \qquad (6)$$

In a perfect compensation, $K_{sxh\_i} = 0$ and $K_{cxh\_i} = 1$ should be achieved so that all the load harmonic current flows into the LC-HAPF ($i_{cxh} = i_{Lxh}$). In order to achieve this objective, K should be a large value.

C. LC-HAPF Harmonic Circuit Model Due to v_{sxh} only

For the load current i_{Lx} does not contain harmonic components ($i_{Lxh} = 0$), the LC-HAPF single-phase harmonic circuit model due to v_{sxh} is shown in Fig. 2(b). From Fig. 2(b), the i_{sxh} and i_{cxh} due to v_{sxh} only can be expressed as:

$$K_{sxh\_v} = \frac{i_{sxh}}{v_{sxh}} = \frac{1}{K + Z_{sh} + Z_{PPFh}} \qquad (7)$$

$$K_{cxh\_v} = \frac{i_{cxh}}{v_{sxh}} = -\frac{1}{K + Z_{sh} + Z_{PPFh}} \qquad (8)$$

Similarly, in order to achieve $K_{sxh\_v} = 0$, K should also be a large value. From (5) – (8), when only the pure PPF part is employed, $K = 0$.

D. Investigation of LC-HAPF Compensation Performances

In the following, the LC-HAPF steady-state compensating performances are studied and discussed with four evaluation indexes, compared with those of the pure PPF part. Table I shows a set of LC-HAPF system parameters for the analyses.

TABLE I. A SET OF LC-HAPF SYSTEM PARAMETERS

System Parameters	Physical Values
v_x	220V$_{rms}$
L_s, L_{c1}	1mH, 5mH
C_{c1}, R_{c1}	80μF, 0Ω
V_{dc1_U}, V_{dc1_L}	50V

1) LC-HAPF Capability to Prevent Parallel Resonance

Fig. 3 shows the K_{sxh_i} (5) diagram with respect to different frequency and L_s when the PPF part or LC-HAPF is utilized. When only pure PPF is used ($K = 0$), the harmonic current amplification phenomenon occurs. When the LC-HAPF is employed, the parallel resonance phenomenon disappears. Fig. 3 shows that the LC-HAPF has the ability to prevent the parallel resonance inherent in the pure PPF part.

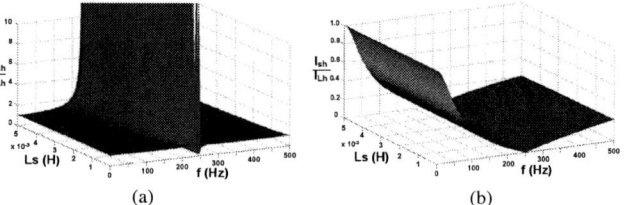

Figure 3.   Capability to prevent parallel resonance: (a) only PPF part is utilized ($K = 0$), (b) LC-HAPF is employed ($K = 50$)

2) LC-HAPF Capability to Prevent Series Resonance

Fig. 4 shows the K_{cxh_v} (8) diagram with respect to different frequency and L_s when the PPF part or LC-HAPF is utilized. When only pure PPF is used ($K = 0$), the harmonic current amplification phenomenon occurs. When the LC-HAPF is employed, the series resonance phenomenon disappears. Fig. 4 shows that the LC-HAPF has the ability to prevent the series resonance inherent in the pure PPF part.

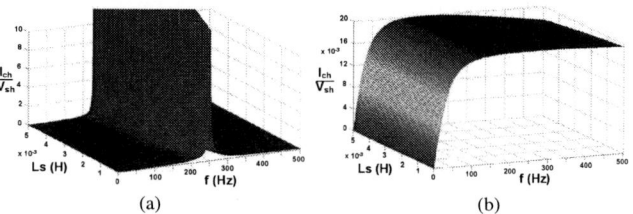

Figure 4.   Capability to prevent series resonance: (a) only PPF part is utilized ($K = 0$), (b) LC-HAPF is employed ($K = 50$)

3) LC-HAPF Capability to Improve Filtering Performances

Fig. 5 shows the bode diagrams of K_{sxh_i} (5) and K_{sxh_v} (7) with respect to different K. When only pure PPF is used ($K = 0$), the harmonic current amplification phenomenon occurs at $\omega = 1120$ rad/s. When the LC-HAPF is employed

($K = 25$ or $K = 50$), K_{sxh_i} and K_{sxh_v} will have a larger attenuation at different harmonic frequencies. Fig. 5 shows the LC-HAPF is capable to improve the filtering performances.

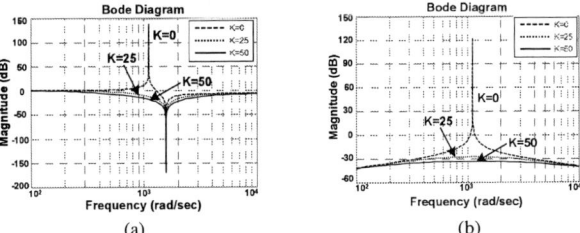

(a)   (b)

Figure 5.   Capability to improve the filtering performances of the PPF part ($K = 0$, $K = 25$, $K = 50$) due to: (a) i_{Lxh} , (b) v_{sxh}

4) LC-HAPF Capability to Enhance System Robustness

Fig. 6 shows the bode diagrams of K_{sxh_i} (5) with respect to different L_s. When only pure PPF is used ($K = 0$), the harmonic current amplification phenomenon will move to the lower frequency side as L_s increases. When the LC-HAPF is used ($K = 50$), the harmonic current amplification effect disappears no matter what L_s value is. Also, its harmonic compensating characteristics does not vary at all. Fig. 6 shows the LC-HAPF is capable to enhance the system robustness.

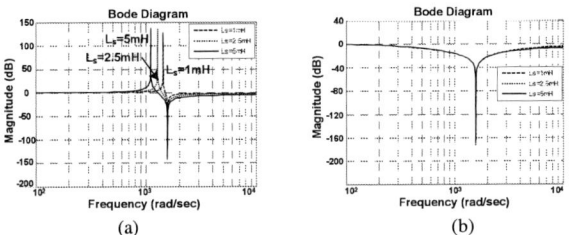

(a)   (b)

Figure 6.   Capability to enhance the system robustness: (a) only PPF part is utilized ($K = 0$), (b) LC-HAPF is employed ($K = 50$)

III. SIMULATION VERIFICATION

In this section, simulation results are included to illustrate and verify the previous LC-HAPF compensation performance analyses. Simulation studies were carried out using PSCAD/EMTDC. Table I shows the LC-HAPF simulated system parameters for balanced loading current quality compensation. Based on the instantaneous power theory [10], the control block diagram for the LC-HAPF is shown in Fig. 7.

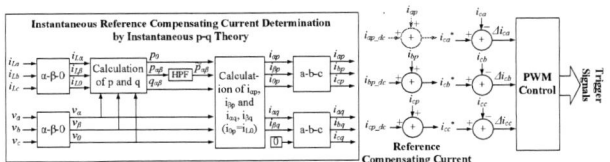

Figure 7.   Control block diagram for the three-phase four-wire LC-HAPF

1) LC-HAPF Capability to Prevent Parallel Resonance

Fig. 8 shows the capability to prevent parallel resonance phenomenon: (a) only PPF part is utilized, (b) LC-HAPF is employed. When L_s increases to 5mH, the parallel resonance will occur close to the 3rd order harmonic frequency as shown in Fig. 3(a). Fig. 8(a) shows that the pure PPF compensation amplifies the 3rd order harmonic from 2.00A$_{rms}$ to 4.59A$_{rms}$. This phenomenon deteriorates the compensation results as shown in Table II, in which $THD_{i_{sx}}$ and THD_{v_x} do not satisfy the international standards [11]–[13]. When the LC-HAPF is employed, Fig. 8(b) and Table II show that the LC-HAPF can obtain good compensation results without parallel resonance, which verified the parallel resonance analysis as in Fig. 3.

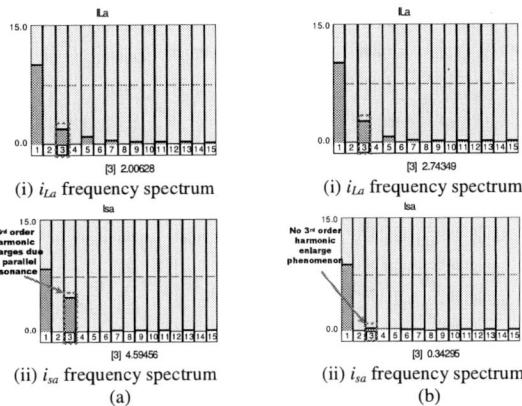

(i) i_{La} frequency spectrum   (i) i_{La} frequency spectrum

(ii) i_{sa} frequency spectrum   (ii) i_{sa} frequency spectrum

(a)   (b)

Figure 8.   Capability to prevent parallel resonance: (a) only PPF part is utilized, (b) LC-HAPF is employed

2) LC-HAPF Capability to Prevent Series Resonance

Fig. 9 shows the capability to prevent series resonance phenomenon: (a) only PPF part is utilized, (b) LC-HAPF is employed. Since L_{c1} and C_{c1} are tuned at 5th order, the series resonance will occur when 5th order system harmonic voltage presents, as shown in Fig. 4(a). When 4% of 5th order system harmonic voltage is added, Fig. 9(a) shows that the pure PPF operation will yield a series resonance with 5th order harmonic content increases from 0.74A$_{rms}$ to 5.94A$_{rms}$. This phenomenon deteriorates the compensation results as illustrated in Table II, in which $THD_{i_{sx}}$ and THD_{v_x} do not satisfy the international standards [11]–[13]. When the LC-HAPF is employed, Fig. 9(b) and Table II show that the LC-HAPF can obtain good compensation results without series resonance, which verified the series resonance analysis as in Fig. 4.

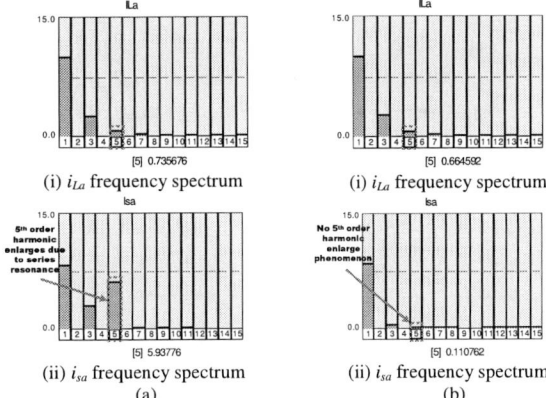

(i) i_{La} frequency spectrum   (i) i_{La} frequency spectrum

(ii) i_{sa} frequency spectrum   (ii) i_{sa} frequency spectrum

(a)   (b)

Figure 9.   Capability to prevent series resonance: (a) only PPF part is utilized, (b) LC-HAPF is employed

978-1-4577-1608-9/11 $26.00 © 2011 IEEE   23

3) LC-HAPF Capability to Improve Filtering Performances

Fig. 10 shows the capability to improve the filtering performances: (a) only PPF part is utilized, (b) LC-HAPF is employed. From Fig. 10(a), when only PPF is employed, the i_{sx} fundamental and 5th order harmonic content have been reduced. But it yields a larger $THD_{i_{sx}}$ of 37.0% as shown in Table II, in which $THD_{i_{sx}}$ does not satisfy the international standards [11]–[13]. When the LC-HAPF is employed, Fig. 10(b) and Table II show that the LC-HAPF can obtain good compensation performances, which verified the filtering improvement analysis as in Fig. 5.

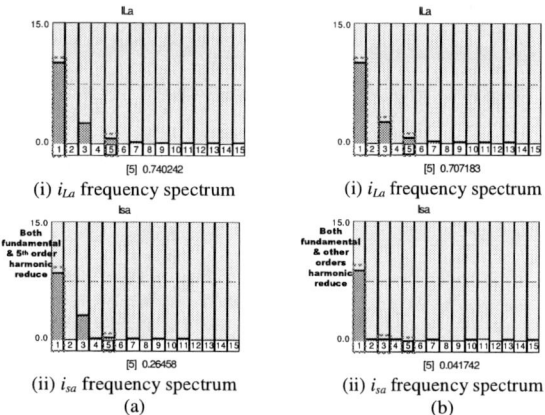

(i) i_{La} frequency spectrum   (i) i_{La} frequency spectrum

(ii) i_{sa} frequency spectrum   (ii) i_{sa} frequency spectrum

(a)   (b)

Figure 10. Capability to improve the filtering performances of the PPF part: (a) only PPF part is utilized, (b) LC-HAPF is employed

4) LC-HAPF Capability to Enhance System Robustness

Fig. 11 shows the capability to enhance the system robustness: (a) only the PPF part is utilized, (b) LC-HAPF is employed. When L_s changes to 3mH, the compensating characteristics are being changed as shown in Fig. 6(a). Fig. 11(a) shows that the PPF part amplifies the i_{sx} 3rd order harmonic content. This deteriorates the compensation results, in which $THD_{i_{sx}}$ and THD_{v_x} do not satisfy the standards [11]– [13]. When the LC-HAPF is employed, Fig. 11(b) and Table II show the LC-HAPF can perform good compensation, which verified the system robustness analysis as in Fig. 6.

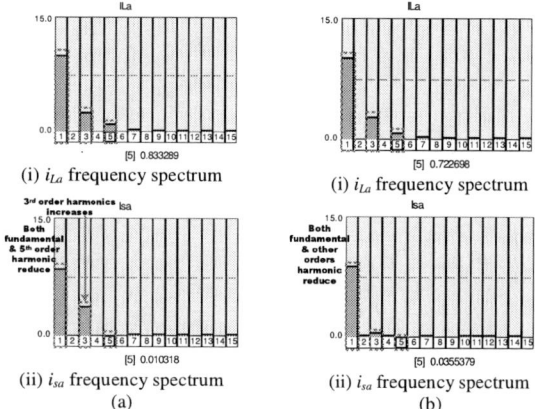

(i) i_{La} frequency spectrum   (i) i_{La} frequency spectrum

(ii) i_{sa} frequency spectrum   (ii) i_{sa} frequency spectrum

(a)   (b)

Figure 11. Capability to enhance the system robustness: (a) only PPF part is utilized, (b) LC-HAPF is employed

TABLE II.   SIMULATION RESULTS BEFORE AND AFTER PPF AND LC-HAPF COMPENSATIONS

Capabilities		Before Compensation			After Compensation			
		$THD_{i_{sx}}$ (%)	i_{sn} (A$_{rms}$)	DPF	$THD_{i_{sx}}$ (%)	i_{sn} (A$_{rms}$)	DPF	THD_{v_x} (%)
Parallel Resonance	PPF	22.0	5.84	0.845	54.5	13.83	0.998	10.5
	LC-HAPF	30.0	8.56	0.838	5.6	1.16	1.000	4.8
Series Resonance	PPF	30.0	8.10	0.838	>75.0	9.25	0.999	5.2
	LC-HAPF	30.0	8.56	0.838	8.5	1.17	1.000	4.6
Improve Filtering	PPF	30.0	8.10	0.838	37.0	9.30	1.000	2.1
	LC-HAPF	30.0	8.56	0.838	5.0	0.90	1.000	2.2
System Robustness	PPF	25.4	6.86	0.837	44.0	10.20	0.999	5.5
	LC-HAPF	30.0	8.56	0.838	5.2	1.05	1.000	4.0

IV. CONCLUSION

In this paper a single-phase harmonic circuit model of a three-phase four-wire center-spilt LC-HAPF is deduced and built. Based on the model, the steady-state compensating performances for pure PPF part and LC-HAPF are discussed and analyzed with four evaluation indexes, capabilities to prevent parallel resonance and series resonance, improve the filtering performances and enhance the system robustness. It is clearly illustrated that the LC-HAPF has the capabilities to prevent parallel and series resonance phenomena inherent in pure PPF part, improve the filtering effects and enhance the system robustness of the PPF part, in which all the deduced and analyzed results are verified by simulations.

REFERENCES

[1] S. T. Senini, and P.J. Wolfs, "Systematic identification and review of hybrid active filter topologies," in *Proc. IEEE 33rd Annual Power Electronics Specialists Conf., PESC. 02*, vol. 1, 2002, pp. 394–399.

[2] H. Fujita, and H. Akagi, "A practical approach to harmonic compensation in power systems – series connection of passive and active filters," *IEEE Trans. Ind. Applicat.*, vol. 27, pp. 1020–1025, Nov./Dec. 1991.

[3] F. Z. Peng, H. Akagi, and A. Nabae, "A new approach to harmonic compensation in power systems – a combined system of shunt passive and series active filters," *IEEE Trans. Ind. Applicat.*, vol. 26, pp. 983–990, Nov./Dec. 1990.

[4] S. Park, J.-H. Sung, and K. Nam, "A new parallel hybrid filter configuration minimizing active filter size," in *Proc. IEEE 30th Annual Power Electronics Specialists Conf., PESC. 99*, vol. 1, 1999, pp. 400–405.

[5] D. Rivas, L. Moran, J.W. Dixon, et al., "Improving passive filter compensation performance with active techniques," *IEEE Trans. Ind. Electron.*, vol. 50, pp. 161–170, Feb. 2003.

[6] S. Srianthumrong, H. Akagi, "A medium-voltage transformerless AC/DC Power conversion system consisting of a diode rectifier and a shunt hybrid filter," *IEEE Trans. Ind. Applicat.*, vol.39, pp.874 – 882, May/Jun. 2003.

[7] W. Tangtheerajaroonwong, T. Hatada, K. Wada, H. Akagi, "Design and performance of a transformerless shunt hybrid filter integrated into a three-phase diode rectifier," *IEEE Trans. Power Electron.*, vol. 22, pp. 1882–1889, Sept. 2007.

[8] H. -L. Jou, K. -D. Wu, J.- C. Wu, C. -H. Li, M. -S. Huang, "Novel power converter topology for three phase four-wire hybrid power filter," *IET Power Electron.*, vol.1, pp. 164 – 173, 2008.

[9] R. Inzunza, H. Akagi, "A 6.6-kV transformerless shunt hybrid active filter for installation on a power distribution system," *IEEE Trans. Power Electron.*, vol. 20, pp. 893 – 900, Jul. 2005.

[10] H. Akagi, S. Ogasawara, Kim Hyosung, "The theory of instantaneous power in three-phase four-wire systems: a comprehensive approach," *in Conf. Rec. IEEE-34th IAS Annu. Meeting*, 1999, vol. 1, pp. 431–439.

[11] IEEE Recommended Practices and Requirements for Harmonic Control in Electrical Power Systems, 1992, IEEE Standard 519-1992.

[12] IEEE Recommended Practice on Monitoring Electric Power Quality, 1995, IEEE Standard 1159:1995.

[13] Electromagnetic Compatibility (EMC), Part 3: Limits, Section 2: Limits for Harmonics Current Emissions (Equipment Input Current ≤16A Per Phase), IEC Standard 61000-3-2, 1997.

978-1-4577-1608-9/11 $26.00 © 2011 IEEE

A FPGA-Based Power Electronics Controller for Hybrid Active Power Filters

Bo Sun[1], U-Fat Chio, Chi-Seng Lam[1], Ning-Yi Dai[1], Man-Chung Wong[1], Chi-Kong Wong[1], Sai-Weng Sin, Seng-Pan U, R. P. Martins[2]

State Key Laboratory of Analog and Mixed Signal VLSI (http://www.fst.umac.mo/en/lab/ans_vlsi/website/index.html)
Faculty of Science and Technology, University of Macau, Macao, China
1 – Also with the Power Electronics Laboratory
2 – On leave from Instituto Superior Técnico/TU of Lisbon, Portugal
E-mail: ma76571@umac.mo

Abstract—**In this paper, a power electronics controller implemented on one field-programmable gate array (FPGA) chip is proposed. The FPGA-based power electronics controller integrates the whole signal processing system including synchronous reference frame (SRF) algorithm, decoupled double synchronous reference frame phase-locked loop (DDSRF-PLL) and hysteresis pulse-width modulation (PWM). It is applicable to three-phase four-wire hybrid active power filters (HAPFs), which use four-leg voltage source inverters (VSIs). Different from the conventional controller using instantaneous reactive power theory, the proposed controller could work when source voltages are unbalanced. The bit width and I/O ports of the FPGA-based controller are user-defined. Hence, the proposed FPGA-based controller is more flexible than those using digital signal processors (DSPs). Parallel processing in FPGA could also achieve a faster control. The prototype of a three-phase HAPF is built and the proposed FPGA-based controller is adopted. Experimental results are provided to show its validity.**

I. INTRODUCTION

Compared with DSP, FPGA has flexibilities in defining bit-widths and I/O ports. Short bit width saves hardware resource and processing time. Long bit width guarantees accuracy. Rich I/O ports mean FPGA-based system can be developed to control power electronics device requires more switching signals since one switching signal is generated by one port. Parallel processing makes FPGA have high process efficiency. The design in FPGA is mapped into actual circuit and the processing is clock triggered. Precision will be better.

The work of this paper is implementing a FPGA-based power electronics controller and applying it in a prototype of three-phase four-wire HAPF. The compensation using FPGA-based power electronics controller is applicable when source voltages are unbalanced, because the compensation current detection algorithm is modified. Many theories have been developed since the concept of instantaneous reactive power

theory for calculating compensation currents was established. Generalized instantaneous reactive power theory [1][2] and compensation by using pq theory [3] are only applicable when source voltages are balanced without harmonics. Compensation by using pqr theory [4] and SRF theory [5] with conventional SRF-PLL can provide accurate compensation currents only when source voltages are balanced with or without harmonics. However, these typical theories cannot calculate accurate compensation currents when source voltages are unbalanced and distorted. DDSRF-PLL [6] is used to replace SRF-PLL in SRF algorithm so that the compensation currents obtained are accurate no matter source voltages are balanced or unbalanced and distorted. In this paper, the operation principles of the control system is introduced in section II. Its implementation on FPGA is discussed in section III. Experimental results in section IV are provided to show the validity.

II. OPERATION PRINCIPLES

Figure 1 shows the configuration of compensation system using three-phase four-wire HAPF. The main circuit of the HAPF is a two-level four-leg VSI. The block diagram of signal processing system in FPGA is given in Figure 2. The source voltages (v_{sa}, v_{sb}, v_{sc}), load currents (i_{la}, i_{lb}, i_{lc}) and injected currents (i_{ca}^*, i_{cb}^*, i_{cc}^*) should be detected from circuit.

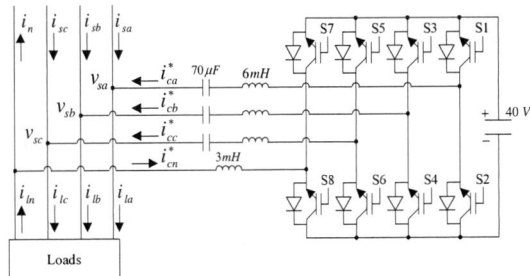

Figure 1. Block diagram of a three-phase four-wire system with HAPF

978-1-4577-1608-9/11 $26.00 © 2011 IEEE

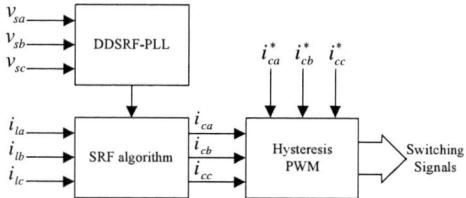

Figure 2. Block diagram of signal processing system in FPGA

SRF algorithm is used for calculating compensation currents. The d and q components of the load currents in the block diagram of SRF algorithm [5] can be obtained by

$$\begin{bmatrix} i_d \\ i_q \end{bmatrix} = [T][C] \begin{bmatrix} i_{la} \\ i_{lb} \\ i_{lc} \end{bmatrix}. \qquad (1)$$

Where

$$[C] = \begin{bmatrix} \sqrt{2/3} & -\sqrt{1/6} & -\sqrt{1/6} \\ 0 & \sqrt{1/2} & -\sqrt{1/2} \end{bmatrix}, \qquad (2)$$

$$[T] = \begin{bmatrix} cos(\theta) & sin(\theta) \\ -sin(\theta) & cos(\theta) \end{bmatrix}. \qquad (3)$$

When the imbalance, reactive power and harmonics in a three-phase system with or without the neutral current should be fully compensated, the three-phase compensation currents are expressed as

$$\begin{bmatrix} i_{ca} \\ i_{cb} \\ i_{cc} \end{bmatrix} = \begin{bmatrix} i_{la} \\ i_{lb} \\ i_{lc} \end{bmatrix} - [C]^T [T] \begin{bmatrix} \bar{i}_d \\ 0 \end{bmatrix}. \qquad (4)$$

The compensation current of the neutral is equal to

$$i_{cn} = -i_{ca} - i_{cb} - i_{cc}. \qquad (5)$$

Based on (1)-(5), the accuracy of the compensation currents are determined by the phase angle provided. DDSRF-PLL can provide instantaneous phase angle of positive sequence under unbalanced and distorted case. It is applied for SRF algorithm. The accuracy of the compensation currents will be not affected by source voltages.

In order to control the inverter output current tracking with reference one, hysteresis PWM [7] is used to obtain the switching signals by comparing the compensation current to the tolerance bands. Thus, the switching signals are obtained.

III. DESIGN AND IMPLEMENTATION

The whole signal processing system is implemented on XILINX XC3SD1800A Spartan-3A DSP FPGA. It has 1800K

system gates or 37440 equivalent logic cells and 519 maximum user I/O ports. The FPGA control board used is XtremeDSP Starter Platform. The clock period adopted is 200 ns which is generated by 125 MHz oscillator. An analog-to-digital converter (ADC) embedded on TDS2812EVMB is employed to detect 9-channel signals from the circuit and transfer the signals to FPGA. The sample period is equal to 50 µs. The FPGA-based power electronics controller adopted 16 input ports to receive signals and 8 output ports to generate switching signals for the HAPF.

A. Communication from DSP to FPGA

The communication from DSP to FPGA is transmitting the 9-channel signals. And serial transmission is adopted. In order to identify the 9-channel signals, 4-bit address is also transmitted. Since each channel signal is expressed by 12 bits, DSP transmits 16 bits every time including 12-bit data and 4-bit address by 16 GPIO ports. The 16 GPIO ports hold each signal 1400 ns. When the transmission is over, the states of 16 GPIO ports hold low voltages. 16 ports of FPGA are defined to receive the 12-bit data and 4-bit address. They sample the signals of GPIO ports every 200 ns. The 12-bit data is ensured to be correct if only the 5 sequential samples of the 4-bit address are identical. FPGA will achieve series to parallel conversion of the 9-channel signals for further parallel processing. Moreover, the 9-channel signals are expressed by 16 bits with Q15 format in FPGA.

B. SRF Algorithm

All the constants or variables in SRF algorithm are expressed by 16 bits with Q15 format. Parallel processing is carried out, so the three-phase and neutral compensation currents are obtained at the same time.

In SRF algorithm, the dc component is extracted by a second-order low-pass filter (LPF). It is cascaded by 2 identical first-order butterworth infinite impulse response (IIR) LPF which cut-off frequency is 40 Hz. Its transfer function is

$$H(z) = \frac{0.0062 + 0.0062 z^{-1}}{1 - 0.9875 z^{-1}}. \qquad (6)$$

C. DDSRF-PLL

FPGA exhibits the flexibility in arranging bit width herein although most constants and variables in DDSRF-PLL can be expressed by 16 bits with Q15 format. LPF, loop filter (LF), voltage-controlled oscillator (VCO) and look-up table (LUT) in the block diagram of DDSRF-PLL shown in Figure 3 are introduced and emphasized.

In the decoupling network of DDSRF-PLL, 4 LPFs are required. And the first-order LPF introduced in SRF algorithm is applied in decoupling network. All filters' parameters are expressed by 16 bits with Q15 format.

LF is a proportional-integral control. The integral operation is approximated by

$$Tz/(z-1). \qquad (7)$$

978-1-4577-1608-9/11 $26.00 © 2011 IEEE   26

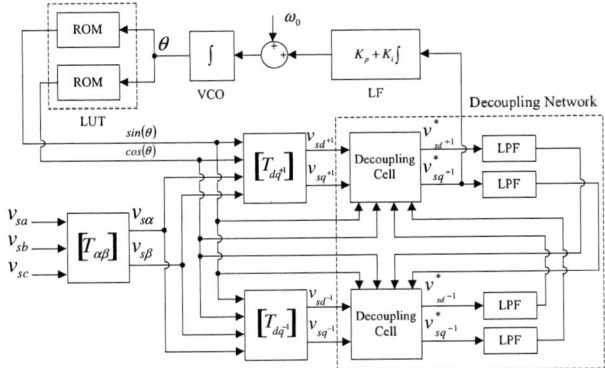

Figure 3.   Block diagram of DDSRF-PLL

T is the sample period. Besides, the constants K_i and K_p in LF satisfies

$$\xi = 0.5 K_p \big/ \sqrt{K_i} \approx 0.707. \qquad (8)$$

Low K_i improves the performance of noise rejection, but dynamic response would be weakened [8]. Constants $K_i=(40\pi)^2$ and $K_p=1.414\times40\pi$ are chosen in the design. In order to avoid overflow happens, the processing result in the parallel path of proportional-integral control is expressed by more than 16 bits. The processing result of proportional-integral control is expressed by 24 bits and Q15 format.

VCO is also an integral operation in digital system. And its output is phase angle. VCO will be reset once its output is equal to or larger than 2π. Thus, overflow is avoided. And its output is expressed by 18 bits and Q15 format.

LUT with 16384 values is adopted to retrieve the corresponding sine or cosine value from a memory address. All the values in LUT are expressed by 16 bits with Q15 format. They are created by MATLAB and stored in the ROMs.

D. Hysteresis PWM

In hysteresis PWM, current is compared with two tolerance bands. If the actual compensation current is larger than the upper band, the upper switch in Figure 1 is turned off. If the actual compensation current is less than the lower band, the upper switch in Figure 1 is turned on. Furthermore, the state of corresponding lower switch is always reverse with the state of the upper switch. It is not difficult to implement hysteresis PWM by comparison operation. And 8 ports are used for generating switching signals. Since FPGA has rich I/O ports, the FPGA-based power electronics controller can be further developed to control multi-level VSIs with more switches.

Dead-time should be provided in practice for protection. The dead-time in the design is 8 μs. The dead-time is held when the state of the switch is changed. Once the 2 sequential states of the same switch are different, dead-time operation is triggered.

IV.   Experimental Results

The main circuit of the HAPF is a four-leg IGBT VSI in experiment. The configuration of compensation system using three-phase four-wire HAPF in experiment is the same as Figure 1.

A. Source Voltages Are Balanced

The waveforms of three-phase source voltages and three-phase source currents before and after compensation in experiment are both shown in Figures 4 and 5, respectively. Experimental results listed in Table I indicate power factor (PF) of three phases (A, B, C), total harmonic distortion (THD) of three phases (A, B, C) and root mean square (RMS) value of neutral current are improved obviously.

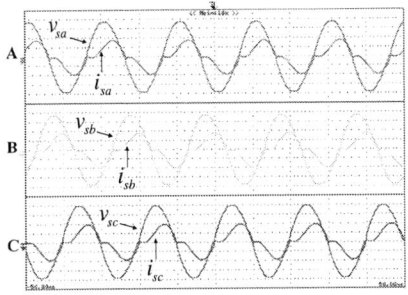

Figure 4.   Waveforms of three-phase source voltages and three-phase source currents before compensatoin

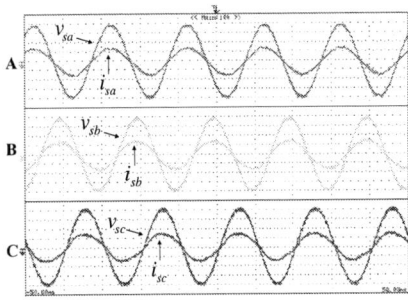

Figure 5.   Waveforms of three-phase source voltages and three-phase source currents after compensatoin

TABLE I.   Experimental Results

		Source Currents Before Compensation	Source Currents After Compensation
PF	A	0.83	1.00
	B	0.86	0.99
	C	0.83	0.99
THD	A	23.4%	7.0%
	B	22.4%	6.0%
	C	23.7%	7.2%
Neutral		1.54443 A	372.544 mA

978-1-4577-1608-9/11 $26.00 © 2011 IEEE

B. Source Voltages Are Unbalanced and Distorted

The imbalance and harmonics result from a resistor cascaded at the source side of phase A. The voltage drop in phase A is 30%. Besides, the maximum THD of source voltages is 9.7% in experiment. The waveforms of three-phase source voltages and three-phase source currents before and after compensation in experiment are shown in Figures 6 and 7, respectively. Experimental results listed in Table II indicate PF of three phases (A, B, C), THD of three phases (A, B, C), current unbalance and RMS value of neutral current are improved.

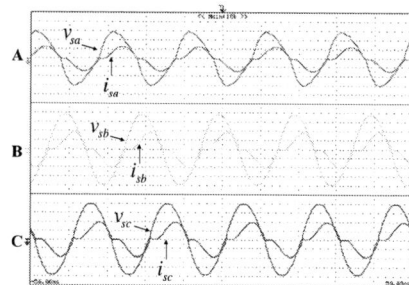

Figure 6.   Waveforms of three-phase source voltages and three-phase source currents before compensatoin

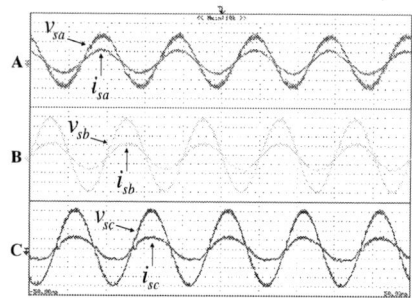

Figure 7.   Waveforms of three-phase source voltages and three-phase source currents after compensatoin

TABLE II.   EXPERIMENTAL RESULTS

		Source Currents Before Compensation	Source Currents After Compensation
PF	A	0.84	0.99
	B	0.86	1.00
	C	0.83	0.99
THD	A	18.6%	5.0%
	B	22.5%	5.7%
	C	23.4%	9.8%
Current Unbalance		18.18%	9.74%
Neutral		1.50406 A	357.348 mA

V.   CONCLUSION

In this paper, a FPGA-based power electronics controller is implemented on XILINX XC3SD1800A Spartan-3A DSP FPGA. Design and implementation are introduced. The application of DDSRF-PLL makes compensation currents obtained by SRF algorithm be accurate no matter source voltages are balanced or unbalanced, with or without harmonics. The validity is proved by experimental results. Moreover, the FPGA-based design possesses many advantages. Bit width for variables or constants is optimized on the basis of accuracy and efficiency. 24 I/O ports are occupied. 16 input ports are used for receiving signals detected from circuit and 8 output ports are used for generating 8 switching signals. Parallel processing is also widely adopted in the design. The proposed controller can generate switching signals within half a sample period 50 μs. FPGA is an advisable choice for power electronics applications.

ACKNOWLEDGMENT

The authors would like to thank the Macao Science and Technology Development Fund (FDCT) and the University of Macau for their financial support.

REFERENCES

[1] Fang Zheng Peng; Ott, G. W., Jr.; Adams, D. J., "Harmonic and reactive power compensation based on the generalized instantaneous reactive power theory for three-phase four-wire systems," IEEE Transactions on Power Electronics, vol. 13, no. 6, pp. 1174–1181, November 1998.

[2] Fang Zheng Peng and Jih-Sheng Lai, "Generalized instantaneous reactive power theory for three-phase power systems," IEEE Transactions on Instrumentation and Measurement, vol. 45, no. 1, pp. 293–297, February 1996.

[3] Aredes, M.; Akagi, H.; Watanabe, E.H.; Vergara Salgado, E.; Encarnacao, L.F., "Comparisons Between the p-q and p-q-r Theories in Three-Phase Four-Wire Systems," IEEE Transactions on Power Electronics, vol. 24, no. 4, pp. 924–933, April 2009.

[4] Hyosung Kim; Blaabjerg, F.; Bak-Jensen, B and Jaeho Choi, "Instantaneous power compensatoin in three-phase systems by using p-q-r theory," IEEE Transactions on Power Electronics, vol. 17, no. 5, pp. 701–710, September 2002.

[5] da Silva, S.A.O.; Donoso-Garcia, P.F.; Cortizo, P.C.; Seixas, P.F., "A Three-Phase Line-Interactive UPS System Implementation With Series-Parallel Active Power-Line Conditioning Capabilities", IEEE Transactions on Industrial Applications, vol. 38, no. 6, pp. 1581-1590, November/December 2002

[6] Rodriguez, P.; Pou, J.; Bergas, J.; Candela, J.I.; Burgos, R.P. and Boroyevich, D., "Decoupled Doubled Synchronous Reference Frame PLL for Power Converters Control," IEEE Transactions on Power Electronics, vol. 22, no. 2, pp. 584–592, March 2007

[7] Le-Huy, H.; Dessaint, L.A., "An Adaptive Current Control Scheme for PWM Synchronous Motor Drives: Analysis and Simulation", IEEE Transactions on Power Electronics, vol. 4, no. 4, pp. 486–495, October 1989

[8] L Rolim, L.G.B.; da Costa, D.R.; Aredes, M., "Analysis and Software Implementation of a Robust Synchronizing PLL Circuit Based on the pq Theory", IEEE Transactions on Industrial Electronics, vol. 53, no. 6, pp. 1919-1926, December 2006

An Algorithm for the Design of
Low Power Linear Phase FIR Filters

Wen Bin Ye *and* Ya Jun Yu
School of Electrical and Electronic Engineering
Nanyang Technological University, Singapore 639798
yewe0003@e.ntu.edu.sg, eleyuyj@pmail.ntu.edu.sg

Abstract— **In this work, a novel algorithm is proposed for the design of low power linear phase finite impulse response (FIR) filters with discrete coefficients. The proposed algorithm finds an adder and shift network using a small logic depth to realize the coefficient multipliers. This is achieved by constraining the synthesis of each coefficient on its minimum adder depth. The new synthesis mechanism tends to use a few more adders to successfully realize the discrete filter coefficients than the current best algorithms, but with lower average logic depth, resulting in a low power implementation. Designs of benchmark filters and simulation results verified the above claims.**

I. INTRODUCTION

Finite impulse response (FIR) filters have many applications in digital signal processing systems for their guaranteed stability and linear phase property. However, compared with infinite impulse response (IIR) filters, FIR filters usually suffer from high computational complexity, and therefore, consume more power, if the same magnitude response is realized. In the synthesis of FIR filters, the transposed direct form as shown in Fig. 1 is a popular structure due to its inherent pipeline realization. Moreover, in the transposed direct form, the input signal is multiplied by a set of constant coefficients, such that the coefficients multipliers can be replaced by a network of adders and shifts (as show in Fig. 1(b)), an operation named as multiple constant multiplications (MCM) [1]–[9]. As the shifts for fixed coefficients can be hardwired, and thus are considered cost-free, achieving the minimum number of adders/subtractors to realize the filters becomes the main objective in the low complexity MCM design [5], [10]–[15]. While low power is considered, the maximum and average logic depths in the MCM network become important concerns on top of the number of adders.

In FIR filter designs the MCM algorithms could either be applied on a given set of discrete coefficients [1]–[9], or be incorporated with the optimization search of the discrete coefficients for a given filter specification [10]–[15]. Researches have shown that the later one generally generates much better results than the former one, in terms of number of adders.

In the techniques incorporating the MCM algorithms and discrete coefficient search, a subexpression search space is proposed in [12], making the design of filters with practical length possible. A subsequent research [14] introduced a dynamically expending subexpression space, achieving more flexibility in the search and synthesis of discrete coefficients. In the most recent published technique [15], an optimum-aware algorithm was proposed, in which the designs synthesized with the minimum number of adders are achieved with high probability; in addition, maximum logic depth constraints could be imposed on the design for the low power consideration. While many MCM algorithm applied on a given set of discrete coefficients have considered the low average logic depth [2], [16], [17], discrete coefficient search and optimization algorithms [10]–[14] so far aimed only at the minimum adder cost. Although the technique in [15] can impose a maximum logic depth constraint to

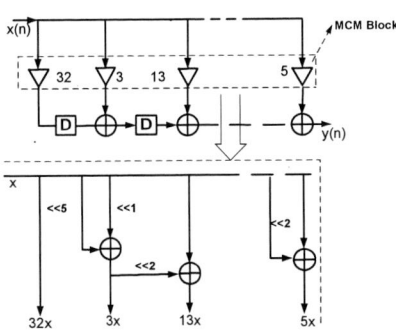

Fig. 1: Diagram of FIR filters with coefficients synthesized as a network of adders and shifts

the design, this technique has three obvious limitations: first, it is not known what is the best tradeoff between the maximum logic depth and the number of adders; second, if a design is imposed a maximum adder depth which is smaller than that of a optimum design without such constraints, the probability of obtaining the assured minimum adder cost decreases; last, even if a design with minimum number of adders is achieved under the maximum logic depth constraint, it does not imply directly the minimum average logic depth of the overall design; the minimum average logic depth is generally believed a more relevant criteria for low power design when the number of adders are comparable.

Therefore, this paper proposes an algorithm aiming at designs with low average logic depth. This is achieved by constraining the synthesis of each coefficient on its minimum adder depth (MinAD) during the tree search of discrete coefficients. With such constraints, our designs tend to use a few more adders to synthesize the filters, but always result in a reduction of the average logic depth and power dissipation compared with the best algorithms currently available.

The remaining of this paper is organized as follows. Section II reviews of the branch and bound depth first tree search algorithm to optimize the FIR filter in discrete space. In Section III, a novel synthesis mechanism which aims at low average logic depth is introduced. Experimental results are presented in Section IV, and Section V concludes the paper.

II. REVIEW OF TREE SEARCH ALGORITHM FOR OPTIMIZATION OF FIR FILTERS WITH MCM IMPLEMENTATION

In this section, the the design of filters with continuous coefficients is first formulated as a linear programming problem to minimize the filter ripples . Secondly, branch and bound depth first algorithm is reviewed for the search of discrete coefficients. The dynamic expending subexpression space for the synthesis of discrete coefficients in MCM block form is also introduced.

A. Problem Formulation for the FIR Filter design

The zero-phase frequency responses of a linear phase FIR filter with N taps can be express as [12]:

$$H(\omega) = \sum_{n=0}^{\frac{N-1}{2}} h(n)Trig(\omega, n) \qquad (1)$$

where $h(n)$ are the filter coefficients and $Trig(\omega, n)$ is an appropriate trigonometric function depending on the parity of N and symmetry of the filter. To find the coefficient set $h(n)$, a linear programing is formulated to [15]

$$Minimize: f = \delta - \delta_p b$$
$$Subject\ to: b - \delta \leq H(\omega) \leq b + \delta, \text{for} \quad w \in [0, \omega_P]$$
$$-(\delta_s \delta)/\delta_p \leq H(\omega) \leq (\delta_s \delta)/\delta_p, \text{for} \quad w \in [\omega_s, \pi]$$
$$b_l \leq b \leq b_u \qquad (2)$$

where δ is the peak ripple and b is a floating passband gain. δ_p, δ_s, ω_p and ω_s are the given passband ripple, stopband ripple, passband edge and stopband edge, respectively. b_l and b_u are two constants, defining the lower bound and upper bound of the passband of gain. In this paper, they are chosen to be 0.7 and 1.4, respectively. Note that the objective function of the linear programming is not the normalized peak ripple (NPR) defined as δ/b; instead, $\delta - \delta_p b$ is minimized. With such formulation, the algorithm may not find the optimum solution ($h(n)$ together with δ and b) in the sense of minimum NPR, but it always manages to find a feasible solution making $\delta - \delta_p b < 0$, if such solution exists [15]. A feasible solution refers to a set of coefficients meeting the ripple specification.

B. Branch and Bound Depth First Algorithm

Branch and bound depth-first search is one of the most frequently used method to solve the mixed integer linear programming (MILP) problem. Basically, it is a tree search algorithm, where the root of the tree is the optimum continuous solution of the FIR filter. The root produces its child nodes by quantizing a select coefficient to its discrete values. All child nodes will further produce their own child nodes by quantizing another coefficient to their discrete values. The process continues until it reaches a leaf which contains only coefficients with discrete values. The nodes having the same parents are called siblings. When a node is generated, the quantized coefficients are synthesized as additions of shifted subexpression basis [13]. In such a manner, the program traverses all the nodes which have a possibility of yielding a feasible solution using less number of adders. The basic steps are:

1) The root is generated by solving the linear programming formulated in (2). The feasible range of each coefficient is obtained as proposed in [15]. The search forwards to its child.

2) A coefficient is selected from its parent node and quantized to a discrete value in its range found in step 1.

3) The quantized discrete value is synthesized by additions of shifted subexpression bases. The subexpression basis set is updated if new bases are generated.

4) A first pruning check: if the total number of adders to synthesize the current node, denoted as N_{total}, is larger than the minimum number of adders so far used to synthesize a leaf node, denoted as N_{best}, this node is cut off, and go to step 7; Otherwise go to step 8;

5) The rest unquantized coefficients are reoptimized to compensate for the loss of frequency response.

6) A second pruning check: it is to check the feasibility of the node. If $f > 0$, the current node already does not meet the ripple specification and is cut off. Go to step 7; otherwise go to step 8.

7) If the node is pruned, the program switches to its siblings and go to step 2. If such siblings do not exist, i.e. the selected coefficient has been fixed to all discrete values, the program backtracks to the siblings of its parent node and go to step 2.

8) If the node is not pruned, it grows further by selecting the next coefficient and fixing it to the next discrete values. Go to step 2. If such coefficients do not exist, a leaf is reached. Go to step 9.

9) N_{best} is updated to N_{total} and this solution is recorded. The search process moves to the siblings of this leaf and go to step 2. If such siblings do not exist, the program backtrackes to the siblings of the parent node and go to step 2.

C. Dynamically Expending Subexpression space

In the above branch and bound algorithm, the MCM coefficients are synthesized in step 3 when the selected coefficient is quantized. The synthesis is based on the subexpression space constructed from a subexpression basis set. The root has a initial zero order set including elements 0 and 1; the order of the set is defined by the number of adders required to synthesize all elements in the set. All other nodes inherit the basis sets from their parents, and the sets are expanded by including the current quantized coefficient (after repeatly divided by 2 until an odd number) and other intermediate bases if additional adders are required to synthesize the current quantized coefficient. In [15], if a child's basis set is more than 1 order higher than its parent, the other intermediate bases are not inserted to the child's basis set; instead, the basis set is flagged with "uncertain". With the introduction of the uncertainty, the optimum is guaranteed if the best leaf node is associated with a certain basis set.

III. A NEW SYNTHESIS MECHANISM

As discussed in Introduction, the synthesis technique reviewed in section II-C may be able to find the optimum solution with minimum number of adders; however, the average adder depth of the MCM blocks is not controlled, resulting in solutions maybe consuming higher powers. In this section, a synthesis technique that constraints the average adder depth is proposed. First, the concept of the MinAD of the adder of a discrete value is introduced. Based on this concept, a synthesis mechanism which ensures that each adder is synthesized on its MinAD is proposed. The MCM block synthesized in such a way is guaranteed with low average adder depth.

A. Minimum Adder Depth of A Coefficient

Canonic signed digit (CSD) format represents a fix point coefficient value using minimum number of non-zero bits. In CSD format, each coefficient is expressed by a string of signed digits, for example $\overline{1}01$ representing -3, where $\overline{1}$ means -1. According to [2], if the number of the non-zero digits of a coefficient expressed in CSD is N_{zero}, irrespective of how this coefficient is decomposed into shift and add operations, the minimum achievable adder depth of the adder network to realize this coefficient is confined as:

$$S = \lceil Log_2 N_{zero} \rceil \qquad (3)$$

This number S is defined as the minimum adder depth (MinAD) of the coefficient. Fig. 2 shows two different adder networks synthesizing coefficient values 3, 11, and 43. In Fig. 2 (a) each coefficient value requires an additional adder resulting in a maximum logic depth

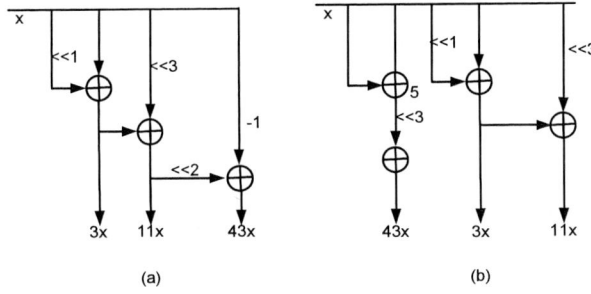

Fig. 2: (a)the discrete coefficient value 3, 11 and 43 are synthesized without MinAD constraint (b) the discrete coefficient value 3, 11 and 43 are synthesized with MinAD constraint

TABLE I: an example of the proposed synthesis procedure and the update of basis sets

Coefficients	main basis Set	auxiliary basis set	basis set certain
h(0)=3	{0,1,3}	null	Y
h(1)=11	{0,1,3,11}	null	Y
h(2)=43	{0,1,3,11,43}	{5}	Y

of 3. According to (3), the MinAD of 3, 11 and 43 are 1, 2 and 2, respectively. This indicates that an adder network with maximum logic depth of 2 could synthesize the three coefficients, though more adders may be required. Fig. 2 (b) shows an implementation with maximum logic depth 2, but one more adder is used than that in Fig. 2 (a). In this paper, if a coefficient is synthesized with the logic depth of MinAD, this coefficient is said to be synthesized on its MinAD. In the next subsection, a synthesis technique that ensures the realization of each coefficients on its MinAD is proposed. In general, the terms of logic depth and adder depth are the same in an MCM block and referring to the number of adders in the longest path of a synthesis. In the following of this paper, to distinct the difference between the synthesis of a particular coefficient and the synthesis of the entire MCM block, we use "adder depth" to refer the former, and "logic depth" to refer the later.

B. A New Synthesis Mechanism Ensuring Coefficient Synthesized on its MinAD

In our proposed technique, besides the expanding subexpression basis set associated with each node, each node is also associated with an expanding auxiliary basis set. The auxiliary set for the root is initialized to be null, and each node inherits the set from its parent; the update of the set will be described incorporating with the coefficient synthesis. When a node is generated in step 3 in the tree search described in section II-B, a selected coefficient $h(k)$ is fixed to a discrete value in its range; the discrete value is denoted as $h^0(k)$. Now this coefficient is to be synthesized based on its associated subexpression basis set. The detailed procedure (as shown in Fig. 3) is given as follows.

1) The MinAD to realize $h^0(k)$ is computed according to (3).

2) If $h^0(k)$ could be synthesized based on the current basis set and the auxiliary set using not more than an adder, and the synthesis is on its MinAD , the synthesis is deemed optimum in the sense of minimum number of adders and the lowest power consumption.

3) If $h^0(k)$ can not be successfully synthesized on its MinAD using not more than one adder, a first order basis is inserted into the auxiliary basis set to help the synthesis of this coefficient on its MinAD. The basis set is also updated to include $h^0(k)$. Thus, the

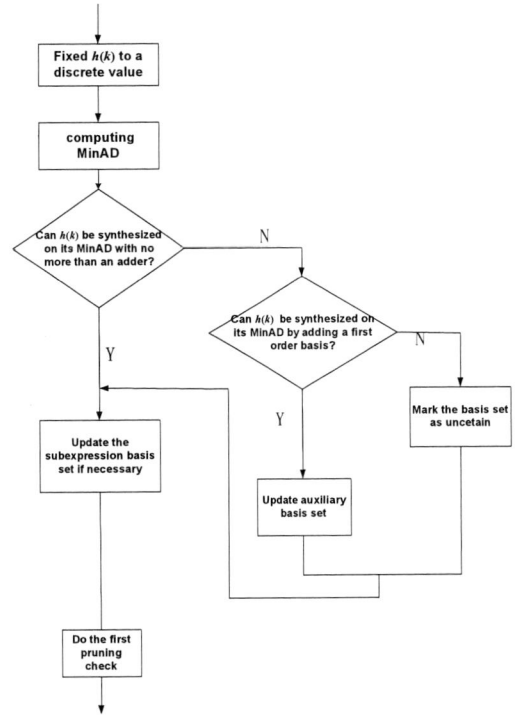

Fig. 3: procedure to synthesize a coefficient $h(k)$ on its MinAD

basis set itself is not certain, but the union of the basis set and auxiliary basis set is still certain. This differentiates the proposed technique from that in [15], where, in such cases, the discrete value is just inserted into the basis set and the set is flagged to be uncertain. With the basis set dynamically expanding, the uncertainty may be subsequently eliminated when more coefficients are synthesized. However, the price paid for that may be the increase of the adder depth for the coefficient. If the maximum logic depth constraint is adopted, it produces uncertain results in most cases. In order to produce certain solution without high power and area penalty, the above auxiliary basis set is introduced. Though this sacrifices the optimality in the sense of minimum number of adders in some cases, experiment results show that the increased number of adders is not significant and power saving is achieved.

4) If even by inserting a first order basis in the auxiliary basis set, the discrete value is still can not be synthesized on its MinAD, this value is just added to the basis set and the basis set is flagged as uncertain without updating the auxiliary set. The reason that not allowing more than one basis to be inserted into the auxiliary set is a tradeoff between the certainty of the results and the guarantee of the minimum number of adders. From our design experiences, adding not more than an auxiliary basis for each node is enough to make the final results be certain, and meanwhile to maintain each coefficient on its MinAD.

A simple example is given to illustrate the synthesis mechanism in details. As shown in Table I, first, $h(0)$ is fixed to 3 and it can be synthesized on it MinAD based on the initial zeroth order basis set. The basis set is updated to include 3, and the auxiliary basis set is Null. Thereafter $h(1)$ is fixed to 11 whose CSD representation is $10\bar{1}0\bar{1}$. According to (3), the MinAD of 11 is 2. Clearly, $11=3 \times 2^2 - 1$ can be realized on its MinAD based on the current basis set.

TABLE II: Results and Comparison of eight benchmark filters

Name	EWL	#tap	A:the method proposed in [15]				B: [15] with adder step constraint				C:our proposed method			
			NA	MaxLD	ALD	Power	NA	MaxLD	ALD	Power	NA	MaxLD	ALD	Power
S1	7	24	23	3	1.78	3.09	the same as the algorithm A				the same as the algorithm A			
G1	7	16	15	3	1.46	1.75	the same as the algorithm A				the same as the algorithm A			
S2	9	60	76	4	2.18	27.30	the same as the algorithm A				77	4	2.02	25.50
A1	10	59	68	4	2.30	23.88	the same as the C				71	4	1.64	21.72
Y1	9	30	30	4	1.93	6.60	31	3	1.83	6.04	31	3	1.83	5.98
Y2	11	34	37	5	2.43	7.01	uncertain				38	4	2.31	6.22
F7	11	40	53	5	2.49	20.24	54	4	2.57	20.22	54	4	2.35	17.80
F11	10	25	35	5	2.83	12.91	uncertain				37	4	2.45	11.20

TABLE III: The specification for F7 and F11

Filters	Filter tap	ω_p	ω_s	δ_p	δ_s
F7	40	0.15π	0.25π	0.011618	0.011764
F11	25	0.25π	0.3π	0.1355	0.1352

Therefore, when $h(2)$ is to be fixed to 43, the basis set is $\{0,1,3,11\}$ with certainty and the auxiliary basis set is still null. $h(2)$ can be synthesized based on the current basis set with one adder as $(43 = 11 \times 2^2 - 1)$, but the adder depth of such synthesis of 43 is 3 which is larger than its MinAD of 2. Thus, a first order basis 5 is inserted into the auxiliary basis set, such that 43 could be synthesized as $5 \times 2^3 + 3$ on its MinAD. Therefore the basis set and the auxiliary set are updated to be $\{0, 1, 3, 11, 43\}$ and $\{5\}$, respectively. Without the auxiliary basis set, the basis set is still certain, but not all members could be synthesized on its MinAD. With the price of one more adder to synthesize the auxiliary basis, the MinAD for each coefficient is achieved.

IV. EXPERIMENT RESULTS

Eight benchmark filters from literature are synthesized to show the superiority of the proposed algorithm in low power implementation. S1,S2, Y1, Y2, G1, A1 are the examples taken from [15], while F7 and F11 are the 7th and 11th examples of [1], respectively. For F7 and F11, the passband ripple and stopband ripple were not explicitly given in the original paper; they are evaluated based on the coefficients provided in [1], and are listed in Table II. The filters are synthesized using 0.18 μm digital libraries, and the Synopsys Tool, Prime Power, is adopted to simulate the power consumption.

The proposed method is compared with the method proposed in [15] without maximal adder depth constraint. Results show, as listed in Table III, for filters with short effective word length (EWL), such as G1 and S1 (both with EWL of 7), the proposed algorithm generates the exactly same solutions as that produced by the algorithm proposed in [15]. For the filters with relatively longer EWL, the proposed algorithm produces results outperforming those produced by the algorithm in [15], in the sense of maximum logic depth, average logic depth and power consumption, although our algorithm generally uses one or two more adders to synthesize the filters.

For a fairer comparison, our algorithm is further compared with that in [15] with maximum logic depth constraints. The maximum logic depth of each filter obtained in our technique is used as the constraints for the algorithm in [15]. Even with such information which is usually not prior known to algorithm in [15], only one filter generated by [15] is as good as that of our proposed algorithm; the other are either poorer in the sense of average logic depth and power consumption, or not certain in the results obtained.

V. CONCLUSION

In this work, an FIR filter optimization technique is proposed for the low power implementation of the MCM coefficients. The proposed method ensures that each adder in the MCM block is synthesized on its MinAD while keeps the number of adders low. Thus, the average logic depth of the resulting MCM block is ensured to be low without any maximum logic depth constraints enforced, and therefore the power consumption is reduced.

REFERENCES

[1] L. Aksoy, E. Gunes, and P. Flores, "An exact breadth-first search algorithm for the multiple constant multiplications problem," Tallinn, Nov. 2008, pp. 41–46.

[2] M. Faust and C. H. Chang, "Minimal logic depth adder tree optimization for multiple constant multiplication," in *Proc. IEEEInt. Symp. Circuits Syst.*, Paris, May 2010, pp. 457–460.

[3] R. I. Hartley, "Subexpression sharing in filters using canonic signed digit multipliers," *IEEE Trans. Circuits Syst. II*, vol. 43, pp. 677–688, Oct. 1996.

[4] D. R. Bull and D. H. Horrocks, "Primitive operator digital filters," *IEE Proc. G*, vol. 138, pp. 401–412, June 1991.

[5] A. G. Dempster and M. D. Macleod, "Use of minimum-adder multiplier blocks in FIR digital filters," *IEEE Trans. Circuits Syst. II*, vol. 42, pp. 569–577, Sept. 1995.

[6] A. G. Dempster, S. S. Demirsoy, and I. Kale, "Designing multiplier blocks with low logic depth," in *Proc. IEEE International Symposium on Circuits and Systems*, Scottsdale, Arizona, May 2002, pp. 773–776.

[7] O. Gustafsson and L. Wanhammar, "A novel approach to multiple constant multiplication using minimum spanning trees," in *Proc. IEEE Midwest Symp. Circuits Syst*, Tulsa, OK, Aug. 2002, pp. 652–655.

[8] O. Gustafsson, H. Ohlsson, and L. Wanhammar, "Improved multiple constant multiplication using minimum spanning trees," in *Proc. IEEE Asilomar Conf. Signals Syst. Comp.*, Pacific Grove, CA, Nov. 2004, pp. 63–66.

[9] O. Gustafsson, "A difference based adder graph heuristic for multiple constant multiplication problems," in *Proc. IEEEInt. Symp. Circuits Syst.*, New Orleans, LA, May 2006, pp. 1097–1100.

[10] O. Gustafsson and L. Wanhammar, "ILP modeling of the common subexpression sharing problem," in *Proc. IEEE ICECS'02*, Dubrovnic, Croatia, 2002, pp. 1171–1174.

[11] J. Yli-Kaakinen and T. Saramäki, "A systematic algorithm for the design of multiplierless FIR filters," in *Proc. IEEE ISCAS'01*, Sydney, Australia, 2001, pp. 185–188.

[12] Y. J. Yu and Y. C. Lim, "Design of linear phase FIR filters in subexpression space using mixed integer linear programming," *IEEE Trans. Circuits Syst. I*, vol. 54, pp. 2330–2338, Oct. 2007.

[13] Y. J. Yu, D. Shi, and Y. C. Lim, "Design of extrapolated impulse response FIR filters with residual compensation in subexpression space," *IEEE Trans. Circuits Syst. I*, vol. 56, pp. 2621–2633, Dec. 2009.

[14] Y. J. Yu and Y. C. Lim, "Optimization of linear phase FIR filters in dynamically expanding subexpression space," *Circuits, Systems, and Signal Processing*, vol. 29, pp. 65–80, June 2009.

[15] D. Shi and Y. J. Yu, "Design of linear phase FIR filters with high probability of achieving minimum number of adders," *IEEE Trans. Circuits Syst. I*, vol. 58, pp. 126–136, Jan. 2011.

[16] A. G. Dempster and M. D. Macleod, "Use of minimum-adder multiplier blocks in FIR digital filter," *IEEE Trans. Circuits and Systems II: Analog and Digital Signal Processing,*, vol. 42, pp. 569–577, Sept. 1995.

[17] K. Johansson, "Low power and low complexity shift-and-add based computations," Ph.D. dissertation, Linköping University, 2008.

A New Motion Compensation Algorithm And New Test Algorithm To Compare The Result Of Motion Compensation

Chunchun chen
School of Communication
and Information Engineering,
Shanghai University,
Shanghai, China
Email:chunchun.chen2010@gmail.com

Xuewei Chen
School of Communication
and Information Engineering,
Shanghai University,
Shanghai, China
Email:xueweigm@163.com

Eryan. Yang
School of Communication
and Information Engineering,
Shanghai University,
Shanghai, China
Email:eryan.yang@gmail.com

Abstract— In this paper, we propose a novel motion compensation (MC) algorithm. Compared with previous MC algorithm using only one interpolation method, the proposed MC algorithm chooses different interpolation methods according to the calculated sum of absolute difference (SAD) in the process of motion estimation (ME). The proposed MC algorithm can improve the quality of the interpolated frames. Based on the SAD calculated in ME, interpolation method is divided into three cases. Then in order to demonstrate the performance of the proposed MC algorithm, we propose a new test algorithm, which compares the results of several MCs. Experimental results indicate that, compared with common MC algorithm and the linear MC algorithm, quality of the image using proposed MC algorithm has been improved.

I. INTRODUCTION

With the rapid development of digital TV and video technology, digital video standards have been developed. There have been many different display formats in our life. For example, the standard frame rate of film is 24fps, while HDTV's is 60fps [1]. In the conversion between various display formats with different frame rates, the frame rate up conversion (FRUC) is necessary. Based on the related two or more video frames, FRUC produces new frames and interpolates the new frames into original frames. The interpolated frame that predicts the missing ones can be either linear or nonlinear. In this way, FRUC achieves the conversion of video formats with different frame rates. It is widely used in HDTV [2], video conference and low bit rate coding.

The FRUC algorithms are mainly divided into two categories: non-motion compensation interpolation and motion compensation interpolation (MCI). Non-motion compensation interpolation uses the combination of video frames without taking the motion of objects into account [3]. It includes the methods such as frame repetition and frame averaging. While MCI uses the advanced conversion technique that employs the motion of objects. By ME, it gets motion vector and then uses it for interpolation. As having taken into account the motion of objects within the frame, the target trajectory can be well reflected by the new frame. As non-motion

compensation interpolation is not suitable for large scene motion video sequences, in practice, we generally use the MCI algorithm [4]. So in this paper the FRUC algorithm we choose is based on the MCI algorithm [5]. There are two main processes to achieve MCI algorithm: ME and MC. ME is an important part of inter prediction [6]. It is used to find the motion vector of images and it can effectively remove the temporal redundancy of image sequences. The accuracy of ME directly determines the quality of interpolated frames. Through motion vectors generated by the movement of previous frame, MC can get a prediction of the frame. The predicted frame is close to current frame. The more perfect the ME algorithm is, the more accurate the estimated motion vector is, the better the performance of MC is, and the smaller the prediction error is. As for MC, it predicts and compensates new frames, the interpolated frames. It is an effective way to reduce redundant information between frame sequences. In order to improve the quality of the interpolated frames, we propose a novel MC algorithm in this paper.

This paper is structured as follows: Section A describes the MC algorithm we choose in this paper and presents details of the proposed algorithm and the test model. In section B [7], we obtain the experimental results and discuss its performance by compared with other MC algorithms. Finally, the conclusion is provided. Compared with the common MC algorithm and the linear MC algorithm, quality of the image using proposed MC algorithm has been improved.

A. PROPOSED ALGORITHM

In MCI, the interpolated frame is based on the motion vector information calculated between the reference frame and current frame. In order to reduce the error of motion vector calculated by ME, we choose the full search (FS) algorithm as its ME algorithm [8]. The best-matched block can be obtained by the FS algorithm through checking all candidate blocks. Firstly, to reduce the operation complexity, every 8*8 pixels in a block are assumed to have the same motion vector. The frame is divided into many non-overlapping macro blocks and

978-1-4577-1608-9/11 $26.00 © 2011 IEEE

all the pixels in the macro blocks do the same translation, so the motion vectors of each macro block can be estimated independently. Then the SAD of each block is calculated, starting from the center of search window in a clockwise spiral direction from near and far, until all the points within the search scope are calculated. The block with the smallest SAD in all candidate blocks is the best-matched block and its relative displacement is the motion vector that is asked for. Because the FS algorithm is reliable and able to get the global best-matched block, it usually is the standard of ME algorithm when we compare the performance of MC algorithms [9]. Finally, we will introduce the test model to verify the performance of the proposed MC algorithm.

1) Motion Estimation And Matching Criteria : In ME, we use SAD of only the luminance signal (Y) as the matching criteria. As the input and output signals are in RGB format, color space conversion is needed before the ME and MC. Among these converted YUV signals [10], we just utilize the Y signal to estimate motion vectors in the SAD criteria. After ME and MC, we have to convert the YUV color space back into RGB color space for outputting the video stream. When it comes to the matching criteria, we generally make use of the concept of SAD [11]. SAD is calculated for pixels between the current block in current frame and the candidate blocks in previous frame [12] and it is expressed as equation (1). [13]

$$SAD(i,j) = \sum_{m=1}^{M} \sum_{n=1}^{N} |f_k(x,y) - f_{k-1}(x+i, y+j)| \quad (1)$$

In this equation, M/N is the size of the block, (i,j) is the displacement of the current block, and f_k/f_{k-1} is the gray level in the k/k-1 frame. Then we use the FS algorithm to get the motion vectors. By using the FS algorithm, we can evaluate the performance of MC algorithms more accurate.

2) Motion Compensation: After the full search motion estimation, we have obtained the motion vectors for all the blocks. Then the interpolated frames are produced by linear combination of the current frame and reference frame using the motion vectors obtained by FS algorithm. This proposed FRUC outperforms other traditional algorithms due to its exploiting of motion vectors between successive frames. Let $(x,y),(x_0,y_0)$, (x_1,y_1) denote the spatial location of pixels in the interpolated frames, the reference frames and the current frames. If we use the common MC algorithm, the pixels of interpolated frames can be expressed as :

$$(x,y) = \frac{1}{2}[(x_0,y_0) + (x_1,y_1)] \quad (2)$$

The method we propose is shown in Fig.1. According to the calculated SAD, the interpolation is divided into three cases. when the SAD is smaller than threshold1, the corresponding blocks in interpolated frames, reference frames and current frames are roughly the same, so we can take the block of reference frames as the block of interpolated frames; While when the SAD is larger than threshold2, there is big difference between the reference frame and interpolated frame, in

this case the current frame can be seen as the interpolated frame. When the SAD is between threshold1 and threshold2, difference between the reference frame and interpolated frame is neither too large nor too small. So the interpolation should be mainly based on one frame of the reference frame and current frame. The interpolated frames can be expressed as:

$$(x,y) = \frac{1}{2}[\alpha_1(x_0,y_0) + \alpha_2(x_1,y_1)] \quad (3)$$

In this paper, α_1 is set to 0.1, and α_2 is set to 0.9. By changing the weighting of reference frame and current frame, it can avoid blur brought by the interpolated frame. Thus the MC algorithm we propose is as shown in Fig.1.

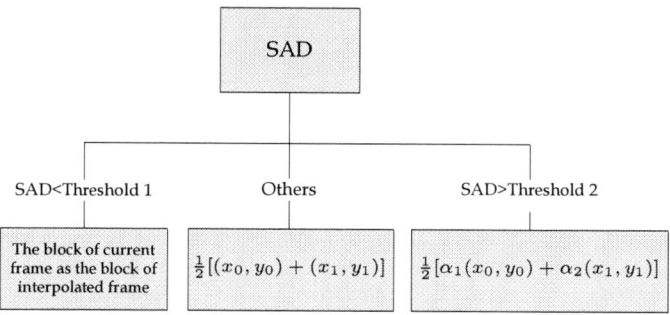

Fig. 1.   New motion compensation algorithm

In the proposed method, interpolated frame is obtained based on the calculated SAD, rather than always taking the average of current frame and reference frame. Therefore, its performance is better than the common method only taking the average of current frame and reference frame. In order to demonstrate the performance of the proposed algorithm, we present a test model, which is shown in Fig.2. In this test model, first we get the interpolated frames by the original frames and current frames using ME and MC method. Then we use the interpolated frames and current frames to get the new original frames by the same method. At last we compare the new original frame with original frame by the Peak Signal to Noise Ratio (PSNR). PSNR is one of the most popular criteria used to evaluate the objective image quality in the field of the image processing and it is expressed as Equation (4) [14].

$$PSNR = 10 * \log 10(\frac{255^2}{MSE}) \quad (4)$$

B. EXPERIMENTAL RESULTS

In this section, results of the proposed MC algorithm are discussed and compared with other MC algorithms. In these experiments, the block size of proposed algorithm is 8*8 pixels and the search coverage in the FS algorithm is 64 pixels. The test sequences we use in this paper are Bouncing balls.

In the first experiment, results of original frame, new original frame generated by the frame that interpolated by the common MC algorithm and by the frame that interpolated by the common MC algorithm are obtained, which are shown in

978-1-4577-1608-9/11 $26.00 © 2011 IEEE

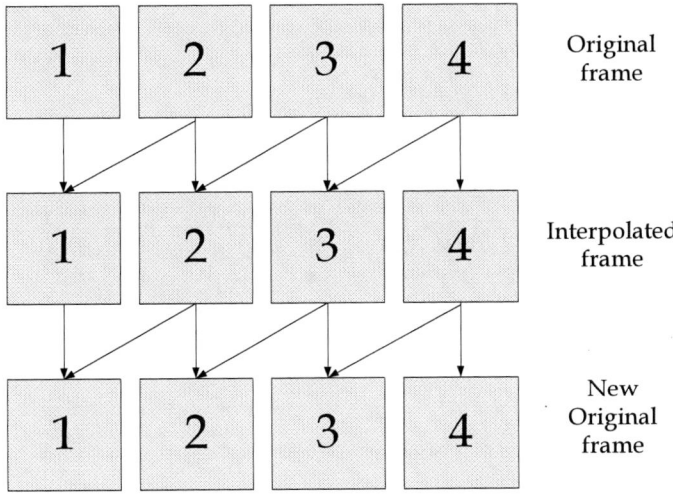

Original frame

Interpolated frame

New Original frame

Fig. 2. Test model

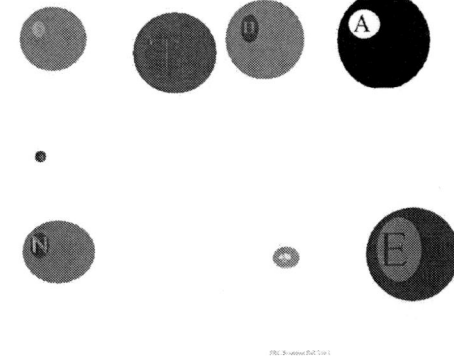

Fig. 4. Generated by the frame interpolated by the common MC algorithm

Fig.4, Fig.5 and Fig.6. Results of original frame are shown in Fig.3. Fig.4 shows another original frame generated by the frame that interpolated by the common MC algorithm. Fig.5 shows another original frame generated by the frame that interpolated by the proposed MC algorithm. Fig.6 shows the other original frame generated by the frame that interpolated by the linear MC algorithm.

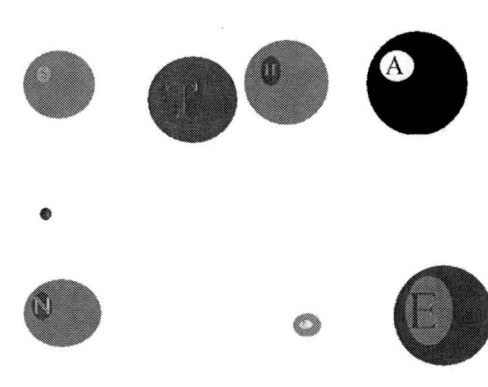

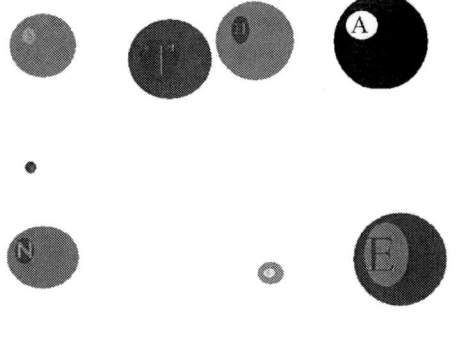

Fig. 3. Original frame

Fig. 5. Generated by the frame interpolated by the proposed MC algorithm

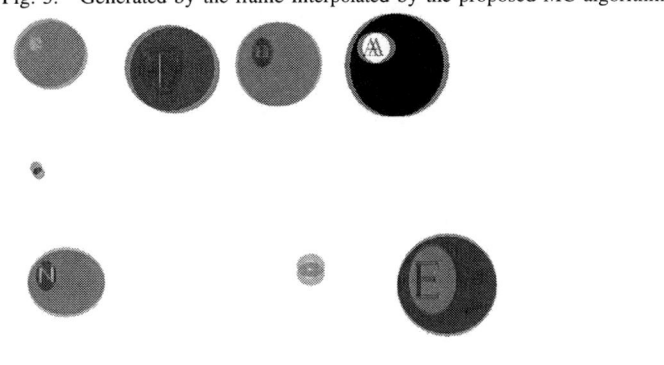

Fig. 6. Generated by the frame interpolated by the linear MC algorithm

Through the circle marked "R", the fast moving ball in the sequence, we can see that the frame interpolated by the proposed MC algorithm is better than the frame interpolated by the common MC algorithm; But the result balls just seem the same when they are very slow. When we compare the frame interpolated by the linear MC algorithm with the original frame, we can see that almost all the balls have shadows in their bridge. So we can conclude that, judged by the comparison modle we propose, result of the proposed MC algorithm is better than the common MC algorithm, and the result of linear MC algorithm is the worst. Then we will use the PSNR to compare the results.

PSNR is one of the most popular criteria used to evaluate the objective image quality in the field of the image processing. So in the second experiment, we compare the PSNR of original

frame with that of frames interpolated by the proposed MC algorithm, common MC algorithm and linear MC algorithm. Fig.7 shows the PSNR of the frame interpolated by the proposed MC algorithm, the common MC algorithm and the linear MC algorithm. From the PSNRs shown in Fig.7, we can conclude that quality of images using proposed algorithm has improved a lot. In these three mentioned MC algorithms, proposed algorithm is the best one of them; The common MC algorithm is better than the linear MC algorithm. And we can also see that results of the first frame in proposed algorithm is the same as that in other two algorithms, because there is almost no motion in the initial frames. Only when there is motion, results of the three mentioned algorithms will be different.

II. CONCLUSION

In this paper, we present a novel MC algorithm and a new test algorithm to compare its performance with other MC algorithms. From the analysis of the experimental results, we can conclude that the performance of proposed MC algorithm is better than other two mentioned MC algorithms. Compared with proposed MC algorithm and common MC algorithm, results of linear MC algorithm using only one interpolation method is the worst, so we usually disuse the linear MC algorithm. And when compared with common MC algorithm, result of proposed MC algorithm is better. In short, in this paper we can work out that the proposed MC algorithm is better than the common MC algorithm, and the common MC algorithm is better than the linear MC algorithm.

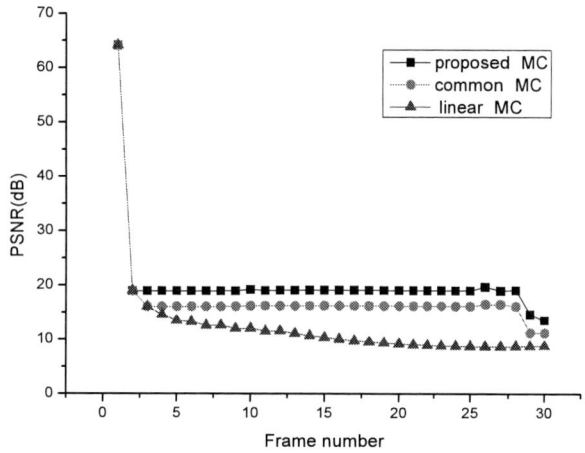

Fig. 7. PSNRs of proposed MC algorithm, common MC algorithm and linear MC algorithm

ACKNOWLEDGMENT

This project is Supported by the Key Project of Chinese Ministry of Education.(NO210069), STCSM NO. 09220502600, and STCSM NO. 09PJ1405100.

REFERENCES

[1] R. Armitano, D. Florencio, and R. Schafer, "The motion transform: a new motion compensation technique," vol. 4, pp. 2295 –2298 vol. 4, May 1996.

[2] H. Jozawa, K. Kamikura, A. Sagata, H. Kotera, and H. Watanabe, "Two-stage motion compensation using adaptive global mc and local affine mc," *Circuits and Systems for Video Technology, IEEE Transactions on*, vol. 7, no. 1, pp. 75 –85, feb 1997.

[3] P. Hsu, K. Liu, and T. Chen, "A low bit-rate video codec based on two-dimensional mesh motion compensation with adaptive interpolation," *Circuits and Systems for Video Technology, IEEE Transactions on*, vol. 11, no. 1, pp. 111 –117, Jan. 2001.

[4] J. Huska and P. Kulla, "A new block based motion estimation with true region motion field," pp. 182 –188, 2007.

[5] D. Kim and K. Sohn, "Static text region detection in video sequences using color and orientation consistencies," *Pattern Recognition, 2008. ICPR 2008. 19th International Conference on*, pp. 1 –4, dec. 2008.

[6] S.-J. Kang, D.-G. Yoo, S.-K. Lee, and Y. H. Kim, "Hardware implementation of motion estimation using a sub-sampled block for frame rate up-conversion," in *SoC Design Conference, 2008. ISOCC '08. International*, vol. 02, nov. 2008, pp. II–101 –II–104.

[7] S.-E. Kim, J.-K. Han, and J.-G. Kim, "An efficient scheme for motion estimation using multireference frames in h.264/avc," *Multimedia, IEEE Transactions on*, vol. 8, no. 3, pp. 457 – 466, june 2006.

[8] Y. Nakaya and H. Harashima, "Motion compensation based on spatial transformations," *Circuits and Systems for Video Technology, IEEE Transactions on*, vol. 4, no. 3, pp. 339 –356, 366–7, jun 1994.

[9] C.-M. Kuo, S.-C. Chung, and P.-Y. Shih, "Kalman filtering based rate-constrained motion estimation for very low bit rate video coding," *Circuits and Systems for Video Technology, IEEE Transactions on*, vol. 16, no. 1, pp. 3 – 18, jan. 2006.

[10] H. Lim, A. Kassim, and P. De With, "Predictive 3d search algorithm for multi-frame motion estimation," *Consumer Electronics, IEEE Transactions on*, vol. 54, no. 4, pp. 1938 –1946, november 2008.

[11] B. Kamolrat, W. Fernando, and M. Mrak, "Adaptive motion-estimation-mode selection for depth video coding," in *Acoustics Speech and Signal Processing (ICASSP), 2010 IEEE International Conference on*, march 2010, pp. 702 –705.

[12] ——, "Adaptive motion-estimation-mode selection for depth video coding," in *Acoustics Speech and Signal Processing (ICASSP), 2010 IEEE International Conference on*, march 2010, pp. 702 –705.

[13] Z. wu Yuan and Y. ming Hu, "Global motion estimation algorithm based on pyramid iteration," in *E-Product E-Service and E-Entertainment (ICEEE), 2010 International Conference on*, nov. 2010, pp. 1 –4.

[14] G. Dane and T. Nguyen, "Analysis of motion vector errors in motion compensated frame rate up conversion," in *Signals, Systems and Computers, 2003. Conference Record of the Thirty-Seventh Asilomar Conference on*, vol. 2, nov. 2003, pp. 1534 – 1538 Vol.2.

On-Chip Process and Temperature Compensation and Self-Adjusting Slew Rate Control for Output Buffer

Ron-Chi Kuo, Hsin-Yuan Tseng, Jen-Wei Liu, and Chua-Chin Wang*, *Senior Member, IEEE*

Department of Electrical Engineering

National Sun Yat-Sen University, Kaohsiung, Taiwan

*Email: ccwang@ee.nsysu.edu.tw

Abstract—A novel process corner detection technique as well as process and temperature compensation method for sub-2×VDD output buffer is proposed. The threshold voltage (Vth) of PMOSs and NMOSs varying with process and temperature deviation could be detected, respectively. By adjusting output currents, the slew rate of output signal could be compensated over 117 %. The maximum data rate with compensation is 120 MHz in contrast with 95 MHz without compensation, which is measured on silicon with an equivalent probe capacitive load of 10 pF.

Index Terms—Process and temperature variation, threshold voltage detection, mixed-voltage-tolerant, I/O buffer, floating N-well circuit, gate-oxide reliability

I. INTRODUCTION

THE sensitivity of modern VLSI circuits to process, and temperature (PT) variation restricts its performance and yield, especially when the technology is evolved toward nanoscale. The performance of VLSI circuit can be interpreted as a function of PT variation, as shown in Fig. 1 [1]. In past ten years, many prior works proposed different techniques to enhanse capability against PT variation and enlarge the acceptable envelop as much as possible to increase the yield. Though the logic delay method has been widely utilized to detect PVT variation [2]-[7], it can only recognize three corners, TT, FF, and SS. Therefore, from a perspective of transistor level, a novel corner detection technique is proposed to detect all process corners, i.e., TT, FF, SS, SF, FS. That is, the process variation of PMOS and NMOS could be examined, respectively. Moreover, for the high-speed interface circuits, the specification of slew rate is varied by different communication systems. Hence, a compensation mechanism with the NMOS and PMOS threshold voltage detectors is needed to self-adjust the slew rate of output buffers, as shown in Fig. 2 (a).

II. 2×VDD OUTPUT BUFFER WITH PT COMPENSATION

Fig. 2 (b) shows the block diagram of the proposed system comprising of a clock generator, a NMOS threshold voltage detector, a PMOS threshold voltage detector, comparators including comparatorN and comparatorP, Digital circuits for NMOS and PMOS, and a sub-2×VDD output buffer.

A. NMOS Threshold Voltage Detector

An output signal, VN, is generated by the NMOS threshold voltage detector, as revealed in Fig. 3 (a). When reset_N is at 1.8 V and CLK is at 0 V to activate the NMOS threshold voltage detector, the voltage of net902 is discharged to 0 V.

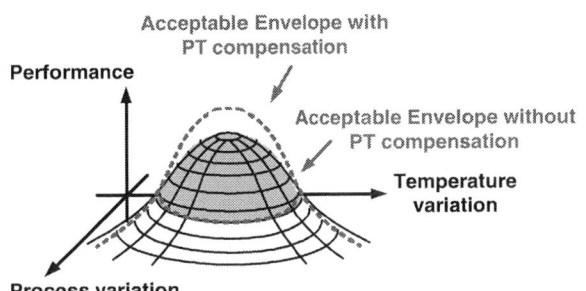

Fig. 1.   Performance envelop as a function of process and temperature varitaion.

At the same time, the voltage of net901 is charged to (1.8-Vthn) V, which equals to the voltage across the capacitor C90. Then, CLK is pulled up from 0 V to 1.8 V, the voltage of net902 is pulled down to (Vthn-1.8) V, while the voltage of net901 is charged to -Vthn V. When CLK is dropped from 1.8 V to 0 V again, the voltage of net902 is pulled up to (1.8-2×Vthn) V. Hence, VN is charged to (1.8-3×Vthn) V after the first charging cycle. When the above operation is repeated in the second charging cycle, VN is pulled up to (3.5-6×Vthn) V. The function of VN can be obviously derived as

$$VN(n) = (VDD - 3 \times Vthn) \times n \qquad (1)$$

where VDD is the supply voltage, Vthn is the threshold voltage of NMOS, and n is the number of clock cycles. The voltage level of VN is rising with the cycle count. Meanwhile, the rising speed is determined by Vthn and VDD. Therefore, the clock count needed for VN to reach the reference voltage level, VREFN, will vary with different corners, as shown in Fig. 4 (a). By recognizing the clock counts, corner variations could be detected precisely.

B. PMOS Threshold Voltage Detector

The design concept is similar with that of the NMOS threshold voltage detector, as shown in Fig. 3 (b). The difference from the NMOS threshold voltage detector is that the voltage level of the output signal, VP, is falling with the clock count. Besides, the falling speed is determined by the threshold of PMOS, Vthp, and VDD. The function of VP can be obtained as follows.

$$VP(m) = 3m \times Vthp - (m - 1) \times VDD \qquad (2)$$

978-1-4577-1608-9/11 $26.00 © 2011 IEEE

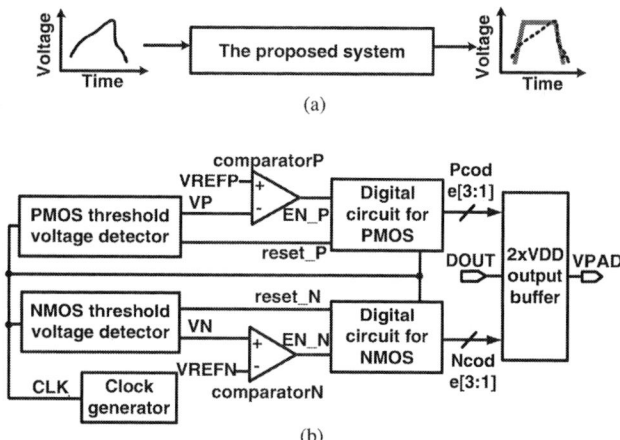

(a)

(b)

Fig. 2. (a)The slew rate is compensated through our system. (b)The block diagram of the proposed system.

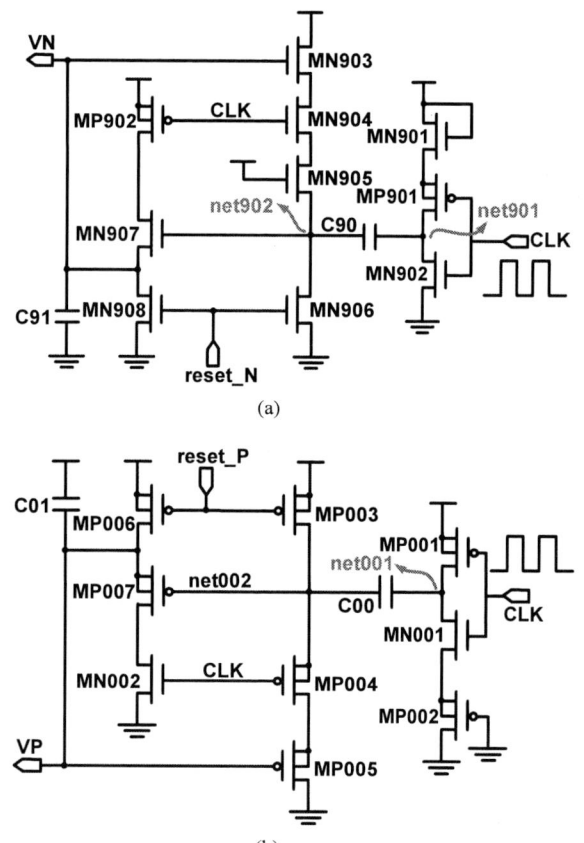

(a)

(b)

Fig. 3. (a) NMOS threshold voltage detector. (b) PMOS threshold voltage detector.

Where m is the number of clock cycle different from n in Eqn. (1). By reading the clock counts when VP meets the reference voltage VREP, different PVT corners could also be detected separately, as shown in Fig. 4 (b).

C. Digital Circuits for NMOS and PMOS

Fig. 5 illustrates the block diagram of Digital circuits for NMOS and PMOS, respectively, where a 4-bit counter, an encoder, and D flip-flops are included in one Digital circuit. According to various corners, the encoder will create a code to be latched in D flip-flops. When VP and VN reach the reference voltage VREFP and VREFN, respectively, comparators deliver, EN_P and EN_N, respectively, to latch D flip-flops. The code loaded into the D flip-flops indicates the required compensation status to control the output currents.

D. Sub-2×VDD Mixed-Voltage Tolerant Output Buffer

The sub-2×VDD mixed-voltage output buffer is composed of a Pre-driver, a Vg1 generator, a VDDIO detector, and an output stage, as shown in Fig. 6. Pre-driver is used to encode three control signals, DOUT, Pcode[3:1], and Ncode[3:1], to adjust output currents for slew rate compensation. VDDIO detector and Vg1 generator can generate appropriate gate drive voltages in different voltage modes without leakage currents and overstress problems [7].

E. Output stage

Since the supply voltage (VDD) of the core circuits is 1.8 V in 0.18 μm CMOS process, the output stage must be realized with two groups of stacked PMOS and NMOS transistors, respectively, for transmitting 2×VDD signals, as depicted in Fig. 6. PMOSs $P_{1a} \sim P_{1c}$ are connected in parallel so that the slew rate of the output signal can be improved by adjusting the currents flowing through $P_{1a} \sim P_{1c}$. According to different process and temperature variation quantity, control signal, Pcode[3:1] and Ncode[3:1], can decide the number of turned on PMOSs of output stage. The operating mechanism of $N_{1a} \sim N_{1c}$ are same as PMOSs mentioned above. $P_{1a} \sim P_{1c}$ and $N_{1a} \sim N_{1c}$ are designed with different sizes to generate different currents, which could successfully achieve coarse and precise adjustment.

III. MEASUREMENT RESULTS

This work is fabricated using TSMC 0.18 μm CMOS technology without thick-oxide devices. Fig. 7 shows the die photo of the proposed system, where the overall chip size is 1064 μm × 586 μm and the compensation circuit area is only 160 μm × 65 μm. Fig. 8 shows the measurement settings for process and temperature compensation. In Fig. 8 (a), the body voltage of PMOSs, Vnewll, are varied from 0.5~0.7 V to generate different process corners. Besides, VDDIO are given 3.3/1.8/0.9 V, respectively, in different transmitting modes while VDD remains at 1.8 V. Fig. 8 (b)(c) show that our chip is heated by a thermo chamber between -40°C~100°C. Referring to Fig. 9 (a)(b)(c), the maximum data rate of VPADs without compensation are measured to be 85/95/75 MHz when VDDIO is at 3.3/1.8/0.9 V given core VDD=1.8 V, respectively. The worst improvement of slew rate compensation, from 1.28 V/ns to 2.79 V/ns, occurs at [TT, 25°C] corner when VDDIO=3.3 V, as shown in Fig. 8 (d). The maximum data rate is measured to 120 MHz with compensation. The performance comparison with prior works is tabulated in Table I. Our design has the

978-1-4577-1608-9/11 $26.00 © 2011 IEEE

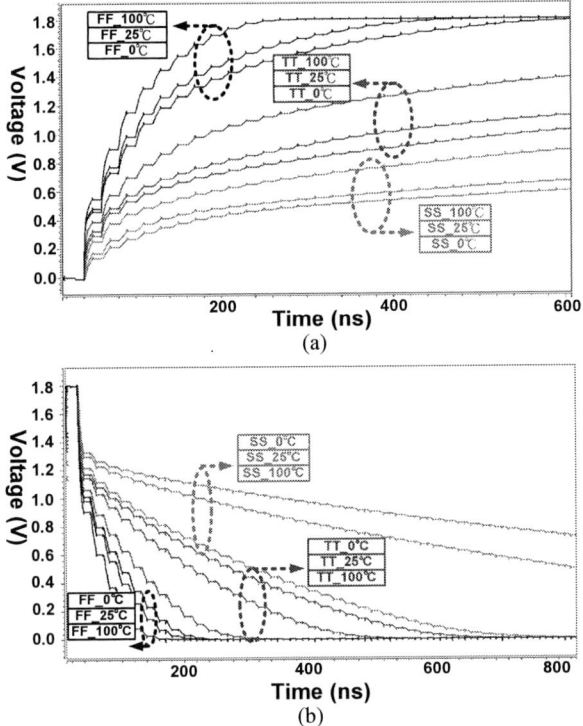

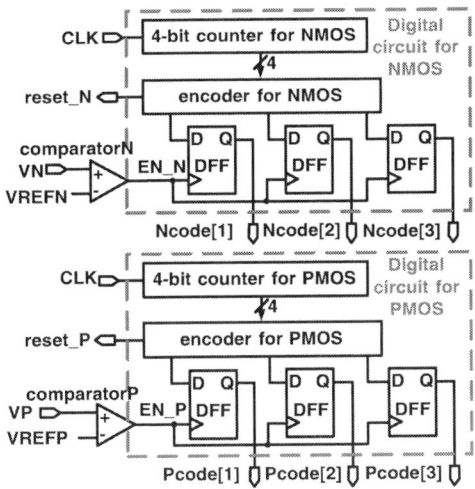

Fig. 4. Simulation waveform of (a) NMOS threshold voltage detector, (b) PMOS threshold voltage detector for different corners.

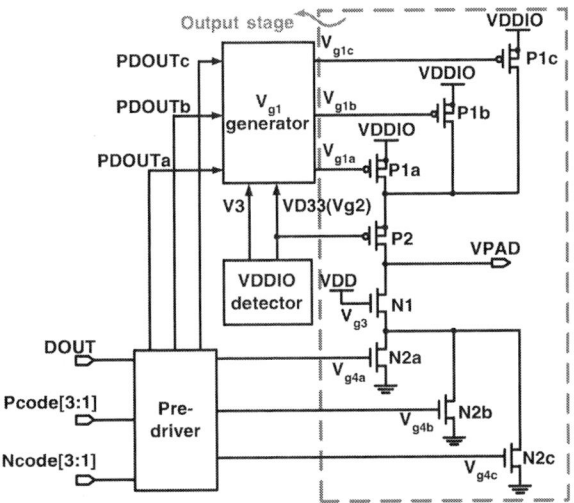

Fig. 6. Sub-2×VDD mixed-voltage tolerant output buffer.

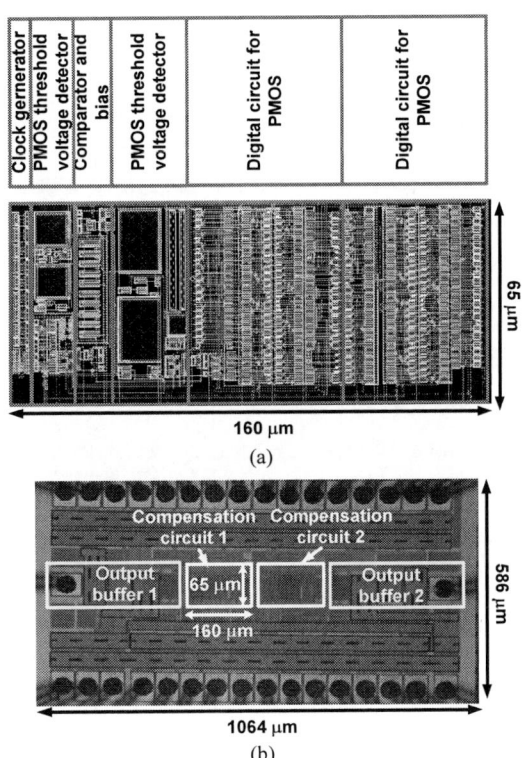

Fig. 7. (a) Layout of the compensation circuit. (b) Die photo of the proposed system.

Fig. 5. Digital circuits for NMOS and PMOS.

edge of maximum slew rate improvement and the capability to detect all corners.

IV. CONCLUSION

An on-chip process and temperature compensation and self-adjusting slew rate control technique for output buffer is proposed in this paper. The maximum slew rate improvement can be achieved over 117 %. Besides, the effects of gate-oxide overstress and the leakage current are both eliminated.

V. ACKNOWLEDGMENT

This investigation is partially supported by Ministry of Economic Affairs, Taiwan, under grant 99-EC-17-A-01-S1-104 and under grant 99-EC-17-A-19-S1-133. It is also partially supported by National Science Council, Taiwan, under grant NSC99-2221-E-110-082-MY3, NSC99-2923-E-110-002-MY2, NSC-99-2221-E-110-081-MY3, NSC-99-2220-E-110-001. It is

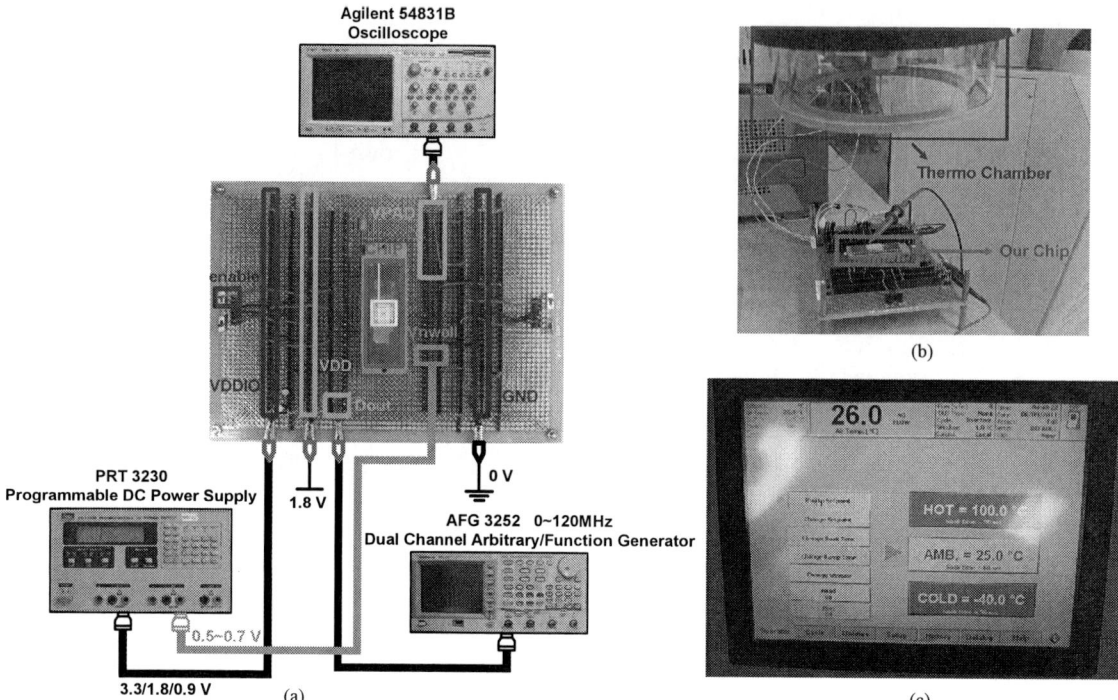

Fig. 8. The measurement settings of the proposed system. (a)In different transmitting modes when VDDIO is at 3.3/1.8/0.9 V. (b)The chip is heated and cooled down by a thermo chamber. (c)Thermo test is between -40°C~100°C.

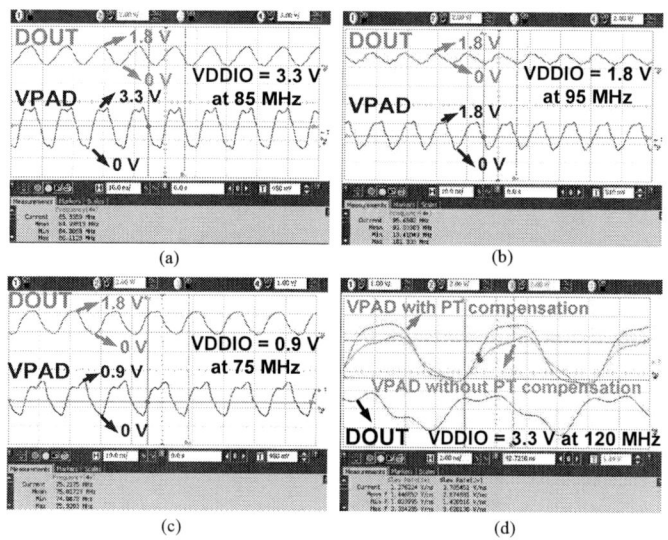

Fig. 9. The maximum data rate of VPAD when (a) VDDIO is at 3.3 V. (b) VDDIO=1.8 V. (c) VDDIO=0.9 V. (d) When VDDIO=3.3 V @120 MHz, the slew rate of VPAD can be compensated over 117 %.

TABLE I
PERFORMANCE CAPARISON OF MIXED-VOLTAGE I/O BUFFER

	Ours	[?] ISSCC	[?] JSSC	[?] JSSC
Year	2010	2007	2003	2003
Process (μm)	0.18	0.18	0.35	0.18
Results	Measured	Post-sim	Measured	Measured
Slew rate (V/ns)	1.28-2.79	2.10-3.58	1.60-2.20	0.40-0.99
Process corners detected	TT, FF, SS FS, SF	TT, FF, SS	TT, FF, SS	TT, FF, SS
Power (mW)	0.427	13.7	N/A	N/A
Slew rate improvement	>117%	N/A	N/A	>32%

also partially supported by the Southern Taiwan Science Park Administration(STSPA), Taiwan, under contract no. EZ-10-09-44-98. The authors would like to express their deepest gratefulness to CIC (Chip Implementation Center) of NARL (National Applied Research Laboratories), Taiwan, for their thoughtful chip fabrication service.

REFERENCES

[1] B. Razavi, "Short-channel effects and device models," *Design of Analog CMOS Integrated Circuits, Mcgraw-Hill*, pp. 599-600, 2001.

[2] Y.-H. Kwak, I. Jung, H.-D. Lee, Y.-J. Choi, Y. Kumar, and C. Kim, "A one cycle lock time slew-rate-controlled output driver," in *IEEE Int. Solid-State Circuits Conf. Dig. Tech. Papers*, pp. 408-611, Feb. 2007.

[3] T. Matano, Y. Takai, T. Takahashi, Y. Sakito, I. Fujii, Takaishi, et al., "A 1-Gb/s/pin 512-Mb DDRII SDRAM using a digital DLL and a slew-rate-controlled output buffer," *IEEE J. Solid-State Circuits*, vol. 38, no. 5, pp. 762-768, May 2003.

[4] S.-K. Shin, W. Yu, Y.-H. Jun, J.-W. Kim, B.-S. Kong, et al., "A slew rate controlled output driver using PLL as compensation circuit," *IEEE J. Solid-State Circuits*, vol. 38, no. 7, pp. 1227-1233, Jul. 2003.

[5] M.-D. Ker, T.-M. Wang, and F.-L. Hu, "Design on mixed-voltage I/O buffers with slew-rate control in low-voltage CMOS process," in *Proc. IEEE Int. Conf. on Electronics, Circuits and Syst.*, pp. 1047-1050, 2008.

[6] V. Narang, B. Arya, and K. Rajagopal, "Novel Low Delay Slew Rate Control I/Os," in *Proc. 1st Asia Symp. on Quality Electronic Design*, pp. 189-193, 2009.

[7] J.-Y. Park, Y. Koo, D.-K.n Jeong, W. Kim, C. Yoo, and C. Kim, "A high-speed memory interface circuit tolerant to PVT variations and channel noise," in *Proc. European Solid-State Circuits Conf*, pp. 293-296, 2001.

[8] C.-C. Wang, R.-C. Kuo, and J.-W. Liu, "0.9 V to 5 V bidirectional mixed-voltage I/O buffer with an ESD protection output stage," *IEEE Trans. Circuits Syst. II, Exp. Briefs*, vol. 57, no. 8, pp. 612-616, Aug. 2010.

Theorems on the Global Convergence of the Nonlinear Homotopy Method for MOS Circuits

Dan Niu[†], Guangming Hu[‡], Yasuaki Inoue[†]

[†]Graduate School of Information, Production and Systems, Waseda University
2-7 Hibikino, Wakamatsu-Ku, Kitakyushu-Shi, 808-0135, Japan
[‡]Toshiba Microelectronics Corporation
25-1 Ekimae-Honcho, Kawasaki-Ku, Kawasaki-Shi, 210-8538, Japan
Email: niudan2010@akane.waseda.jp

Abstract—**Finding DC operating points of nonlinear circuits is an important and difficult task. The Newton-Raphson method adopted in the SPICE-like simulators often fails to converge to a solution. To overcome this convergence problem, homotopy methods have been studied from various viewpoints. However most previous studies are mainly focused on the bipolar transistor circuits and no paper presents the global convergence of the homotopy method for MOS circuits. This paper extends the nonlinear homotopy method to MOS transistor circuits and presents the global convergence theorems of the homotopy method for MOS circuits.**

I. Introduction

With the development of semiconductor technologies, IC circuit designs become more and more computer-dependent. It is still an important and difficult task to find DC operating points of nonlinear circuits in circuit simulation. SPICE-like circuit simulators [1], widely used for designing LSI's, employ the Newton-Raphson (NR) method for solving modified nodal (MN) equations [2]. However, the NR method or its variants often fail to converge to a solution unless the initial estimation point is close enough to the solution. To overcome this non-convergence problem, globally convergent homotopy methods have been studied by many researchers from various viewpoints [3]-[13].

These various researches include how to construct a homotopy function [3] - [5] [11], how to numerically trace a solution curve [5] [7], how to set an initial solution [6] [10] and so on. However, most previous studies for homotopy methods are mainly focused on the bipolar transistor circuits. Since MOS transistor circuits play a main role in the current analog circuit designs, it is urgent to extend homotopy method research and implementation to MOS transistor circuits.

Recently some papers applying the homotopy methods for MOS circuits can be found. In [8], a homotopy method based on ATANSH (Arc-tangent Schichman-Hodges) is proposed for MOS-based mixed-signal circuits. In [7] [13], the nonlinear homotopy method is proposed for MOS circuits. However, all the previous papers do not present the global convergence theorems of the homotopy method for MOS circuits. In this paper, we further extend the MOS nonlinear

homotopy method proposed in [7] [13] by considering the body effect. Moreover, it is significant that this paper presents the global convergence theorems of the homotopy method for MOS transistor circuits by using the extended MOS homotopy method.

This paper is organized as follows. In Section II, The nonlinear homotopy method for MOS transistor circuits is presented, followed by the global convergence theorems of the nonlinear homotopy method for MOS circuits in Section III. Finally, the conclusions are summarized.

II. Nonlinear Homotoy Method for MOS Circuits

A. MN Equation

We review homotopy methods for solving systems of equations of the form

$$f(x) = 0, f(\cdot) : R^n \to R^n. \qquad (1)$$

In the MN equation, Eq. (1) is rewritten as follows [6]:

$$f_g(v, i) \triangleq D_g g(D_g^T v) + D_E i + J = 0 \qquad (2a)$$

$$f_E(v) \triangleq D_E^T v - E = 0 , \qquad (2b)$$

Where $f = (f_g, f_E)^T$, $f_g : R^n \to R^N$, $f_E : R^N \to R^M$,

$x = (v, i)^T \in R^n$, and $n = N + M$. The variable vector $v \in R^N$ denotes the node voltages to the datum node and the variable vector $i \in R^M$ denotes the branch currents of the independent voltage sources. Also, the continuous function $g : R^K \to R^K$ is a VCCS (voltage-controlled current source) type. In addition, D_g is an $N \times K$ reduced incidence matrix for the g branches and D_E is an $N \times M$ reduced incidence matrix for the independent voltage source branches. Moreover, $J \in R^N$ is the current vector of the independent current sources and $E \in R^M$ is the voltage vector of the independent voltage sources.

978-1-4577-1608-9/11 $26.00 © 2011 IEEE

In homotopy methods, we consider an auxiliary equation

$$f^0(x) = 0, \quad f^0 : R^n \to R^n \qquad (3)$$

with a known solution x^0 and construct a homotopy

$$h(x,t) = tf(x) + (1-t)f^0(x), \qquad (4)$$

where $h : R^{n+1} \to R^n$ and $t \in [0,1]$ is the homotopy parameter. Then the solution curve of the homotopy equation

$$h(x,t) = 0 \qquad (5)$$

is traced from the known initial solution $(x^0, 0)$ at $t = 0$. If the solution curve reaches the $t = 1$ hyperplane at $(x^*, 1)$, then a solution x^* to Eq. (1) is obtained [3].

In order to prove the global convergence of homotopy method for MOS circuits, a proper model must be considered.

B. EKV model

In this paper, the EKV model [10] is used for its simplicity. In this model, all the weak, strong and moderate inversion can be expressed in just one equation. What's more, in this model the body node rather than the source node is used as the voltage reference point. The pinch-off voltage (V_p) is used to model the transition regions of the devices.

The drain current of this model is split into a forward current component I_F and a reverse current component I_R. In order to make the strong inversion and weak inversion current expressions been smoothly connected, the drain current normalization and interpolation between weak and strong inversion are made. The voltages and currents are simply normalized using two constant parameters (the thermal voltage V_T and the specific current I_s) [10].

$$v_{p,s,d} = \frac{V_{p,s,d}}{V_T}, \quad i_{d,f,r} = \frac{I_{D,F,R}}{I_s}, \qquad (6)$$

where $V_{p,s,d}$ are the pinch-off, source and drain to body voltage respectively, $v_{p,s,d}$ are the corresponding normalized voltages. $I_{D,F,R}$ are the drain, forward and reverse current respectively and $i_{d,f,r}$ are the normalized currents.

In conclusion, a generalized expression for the drain current is as follows:

$$i_d = i_f - i_r = F(v_p - v_s) - F(v_p - v_d), \qquad (7a)$$

$$F(v) = [\ln(1 + e^{v/2})]^2. \qquad (7b)$$

C. NonlinearMOS Homotopy Method

In this paper, we extend the MOS nonlinear homotopy method [7] [13] and present its global convergence theorems. This homotopy method is constructed using a nonlinear auxiliary function closely related to the original circuit equation to be solved. The auxiliary function is as follows:

$$f^0(x) = f(x) - f(x^0) + \tilde{f}(x) - \tilde{f}(x^0), \qquad (8)$$

Where

$$\tilde{f}(x) = \tilde{f}(v,i) = \begin{bmatrix} D_g \, \tilde{g}(D_g^T v) \\ -Ri \end{bmatrix}. \qquad (9)$$

In Eq. (9), $\tilde{g} : R^K \to R^K$ is a nonlinear function corresponding to the branch $g = (g_1, g_2, \ldots g_K)^T$ in the original circuit equation. R is a non-negative linear resistance. Then the homotopy function is expressed as

$$h(x,t) = f(x) - (1-t)f(x^0) + (1-t)(\tilde{f}(x) - \tilde{f}(x^0)). \qquad (10)$$

In previous nonlinear MOS homotopy method [7] [13], only two MOS diodes are added to the initial homotopy circuit. The voltage vector is denoted as $\begin{bmatrix} V_{gs}, V_{gd} \end{bmatrix}^T$ and the current vector is $\begin{bmatrix} i_s, i_d \end{bmatrix}^T$. This method neglects the body node and does not present the globally convergent property. Since the MOS transistor is a four terminal device, in this paper the current vector i is denoted as $\begin{bmatrix} i_g, i_s, i_d \end{bmatrix}^T$. Three MOS diodes are added and then the initial homotopy circuit can be obtained for a MOS transistor by the circuit interpretation of the homotopy equation at $t = 0$, which is shown in Fig.1. The composite branch (the initial homotopy circuit) $g^{c0} = g + \tilde{g}$ consisting of M, MDgs, MDgd and MDgb is called the composite MOS transistor.

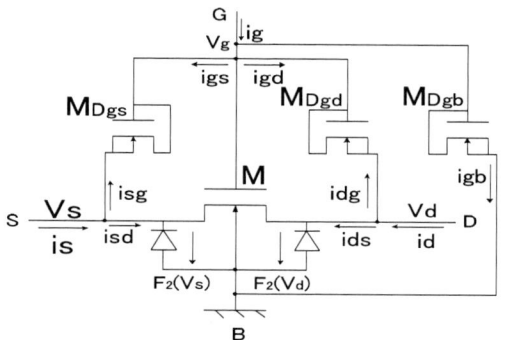

Fig. 1. Composite MOS transistor.

Furthermore, it is clearly seen from Eq. (10) and Fig. 1 that the nonlinear MOS homotopy method is Type I [7] and it is easy to implement on the SPICE-like simulators with no modifications to the existing model subroutines.

III. GLOBAL CONVERGENCE OF NONLINEAR HOMOTOPY METHOD FOR MOS CIRCUITS

A. Theorem of NLH for BJT circuits

Concerning the global convergence of homotopy methods, in order that the solution curve of $h(x,t) = 0$ is guaranteed to reach the $t = 1$ hyperplane, the uniqueness condition and the boundary free condition have to hold [3] [10]. Some homotopy methods for BJT circuits including nonlinear homotopy method (NLH) are proved to be globally convergent for MN equation [3] [5] [11]. For the NLH applying for BJT circuits, we have the following theorem [3].

Theorem 1: Considering the NLH for BJT circuits, assume that υ_e and υ_c (two constant parameters) are sufficiently large, then for any initial point $x^0 \in R^n$ the solution curve of $h(x,t) = 0$ starting from $(x^0, 0)$ reaches $t = 1$. ☐

The condition that υ_e and υ_c are sufficiently large (or the initial homotopy circuit satisfies the uniform passivity) can guarantee the uniqueness condition. The boundary free condition requires that the composite branch $g^c = g + (1-t)\tilde{g}$ satisfies the uniform passivity for all $t \in [0,1)$ on a point. Therefore similarly with the BJT case, the uniform passivity of the composite MOS transistor is also much important for the global convergence theorems.

Firstly, the definition of the terminology uniform passivity is given as follows.

Definition: A continuous function $g: R^K \to R^K$ is said to be uniformly passive on V_b^0 if there exists a $\gamma > 0$ such that $(V_b - V_b^0)^T (g(V_b) - g(V_b^0)) \geq \gamma \|V_b - V_b^0\|^2$ holds for all $V_b^0 \in R^n$. ☐

In this paper, the branch voltage vector V_b for the MOS transistor is denoted as $\begin{bmatrix} V_g, V_s, V_d \end{bmatrix}^T$ ($\begin{bmatrix} V_{gb}, V_{sb}, V_{db} \end{bmatrix}^T$) and $g(V_b)$ is denoted as $\begin{bmatrix} i_g, i_s, i_d \end{bmatrix}^T$.

In order to prove the uniform passivity of the composite MOS transistor, we firstly present the following two lemmas for the single MOS transistor and MOS diode, which are utilized in the subsequent theorems.

Lemma 1: The inequality $V_{ds} i_{ds} \geq 0$ is satisfied for a single MOS transistor on any point V_b. ☐

Proof:

From the EKV model [10], we obtain the drain current equation

$$i_{ds} = F(v_p - v_s) - F(v_p - v_d)$$

$$= \left\{ \ln[1 + \exp(\frac{v_p - v_s}{2})] + \ln[1 + \exp(\frac{v_p - v_d}{2})] \right\} \quad (11)$$

$$\cdot \left\{ \ln[1 + \exp(\frac{v_p - v_s}{2})] - \ln[1 + \exp(\frac{v_p - v_d}{2})] \right\}.$$

From the property of the exponential function and logarithm function, the following inequalities are obtained

$$\ln[1 + \exp(\frac{v_p - v_s}{2})] > 0, \ \ln[1 + \exp(\frac{v_p - v_d}{2})] > 0. \quad (12)$$

In order to prove the lemma 1, we divide the discussion into three parts.

When $V_{ds} > 0$, which implies $V_d > V_s$, therefore $v_d > v_s$ is satisfied. Then the following inequality can be obtained.

$$\ln[1 + \exp(\frac{v_p - v_s}{2})] - \ln[1 + \exp(\frac{v_p - v_d}{2})] > 0. \quad (13)$$

From Eqs. (11) – (13) we obtain $i_{ds} > 0$, which implies that $V_{ds} i_{ds} > 0$ is satisfied.

When $V_{ds} = 0$, we obtain

$$i_{ds} = F(v_p - v_s) - F(v_p - v_d) = 0. \quad (14)$$

In this case, the equation $V_{ds} i_{ds} = 0$ holds.

When $V_{ds} < 0$, in the similar way as $V_{ds} > 0$, we obtain $i_{ds} < 0$, which implies that $V_{ds} i_{ds} > 0$ is also satisfied.

From all the three parts, the Lemma 1 is proved. ☐

What`s more, the MOS diode satisfies the uniform passivity on any point [5] [12]. Then the following lemma can be obtained from the passivity and no-gain properties of the diodes [12].

Lemma 2: The inequality $(V_{ds} - V_{ds}^0)(i_{ds} - i_{ds}^0) \geq 0$ is satisfied on any point V_b^0 for a single MOS diode. ☐

B.   Global convergence of NLH for MOS Circuits

From the lemma 1 and 2, we obtain the following theorem.

Theorem 2: The composite MOS transistor $g^{c0} = g + \tilde{g}$ (initial homotopy circuit) satisfies the uniform passivity on the points V_b^0 satisfying $V_s^0 = V_d^0$. ☐

Proof:

From Fig. 1, the currents of the source, the drain and the gate of the composite MOS transistor are as follows.

$$i_s = i_{sd} + i_{sg} + F_2(V_s), F_2(V_s) = I_s(e^{-V_s/V_T} - 1) \quad (15a)$$

$$i_d = i_{ds} + i_{dg} + F_2(V_d), F_2(V_d) = I_s(e^{-V_d/V_T} - 1) \quad (15b)$$

$$i_g = i_{gs} + i_{gd} + F_2(-V_g), F_2(-V_g) = I_s(e^{V_g/V_T} - 1), \quad (15c)$$

where I_s is the diode saturation current constant, V_T is the thermal voltage.

Therefore, the Inner Product (IP) can be obtained.

$$IP = (V_b - V_b^0)^T (g^{c0}(V_b) - g^{c0}(V_b^0)) = IP_g + IP_d + IP_s$$
$$= (V_{gs} - V_{gs}^0)(i_{gs} - i_{gs}^0) + (V_{gd} - V_{gd}^0)(i_{gd} - i_{gd}^0)$$
$$+ (V_{ds} - V_{ds}^0)(i_{ds} - i_{ds}^0) + (V_g - V_g^0)^T[F_2(-V_g) - F_2(-V_g^0)] + \qquad (16)$$
$$(V_d - V_d^0)^T[F_2(V_d) - F_2(V_d^0)] + (V_s - V_s^0)^T[F_2(V_s) - F_2(V_s^0)]$$

From the assumption $V_d^0 = V_s^0$ and Eq. (14), $i_{ds}^0 = 0$ holds.

From Lemma 1, we obtain that $V_{ds}i_{ds} \geq 0$ holds.

From Lemma 2, we obtain

$$(V_{gs} - V_{gs}^0)(i_{gs} - i_{gs}^0) \geq 0, (V_{gd} - V_{gd}^0)(i_{gd} - i_{gd}^0) \geq 0. \quad (17)$$

Moreover, $F_2(x)$ is uniformly monotone increasing, therefore it is uniformly passive. Then there exist three positive numbers $r_1, r_2, r_3 > 0$ such that

$$(V_g - V_g^0)^T[F_2(-V_g) - F_2(-V_g^0)] \geq r_1 \left\| V_g - V_g^0 \right\|^2, \quad (18a)$$

$$(V_d - V_d^0)^T[F_2(V_d) - F_2(V_d^0)] \geq r_2 \left\| V_d - V_d^0 \right\|^2, \quad (18b)$$

$$(V_s - V_s^0)^T[F_2(V_s) - F_2(V_s^0)] \geq r_3 \left\| V_s - V_s^0 \right\|^2. \quad (18c)$$

Let $\gamma = \min\{r_1, r_2, r_3\}$, we can obtain

$$IP \geq \gamma(\left\| V_g - V_g^0 \right\|^2 + \left\| V_d - V_d^0 \right\|^2 + \left\| V_s - V_s^0 \right\|^2)$$
$$\geq \gamma \left\| V_b - V_b^0 \right\|^2 \qquad (19)$$

holds under the assumption $V_s^0 = V_d^0$, therefore the Theorem 2 is proved. □

The theorem 2 is used for proving the uniqueness condition of the initial solution. For the boundary free condition [5], we have the following Theorem 3.

Theorem 3: The composite branch $g^c = g + (1-t)\tilde{g}$ satisfies the uniform passivity for all $t \in [0,1)$ on the points satisfying $V_g^0 = V_s^0 = V_d^0$. □

The proof of Theorem 3 is similar to the proof of Theorem 2, therefore we omit it for the simplicity.

Under the Theorem 2 and Theorem 3, considering the globally convergent property of the nonlinear homotopy method for MOS circuits, the following Theorem 4 holds.

Theorem 4: Consider the nonlinear homotopy for MOS circuits defined by Eqs. (8) - (10), for any initial solution x^0 satisfying $V_d^0 = V_s^0$, then the solution curve of $h(x,t) = 0$ starting from $(x^0, 0)$ reaches $t = 1$. □

This theorem can be proved by using the proofs of the global convergence theorems of the FPH method [5] [11] and the NLH method [3] for BJT circuits with minor modifications.

The Theorem 5 guarantees the global convergence of the extended MOS homotopy method for any initial solution satisfying $V_d^0 = V_s^0$.

IV. CONCLUSIONS

In this paper, we further extend the previously proposed MOS homotopy method by considering the body effect. Moreover, this paper presents the global convergence theorems of the homotopy method for MOS circuits, which plays a much important role in extending the homotopy methods to MOS circuits.

REFERENCES

[1] L. W. Nagel, "SPICE2: A computer program to simulate semiconductor circuits," Univ. California, Berkeley, CA, ERL-M520, May 1975.

[2] C. W. Ho, A. E. Ruehli and P. A. Brennan, "The modified nodal approach to network analysis," IEEE Trans. Circuits & Syst., vol.CAS-22, no.6, pp.504-509, June 1975.

[3] Y. Inoue, S. Kusanobu, K. Yamamura and M. Ando, "A homotopy method using a nonlinear auxiliary function for solving transistor circuits," IEICE Trans. Inf.&Syst., vol.E88-D, no.7, pp.1401-1408, July 2005.

[4] R. C. Melville, L. TrajkoviC, S.-C. Fang. and L. T. Watson, "Artificial parameter homotopy methods for the DC operating point problem," IEEE Trans. Cornput.-Aided Des. Integrated Circuits & Syst., vol.CAD-12, no.6, pp.861-877, June 1993.

[5] Y. Inoue, S Kusanobu and K. Yamamura, "A practical approach for the fixed-point homotopy method using a solution-tracing circuit," IEICE Trans. Fundamentals, vol.E85-A, no.1, pp.287-298, Jan. 2002.

[6] Y. Inoue, S. Kusanobu, K. Yamamura and M. Ando, "An initial solution algorithm for globally convergent homotopy methods," IEICE Trans. Fundamentals, vol.E87-A, no.4, April 2004.

[7] Kazutoshi SAKO, Hong Yu and Yasuaki INOUE, "A Globally Convergent Method for Finding DC Solutions of MOS Transistor Circuits," IEEJ International Analog VLSI Workshop 2006, Hangzhou China, Nov. 2006.

[8] J.Roychowdhury and R. Melville, "Delivering global DC convergence for large mixed-signal circuits via homotopy/continuation methods," IEEE Trans. Comput.-Aided Des. Integr. Circuits Syst., vol.25, no.1, pp.66-78, Jan. 2006.

[9] C. C. Enz, F. Krummenacher and E. A. Vittoz, "An Analytical MOS Transistor Model Valid in All Regions of Operation and Dedicated to Low-Voltage and Low-Current Applications," Special issue of the Analog Integrated Circuits and Systems Processing Joumal on Low-Voltage and Low-Power Circuits, vol. 8, pp. 83-114, July 1995.

[10] Y. Inoue, S Kusanobu, "Theorems on the Unique Initial Solution for Globally Convergent Homotopy Methods," IEICE Trans. Fundamentals, vol.E86-A, no.9, Sept. 2003.

[11] K.Yamamura and S.Takahashi, "Globally convergent algorithms using the fixed-point homotopy for solving modified nodal equations," IEICE Trans., vol.J81-A, no.7, pp.1094-1098, July 1998.

[12] L. Trajkovi'c, R. C. Melville, and S. C. Fang, "Passivity and no-gain properties establish global convergence of a homotopy method for DC operating points," Proc. IEEE Int. Symp. on Circuits and Systems, New Orleans, LA, vol. 2, pp.914-917, May 1990.

[13] Hong Yu, Inoue, Y., Sako, K., Guangming Hu, Xiaochuan Hu, "An Effective Implementation of the NonlinearHomotopy Method for MOS Transistor Circuits Based on SPICE3," Proc. ICCCAS, pp. 1086 - 1089, July 2007.

Stability Analysis of MEMS Gyroscope Drive Loop Based on CPPLL

Huan-ming Wu[1,2], Hai-gang Yang[1,*], Tao Yin[1], Hui Zhang[1,2]

[1]Institute of Electronics, Chinese Academy of Sciences (CAS)
[2]Graduate University of CAS
Beijing, China
*yanghg@mail.ie.ac.cn

Abstract—Stability analysis of self-oscillation drive loop for MEMS gyroscope based on charge-pump PLL (CPPLL) is presented in this paper. The CPPLL-based driving loop could be realized at a regular voltage CMOS process by maximizing the oscillation amplitude. However, the driving loop has an unstable state at the beginning of the oscillation introduced by the harmonics of the square wave in driving stage. In order to solve this problem, a linear model of the driving loop is established for stability analysis. Based on the analysis, methods of pre-setting the control voltage of VCO and using a hysteresis comparator are proposed to prevent the unstable state occurring. The simulation results verified the validity of the theoretical stability analysis and the effectiveness of the two proposed methods.

I. INTRODUCTION

A driving loop oscillating with large and constant amplitude at a stable frequency is critical for a high performance gyroscope. In most of literatures, an automatic gain control (AGC) or a phase locked loop (PLL) approach is applied in the driving loop to maintain the oscillation [1-3]. To achieve large oscillation amplitude, high driving voltage is commonly used according to these literatures, demanding a high voltage CMOS process implementation. Fortunately, the driving loop based on CPPLL in [4] can be realized at a regular voltage CMOS process as following reasons. First, instead of a sine wave, the square wave is used in the driving stage to achieve the maximum energy transformed to the spring mass in a specific CMOS process. Second, the PLL tracks the resonance frequency of the sensor element to achieve the maximum spring mass displacement with minimum energy. However, the harmonics of the square wave would introduce an unstable state at the beginning of the oscillation. Therefore, stability analysis of the driving loop is necessary. Some effort has been made on the stability analysis of the nonlinear driving loop using the method of time averaging [5-7]. But all these analysis is based on the complicated nonlinear control system model with respect to the amplitude of the oscillation. It is not suitable for the CPPLL-based driving loop with fixed driving voltage, since its oscillation amplitude is also fixed. In this paper, a simple linear model for CPPLL-Based drive loop is established and stability analysis with respect to the phase is presented. Based

on the analysis, two effective circuit design methods are proposed to prevent the unstable state occurring at the beginning of oscillation. Simulation results to verify the theoretical analysis and the proposed methods are also given.

II. BASIS OF CPPLL AND DRIVING LOOP

A CPPLL consists of five functional blocks: phase/frequency detector (PFD), charge pump (CP), low pass filter (LPF), voltage-controlled oscillator (VCO) and divider, as illustrated in Fig. 1. The PFD detects phase or frequency differences between reference and output to activate the charge pump. The charge pump charges or discharges the capacitors in LPF to adjust the control voltage of the VCO, and the oscillating frequency of VCO changes accordingly. The loop locks when the phase and frequency difference drops to zero and the charge pump remains relatively idle.

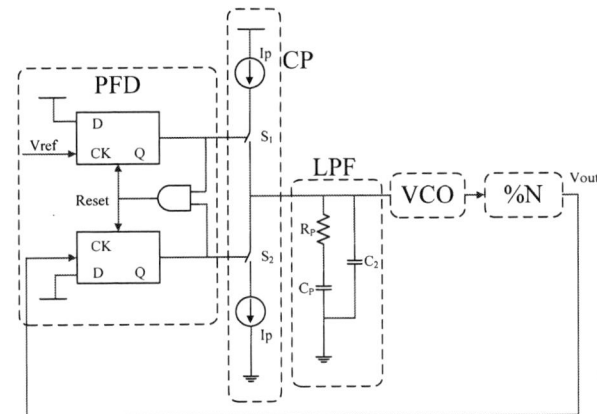

Figure 1. Block diagram of Charge-Pump PLL

A CPPLL can be considered as a linear system by approximating a discrete-time system by a continuous-time model [8]. The CPPLL close-loop transfer function is equal to

$$\frac{\Delta\psi_{out}}{\Delta\psi_{ref}}(s)\Big|_{close} = \frac{\dfrac{I_P K_{VCO}}{2\pi C_2 N}\left(\dfrac{1}{R_P C_P} + s\right)}{s^2\left(\dfrac{1}{R_P C_P}(1+\dfrac{C_P}{C_2}) + s\right) + \dfrac{I_P K_{VCO}}{2\pi C_2 N}\left(\dfrac{1}{R_P C_P} + s\right)}, (1)$$

where $\Delta\psi_{ref}$ and $\Delta\psi_{out}$ are the excess phases of reference and output, I_P is the current of charge pump, R_P is the resistor of LPF, C_P and C_2 are the capacitors of LPF, K_{VCO} is the gain of VCO, and N is division factor of divider.

In Fig.2, gyroscope driving loop based on CPPLL is shown, which consists of MEMS gyroscope, capacitor-to-voltage (C/V) converter, comparator, and CPPLL. The primary mode of MEMS is modeled as a second order spring-damper-system with a dynamic behavior described by

$$\frac{d^2x}{dt^2}+\frac{\omega_x}{Q}\frac{dx}{dt}+\omega_x^2 x=\frac{F}{M}, \qquad (2)$$

where x is the position of the spring mass, ω_x is the natural frequency, Q is the quality factor, F is the electrostatic driving force, and M is the spring mass of the MEMS. Compared to traditional sine wave driving voltage, a square wave V_{out} from divider is applied to the MEMS driving electrodes directly, with fundamental frequency ω_d. Because when the quality factor Q of MEMS is equal to 100, the third harmonic is only 1/2400 of fundamental frequency in position x [3]. Supposing the quality factor Q of MEMS in primary mode is large enough, more than 100, effect of the harmonics in the square wave can be omitted and position x can be expressed as

$$x_{(t)}=A_x\sin(\omega_d t+\varphi_x), \qquad (3)$$

where A_x is amplitude of position x, and φ_x is the phase shift generated from electrostatic driving force F to position x.

A trans-resistance amplifier is used as the C/V converter to generate $90°$ phase shift to ensure the loop phase equal to $0°$. The comparator is used to transfer the sine wave to square wave for PFD.

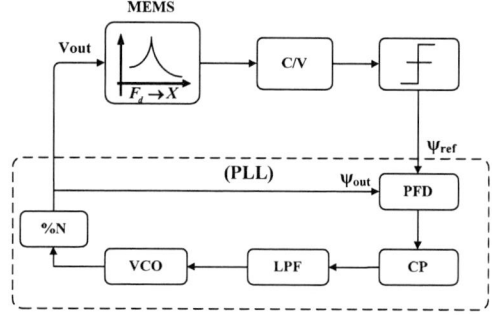

Figure 2. Block diagram of driving loop based on CPPLL

III. ESTABLISHING AND ANALYZING LINEAR MODEL OF DRIVING LOOP BASED ON CPPLL

A linear model of driving loop established in this section is based on the two assumptions as mentioned in section II.

1). the reference frequency of CPPLL is relatively high (about 10 times is preferred) with respect to the CPPLL's closed loop bandwidth, so that the CPPLL behavior can be approximated as a continuous-time system.

2). the quality factor Q of MEMS gyroscope is large enough so that effect of the harmonics of the square wave can be omitted.

The first assumption can be satisfied by carefully selecting the parameters of components in CPPLL. The second assumption is easy to be satisfied, because most of MEMS gyroscope's Quality factor in primary mode is more than 100 even under the atmosphere pressure.

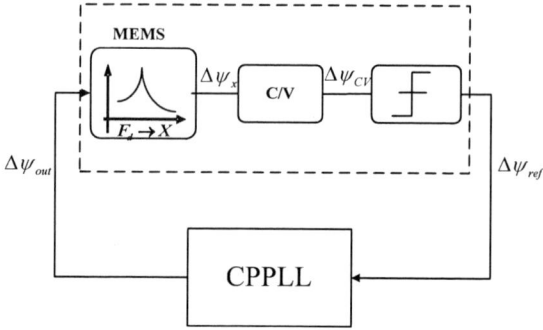

Figure 3. Linear models need to be built

The linear model of CPPLL has been given in (1) with variables $\Delta\psi_{ref}$ and $\Delta\psi_{out}$ being the excess the phases of reference and output signals respectively. Therefore, linear models for MEMS gyroscope, C/V converter and comparator should be built in order to facilitate establishing a linear model for the whole driving loop as illustrated in Fig.3. In (3), φ_x can be expressed as

$$\varphi_x = \arctan\left(-\frac{\omega_x\omega_d}{Q(\omega_x^2-\omega_d^2)}\right). \qquad (4)$$

From (4), a frequency-to-phase response is plotted in Fig. 4. Linearizing the curve at point $(\omega_x, -90°)$, we obtain the slope of the tangent

$$\frac{d\varphi_x}{d\omega_d}\Big|_{\omega_d=\omega_x} = -\frac{2Q}{\omega_x}, \qquad (5)$$

and the tangent equation

$$\varphi_x = -\frac{2Q}{\omega_x}(\omega_d-\omega_x)-\frac{\pi}{2} = -\frac{2Q}{\omega_x}(\frac{d\psi_{out}}{dt}-\omega_x)-\frac{\pi}{2}, \quad (6)$$

where ψ_{out} is the phase of the square wave V_{out}.

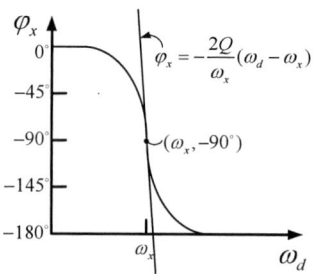

Figure 4. Frequency to phase response of MEMS

Supposing the initial frequency of V_{out} is ω_0, then the phase of the electrostatic driving force F is $\psi_{out}=\omega_0 t+\Delta\psi_{out}$, and the phase of the position x is $\psi_x=\omega_0 t+\Delta\psi_x$. Because φ_x is the phase shift generated from electrostatic driving force F to position x, an equation is obtained as

978-1-4577-1608-9/11 $26.00 © 2011 IEEE

$$\psi_x = \psi_{out} + \varphi_x \Rightarrow \Delta\psi_x = \Delta\psi_{out} + \varphi_x. \qquad (7)$$

Substituting (6) into (7) yields the MEMS linear model in phase domain.

$$\Delta\psi_x = -\frac{2Q}{\omega_x}\frac{d\Delta\psi_{out}}{dt} + \Delta\psi_{out} + 2Q - \frac{2Q\omega_0}{\omega_x} - \frac{\pi}{2}. \qquad (8)$$

Due to the low resonant frequency of MEMS, the bandwidth of C/V converter is large enough to promise a positive 90° phase shift for the signals from MEMS. And the comparator provides 0° phase shift. Therefore, an equation between $\Delta\psi_x$ and $\Delta\psi_{ref}$ is obtained as

$$\Delta\psi_{ref} = \Delta\psi_x + \frac{\pi}{2}. \qquad (9)$$

Combining (8) and (9), a linear model for MEMS, C/V converter and comparator can be expressed as

$$\Delta\psi_{ref} = -\frac{2Q}{\omega_x}\frac{d\Delta\psi_{out}}{dt} + \Delta\psi_{out} + 2Q - \frac{2Q\omega_0}{\omega_x}. \qquad (10)$$

Rewriting (1) in time domain yields

$$KZ\Delta\psi_{ref} + K\frac{d\Delta\psi_{ref}}{dt}$$
$$= \frac{d^3\Delta\psi_{out}}{dt^3} + ZP\frac{d^2\Delta\psi_{out}}{dt^2} + K\frac{d\Delta\psi_{out}}{dt} + KZ\Delta\psi_{out}, \qquad (11)$$

where

$$K = \frac{I_P K_{VCO}}{2\pi C_2 N}, \quad Z = \frac{1}{R_P C_P}, \quad P = 1 + \frac{C_P}{C_2}. \qquad (12)$$

Substituting (10) into (11), an expression of linear model for driving loop is illustrated as

$$\frac{d^3\Delta\psi_{out}}{dt^3} + (ZP + K\frac{2Q}{\omega_x})\frac{d^2\Delta\psi_{out}}{dt^2} + KZ\frac{2Q}{\omega_x}\frac{d\Delta\psi_{out}}{dt}$$
$$= 2QKZ(1 - \frac{\omega_0}{\omega_x}). \qquad (13)$$

Due to

$$\Delta\omega_d = \frac{d\Delta\psi_{out}}{dt}, \qquad (14)$$

we rewrite (13) and get equation as

$$\frac{d^2\Delta\omega_d}{dt^2} + (ZP + K\frac{2Q}{\omega_x})\frac{d\Delta\omega_d}{dt} + KZ\frac{2Q}{\omega_x}\Delta\omega_d$$
$$= KZ2Q(1 - \frac{\omega_0}{\omega_x}) \qquad (15)$$

Solving (15), yields

$$\Delta\omega_d = \omega_x - \omega_0 + C_1 e^{s_1 t} + C_2 e^{s_2 t}, \qquad (16)$$

where C_1 and C_2 are constant value determined by initial state of driving loop, s_1 and s_2 are characteristic roots of (15). Because both of the two roots are negative value, the last two terms in (16) diminish to zero and $\Delta\omega_d$ equals to a stable value $\omega_x - \omega_0$ eventually. In other words, the driving loop based on CPPLL is a stable system and oscillates at the resonant frequency ω_x.

However, it should be noticed that at the beginning of oscillation, if the initial frequency of V_{out} ω_0 is far from the natural frequency ω_x, the linearization of the ω_d to φ_x curve in Fig. 4 was failed and so was the stability analysis. For a square wave with 50% duty cycle, the harmonics is significant large at the odd times of fundamental frequency. If ω_0 is low and the third order harmonic of V_{out} is located on ω_x, the third order harmonic would be significantly amplified by MEMS and introduce another main frequency in position x as shown in Fig. 5. The two main frequencies in position x would cause misjudging of comparator on inversions of output and then lead to the failure of oscillation.

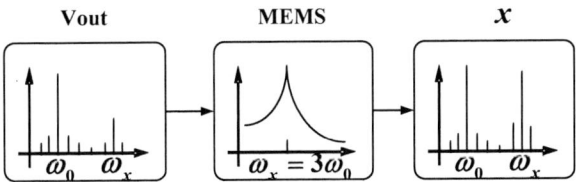

Figure 5. High order harmonics of V_{out} amplified by MEMS

The unstable state of oscillation at the beginning can be eliminated by reducing the effect of the harmonics. Pre-setting the control voltage of VCO to generate V_{out} with initial frequency closed to ω_x can prevent the main high order harmonics located on ω_x. As a result, the effect of the harmonics is significantly reduced. Moreover, a hysteresis comparator can also be used to diminish the effect of the harmonics by filtering the harmonics of its input and make correct inversions. The theoretical analysis and the two proposed methods in circuit design are verified in the following simulations.

IV. SIMULATION RESULTS

A Verilog-A model of the driving loop system is built and parameters of MEMS gyroscope in [5] are adopted for simulation. The numerical values of the parameters of CPPLL used in simulation are listed in table I. The bandwidth of CPPLL is designed as 450 Hz for MEMS gyroscope with 5 kHz natural frequency.

TABLE I
VALUES OF CPPLL'S PARAMETERS USED IN SIMULATION

Parameter	Value
I_P	40 μ A
K_{VCO}	15 M rad/s
C_P	33 n F
C_2	2.2 n F
R_P	30 k Ω

The simulation result of the driving loop without pre-setting the control voltage of VCO (V_{ctrl}) is shown in Fig. 6, where V_{ref} is the reference signal, V_{out} is the output signal from divider, and pos_x is the position. As shown in Fig. 6, V_{ctrl} is continuously raising so are the frequencies of V_{ref} and V_{out}. Therefore, the CPPLL is unlocked at the natural frequency ω_x and the amplitude of pos_x is damping continuously.

The waveform in Fig. 6 is enlarged as shown in Fig. 7 to find the reason of the failure of oscillation. Waveform pos_x on the left side is an irregular sine waveform and its spectrum is shown on the right side, which is caused by the harmonics of the driving wave. The small notches on pos_x is transported to the input of comparator—sig_p - sig_n, and lead to the unwanted inversions of V_{ref}. Therefore, the CPPLL forces the V_{out} tracking the V_{ref} and misses the natural frequency ω_x.

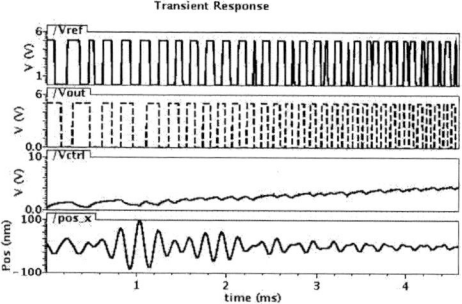

Figure 6. Simulation result of driving loop with initial value of $V_{ctrl} = 0$ V

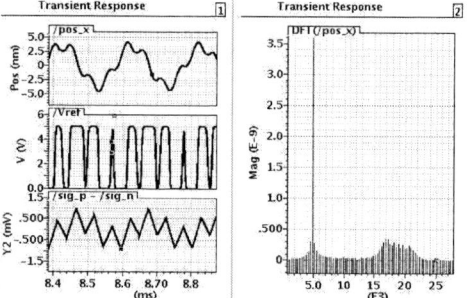

Figure 7. Details of waveforms in Figure 6

The simulation result of pre-setting V_{ctrl} at 1 V initially is shown in Fig. 8. A continuously increasing pos_x reveals that the unstable state of the oscillation is avoided. The V_{ctrl} keeps a constant value after 1 ms and CPPLL locks.

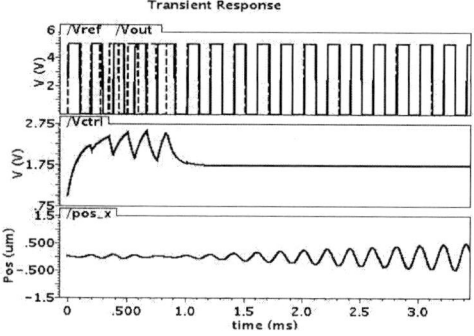

Figure 8. Simulation result of driving loop with initial value of $V_{ctrl} = 1$ V

A driving loop with a hysteresis comparator is also simulated. As shown in Fig. 9, when 2 mV hysteresis voltage is applied in comparator to prevent the unwanted inversion of V_{ref}, the CPPLL locks after 9 ms and the driving loop oscillates successfully. Carefully selection of hysteresis voltage based on the parameters of MEMS and the driving force is important, because overlarge hystereis voltage would filter all signals from C/V and prevent the oscillation.

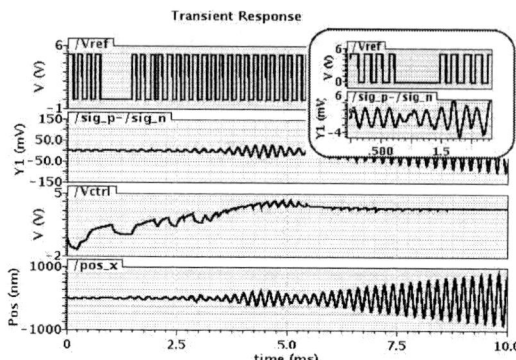

Figure 9. Simulation result of driving loop with a hysteresis comparator

V. CONCLUSION

A linear model for CPPLL-Based drive loop is established by linearizing the MEMS gyroscope, C/V converter and comparator with respect to the phase. Careful analysis on this linear model reveals the existence of an unstable state in the driving loop. Two effective methods of pre-setting the control voltage of VCO and using a hysteresis comparator are proposed to prevent such unstable state occurring. For different MEMS gyroscopes, both the pre-setting voltage and the hysteresis voltage should be carefully chosen based on the parameters of MEMS gyroscopes and driving stage.

REFERENCES

[1] L. Aaltonen, and K. A. I. Halonen, "An analog drive loop for a capacitive MEMS gyroscope," Analog Integrated Circuits and Signal Processing, 63(3): 465-476, 2010.

[2] D. Xia, S. Chen, and S. Wang, "Development of a Prototype Miniature Silicon Microgyroscope," IEEE Sensors Journal, 2009.

[3] B. MO, X. Liu, X. Ding, and X. Tan, "A Novel Closed-loop Drive Circuit for the Micromachined Gyroscope," in Proc. International Conference on Mechatronics and Automation ICMA 2007, Aug. 5-8, 2007, pp.3384-3389.

[4] X. Liu, B. Mo, X. Tan, W. Chen, "Closed-Loop Drive Circuit for the Micromachined Gyroscope Based on the Phase-Locked Technology," Chinese Journal of Nanotechnology and Precision Engineering. 2008, Vol.6, No.6

[5] Z. Wang, Z. Li, and W. Lu, "A new self-oscillation loop for MEMS vibratory gyroscopes," in Proc. 7th International Conference on ASIC ASICON'07, Oct.22-25, 2007, pp.1046-1049.

[6] X. Sun, R. Horowitz, K. Komvopoulos, "Stability and resolution analysis of a phase-locked loop natural frequency tracking system for MEMS fatigue testing," Journal of Dynamic Systems, Measurement, and Control. 2002, 124, 599-605.

[7] P.W. Loveday, C.A. Rogers, "The influence of control system design on the performance of vibratory gyroscopes," Journal of Sound and Vibration. 2002, 255(3), 417-432.

[8] B. Razavi, "Design of Analog CMOS Integrated Circuits", New York: McGraw-Hill Companies, Inc., 2001.

Electromechanical Closed-Loop with High-Q Capacitive Micro-Accelerometers and Pulse Width Modulation Force Feedback

Zhen-hua Ye[1,2], Hai-gang Yang[1], *Member, IEEE*, Fei Liu[1], Tao Yin[1], and Qi-song Wu[1]

1 Institute of Electronics, Chinese Academy of Sciences, Beijing, China
2 Graduate University of the Chinese Academy of Sciences, Beijing, China

Abstract- **This paper presents a closed-loop interface circuit for high quality factor (Q) capacitive micro-accelerometers. High-Q sensing element is desirable for high resolution, but makes the loop control a great challenge. Considering this, closed-loop implementation utilizing analog force feedback is preferred rather than electromechanical $\Delta\Sigma$, which is very popular in recent years. To overcome the non-linearity problem of traditional analogue force feedback scheme, pulse width modulation (PWM) force feedback is developed. The chip measures $2.5\times2.5\text{mm}^2$ in a commercial $0.35\mu\text{m}$ CMOS process. Using a high-Q capacitive micro-accelerometer, the interface circuit attains an input range of $\pm1.25g$, a white noise equivalent acceleration (NEA) of $17\mu g/\sqrt{Hz}$, a less than 0.1% non-linearity from a single 5V supply, and a good degree of robustness.**

I. Introduction

Micro-electro-mechanical-system (MEMS) inertial sensors are becoming increasingly attractive in providing micro-scale, low power, low cost, and high performance solutions for a broad range of applications, including aerospace, military, automotive, and numerous consumer electronic industries. Capacitive micromachined accelerometers, as a representative, gain growing popularity recently, and may overwhelm the conventional macromachined accelerometers in the future [1].

Micromachined sensing system can be simply realized by cascading the micromahined sensor and the readout circuitry. However, constrained by the present MEMS fabrication technology, the open-loop sensing scheme often suffers unacceptable performance degradation due to the non-idealities of micromachined sensor [2]. Electromechanical closed-loop is widely used in high-performance sensing system design. By applying the electrostatic force on the micromachined sensor, negative feedback is established to elaborately control the behavior of the sensing element and minimize the influence of micromachining non-idealities. As a result, electromechanical closed-loop provides better linearity, higher resolution, and larger signal bandwidth.

High-Q sensing element exhibits very low Brownian noise, which is imperative for high-performance applications. Unfortunately, a high-Q component brings the closed-loop into the risk of instability, thus specific care on loop compensation is required. Loop stability can be ensured by using a lead compensator in the feedback path [3], or phase compensator in the forward path [4-6]. Method using positive feedback, together with lag compensation and loop gain controlling, was also developed [7]. All these schemes try to introduce extra left-plane zeros into the loop transfer function

to compensate the effect of the complex poles arising from the high-Q sensing element. Similarly, proportional-derivative (PD) compensation is used in this work.

Electromechanical $\Delta\Sigma$, a very popular closed-loop topology, has many advantages such as direct digit output and high linearity, etc., but faces many challenges in the presence of high-Q sensing element, such as the dilemma between the noise shaping and the loop stabilization [5] and the limitation on input range [6]. This mainly originates from the magnitude discreteness of digital force feedback, which makes the reaction force not an in-time and accurate rebalancing to the external stimulus, but just a time-average and rough one. In fact, the implementation of high-order electromechanical $\Delta\Sigma$ for high-Q sensor is very difficult and complex [4-6].

On the contrary, analogue force feedback behaves well on precisely responding to the external acceleration, with which the close-loop shows excellent controllability and can be easily stabilized by the foregoing compensation methods when high-Q sensor is used. Meanwhile, the noise characteristic is irrelevant with the loop stability and can be greatly improved by traditional low noise circuit design. Nevertheless, there are two main problems for analogue feedback approach. One is the potential threat of "pull-in" in a shock condition, which can be solved to a large extent by utilizing mechanical stoppers [8]. Another one is the poor linearity of voltage-to-force transduction due to the mismatch between the differential actuation capacitors [9]. Without extra linearization, the signal distortion ratio (SDR) is typically limited to lower than 60 dB [10]. To overcome this problem, pulse width modulation (PWM) is developed to achieve high linearity without extra linearization circuit.

This paper describes an electromechanical closed-loop design for high-Q capacitive micro-accelerometer using pulse width modulation force feedback. Section II gives a summarized description of the micro-accelerometer. Section III shows the circuit design details, followed by the simulation and experiment results in section IV.

II. Accelerometer Description

The proposed interface circuit is designed to serve a bulk micromachined capacitive accelerometer [11]. Such an accelerometer consists of a cantilever structure proof mass, as shown in Fig. 1a. It has a fundamental resonant frequency of about 500 Hz, a quality factor of nearly 35, and a differential sensitivity of roughly 20pF/g (g: gravity). The proof mass

---

This work was supported by the Nation High Technology Research and Development Program of China (863 Program) No. 2008AA010701.

978-1-4577-1608-9/11 $26.00 © 2011 IEEE

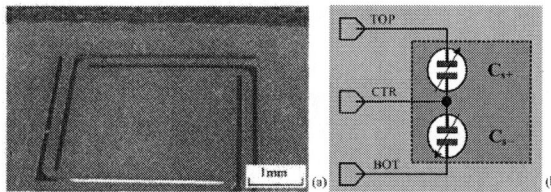

Figure 1. (a) SEM of the cantilever structure proof mass. (b) Equivalent model of the accelerometer.

deviates from the neutral position by a certain displacement according to the external acceleration, making the sensing capacitors vary differentially. By measuring the capacitance difference, the external acceleration is obtained. Fig. 1b is the equivalent model of the accelerometer as a passive three-terminal device. The differential capacitors C_{s+} and C_{s-} are time-multiplexed for both sensing and actuation, as shown in section III.

The prior wok [5] has given the transfer function between the proof mass displacement and the external acceleration:

$$H_m(s) = \frac{x(s)}{a(s)} = \frac{1}{s^2 + \frac{B}{m}s + \frac{k}{m}} = \frac{1}{s^2 + \frac{\omega_n}{Q}s + \omega_n^2} \quad (1)$$

Where B is the mechanical damping coefficient, k is the suspension spring constant, m is the proof mass, ω_n is the resonant frequency, and Q is the quality factor.

III. CIRCUIT DESIGN

There are two fundamental tasks for the interface circuit in an electromechanical closed-loop with high-Q micro-accelerometer. Firstly, compensation is needed for loop stabilization and dynamic characteristic improvement. Secondly, sufficient loop gain is strongly required so as to minimize the influence of practical non-idealities, such as appreciable proof mass displacement, sensing capacitors mismatch, and so on.

The principle of system design is referred to Fig. 2. The micro-accelerometer performs as a one-dimension mass-spring-damper system. High-Q means very small mechanical damping coefficient, which results in insufficient mechanical damping force, and consequently an oscillating time-domain behavior. Since the loop instability originates from the insufficiency of mechanical damping, which is mathematically equal to the product of the mechanical damping coefficient and the proof mass velocity, an "electrical damping" force is raised by the interface circuit to assist the mechanical damper. In detail, the derivative circuit detects the proof mass velocity, and converts it to electrostatic force on the proof mass through the feedback controller. With the help of the "electrical damper", the system settling is greatly accelerated. Meanwhile, "electrical spring" force is provided by the proportional circuit to remarkably enhance the loop gain, thus the proof mass displacement is greatly decreased.

The block diagram of the interface circuit is shown in Fig. 3. The circuit consists of charge sensitive amplifier (CSA), proportional derivative compensator (PD), sample and hold (S&H), and feedback controller (FB). The sensor element

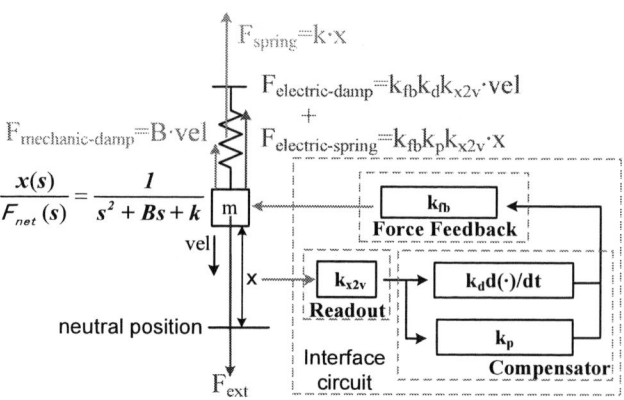

Figure 2. The principle of system design.

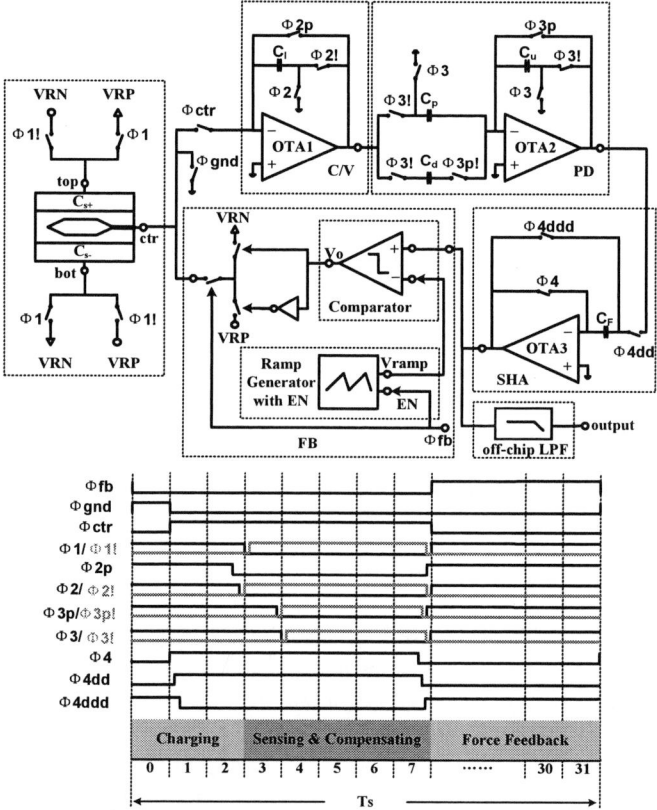

Figure 3. The simplified schematic of the interface circuit and the corresponding clock signals.

interacts with the circuit through a three-terminal interface. An off-chip low pass filter (LPF) is assembled on PCB following the S&H.

The closed-loop system is designed as a switched-capacitor network. As shown in Fig. 3, a single period T_s is divided into 32 units. The first 8 ones are for 3 simultaneous operations: the capacitance-to-voltage converting by CSA, the loop compensation by PD, and the signal sampling by S&H. The next 24 ones are for force feedback by FB. Correlated double sampling (CDS) is employed to depress the electric noise. The clock $\phi2/\phi3$ is a slightly delayed replica of the clock $\phi2p/\phi3p$, so that at the end of resetting phase, the offset

978-1-4577-1608-9/11 $26.00 © 2011 IEEE

and low frequency noise of the operational transconductance amplifiers (OTA's) are stored in the integrating capacitors, and diminished in the next amplifying phase.

A. Charge Sensitive Amplifier

The charge sensitive amplifier reads out the variance of the differential sensing capacitors. The total capacitance (sensing plus parasitic one) is over 400pF, as an expense of high sensitivity. Such a heavy load demands large driving capability and high dc gain of the OTA for accurate readout. In this work, the OTA is of current-mirror structure, offering dc gain of 120dB and gain-bandwidth product of 5M Hz under a 400pF capacitive load.

Denotes the input-referred offset of the OTA as V_{off}, then the CSA output voltage "V_o" can be derived from the charge conservation law:

$$(\frac{V_{dd}}{2}-V_{off})C_{s+}+(V_{off}-\frac{V_{dd}}{2})C_{s-}-V_{off}C_I =$$

$$(-\frac{V_{dd}}{2}-V_{off})C_{s+}+(-V_{off}-\frac{V_{dd}}{2})C_{s-}+(V_o-V_{off})C_I$$

$$\Rightarrow V_o=\frac{V_{dd}}{C_I}(C_{s+}-C_{s-}) \qquad (2)$$

Equation (2) indicates that the capacitance-to-voltage converting is highly linear and immune to the offset of the OTA. Similar analysis shows that the low frequency noise of the OTA is also notably depressed by CDS.

B. Proportional Derivative and S&H

The proportional derivative controller provides sufficient phase lead to stabilize the loop. The transfer function can also be derived from the charge conservation law:

$$\frac{V_o(z)}{V_i(z)}=-(\frac{C_p}{C_u})-(\frac{C_d}{C_u})(1-z^{-1})=-k_p-k_d(1-z^{-1}) \qquad (3)$$

The first item of the equation (3) represents a multiply-by-k effect, and the second item functions as a differentiator. At the end of every compensation phase, the feedback voltage is sampled and held by S&H for the coming force feedback segment.

C. PWM Force Feedback

Highly linear feedback is the critical factor of high linearity closed-loop design. Electromechanical closed-loop faces much more difficulties on this issue than fully-electric one. There are two main non-linearity sources for traditional analogue force feedback: one is the mismatch between the actuation capacitors, another is the non-zero proof mass displacement due to the finite dc loop gain. Usually, the mismatch is on the order of several percent and dominates the nonlinearity. This problem inherently results from the square law relationship between the electrostatic actuation force and the applied voltage [9]. Thereby, time-control rather than voltage-control actuation is developed in this work, named "pulse width modulation (PWM) force feedback".

Fig. 4 shows the principle of PWM force feedback. The potential of the proof mass is switched between V_{act} and V_{gnd}

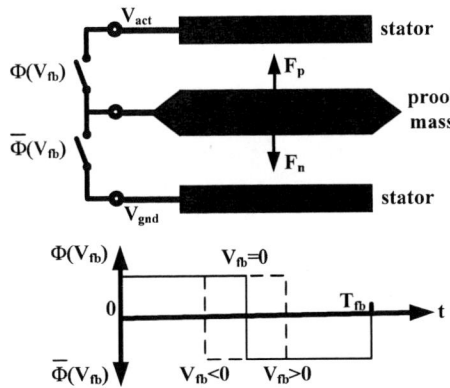

Figure 4. The principle of PWM force feedback.

according to the feedback voltage V_{fb}, so as to determine the polarity and duration time of the electrostatic force applied on it. Because the magnitude of the actuation voltage is constant ($V_{act}-V_{gnd}$), the effective actuation force is proportional to the pulse width of switching signal $\Phi(V_{fb})$. Thus, highly linear force feedback can be realized by precisely-controlled pulse width modulation. The modulator consists of a high linearity ramp generator and a hysterisis comparator, as shown is Fig. 3. By comparing the magnitudes of the ramp signal and V_{fb}, PWM signal is generated.

Denotes F_p and F_n as the force respectively given by the top and bottom electrode set, T_{fb} and T_s as the feedback segment and sampling period, the effective actuation force of PWM force feedback is given as:

$$F_{eff}(V_{fb})=\frac{T_{fb}-\Phi(V_{fb})}{T_s}F_p-\frac{\Phi(V_{fb})}{T_s}F_n \qquad (4)$$

Equation (4) indicates that the mismatch between the actuation capacitors ($F_p\neq F_n$) no longer degrades the feedback linearity, but just results in a dc offset. However, the displacement-dependent nonlinearity still exists, thus sufficient loop gain is necessary. A verilog-A model of the micro-accelerometer was developed for system design and simulation in Cadence analog design environment. Fig. 5 shows the simulated results of system integrated non-linearity (INL) versus the actuator mismatch and loop gain. It is clear that the system SDR keeps better than 60 dB with a loop gain of 42, even when the mismatch rises up to 20%.

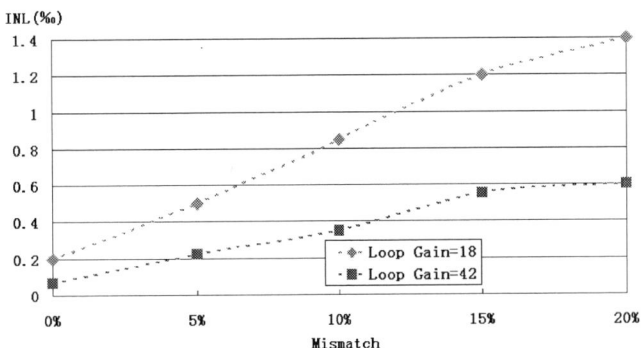

Figure 5. System INL v.s. actuator mismatch and loop gain.

978-1-4577-1608-9/11 $26.00 © 2011 IEEE

IV. EXPERIMENT RESULTS

The prototype is fabricated in a commercial 0.35μm CMOS technology. The chip measures 2.5×2.5mm² and consumes 10mA quiescent current with a single 5V supply. Fig. 6 shows the micrograph of interface ASIC. The system sampling frequency is 125 kHz. Due to the lack of high precision vibration stage, the system sensitivity is characterized by 3 static tests using gravity, which is +1g, -1g, and zero acceleration, and the non-linearity is roughly calibrated as less than 0.1%, as shown in Fig. 7. Fig. 8 shows the measured output power spectra density with zero input. The in-band output white noise floor is 10μV/√Hz, which,

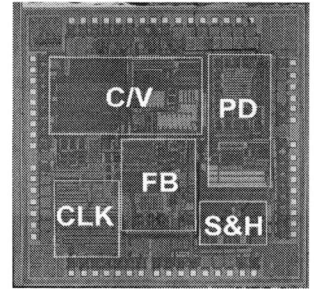

Figure 6. The micrograph of interface ASIC.

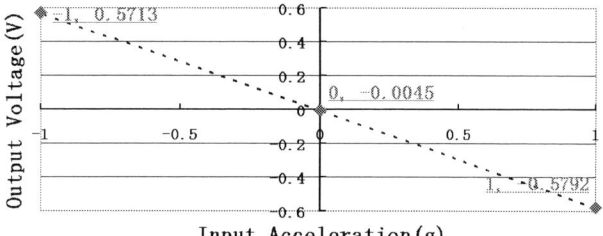

Figure 7. Measured output voltage of 3 static tests.

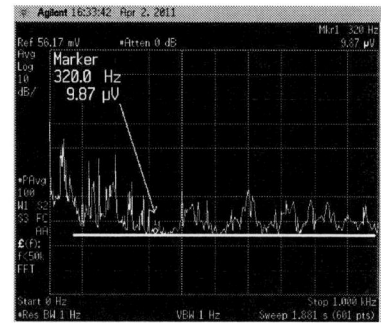

Figure 8. Measured output power spectra density.

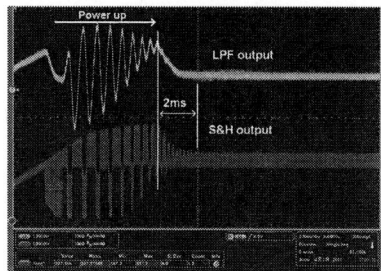

Figure 9. Measured startup.

TABLE 1. PERFORMANCE SUMMARY.

Micro-mechanical Accelerometer	
Resonant Frequancy	489Hz
Quality Factor	35
Whole System	
Sensitivity	0.575V/g (LPF output)
Resolution	17μg/√Hz
Input Range	±1.25g
Non-linearity	< 0.1%

according to the measured sensitivity, corresponds to an NEA of 17μg/√Hz. The numerous unexpected harmonics are mainly caused by the vibrations of the environment. Fig. 9 shows the measured output voltage during startup. The system performance is summarized Table 1.

V. CONCLUSION

This paper demonstrates a comprehensive research on designing electromechanical closed-loop with high-Q capacitive micro-accelerometers. The principle of system design is described. To maintain the closed-loop stable, proportional-derivative compensation is applied. Pulse width modulation force feedback is preferred to traditional analog force feedback for high linearity applications.

REFERENCES

[1] N. Yazdi, F. Ayazi, et al., "Micromachined inertial sensors," Proceedings of the IEEE, vol. 86, pp. 1640-1659, 1998.

[2] W. T. Ang, P. K. Khosla, et al., "Nonlinear Regression Model of a Low-g MEMS Accelerometer," Sensors Journal, IEEE, vol. 7, pp. 81-88, 2007.

[3] Y. Weijie, R. T. Howe, et al., "Surface micromachined, digitally force-balanced accelerometer with integrated CMOS detection circuitry," in Solid-State Sensor and Actuator Workshop, 1992. 5th Technical Digest., IEEE, 1992, pp. 126-131.

[4] V. P. Petkov and B. E. Boser, "A fourth-order Sigma Delta interface for micromachined inertial sensors," Solid-State Circuits, IEEE Journal of, vol. 40, pp. 1602-1609, 2005.

[5] W. Jiangfeng and L. R. Carley, "Electromechanical Delta Sigma modulation with high-Q micromechanical accelerometers and pulse density modulated force feedback," Circuits and Systems I: Regular Papers, IEEE Transactions on, vol. 53, pp. 274-287, 2006.

[6] Y. Dong, M. Kraft, et al., "Higher Order Noise-Shaping Filters for High-Performance Micromachined Accelerometers," Instrumentation and Measurement, IEEE Transactions on, vol. 56, pp. 1666-1674, 2007.

[7] C. D. Ezekwe and B. E. Boser, "A Mode-Matching SD Closed-Loop Vibratory Gyroscope Readout Interface with a 0.004o/s/vHz Noise Floor over a 50 Hz Band," in Solid-State Circuits Conference, 2008. ISSCC 2008. Digest of Technical Papers. IEEE International, 2008, pp. 580-637.

[8] B. V. Amini, R. Abdolvand, et al., "A 4.5-mW Closed-Loop Delta Sigma Micro-Gravity CMOS SOI Accelerometer," Solid-State Circuits, IEEE Journal of, vol. 41, pp. 2983-2991, 2006.

[9] L. Aaltonen, P. Rahikkala, et al., "High-resolution continuous-time interface for micromachined capacitive accelerometer," Int. J. Circuit Theory Appl., vol. 37, pp. 333-349, 2009.

[10] M. Yucetas, J. Salomaa, et al., "A closed-loop SC interface for a +/-1.4g accelerometer with 0.33% nonlinearity and 2mg/vHz input noise density," Solid-State Circuits Conference Digest of Technical Papers (ISSCC), 2010 IEEE International, pp. 320-321, Feb 2010.

[11] W. H. Xu, L. F. Che, Y. F. Li, B. Xiong, and Y. L. Wang, "A Highly Symmetrical Capacitive Accelerometer by Silicon Four-Layer Bonding," Chinese J. Semiconductors, vol. 28, No. 10, pp. 1620–1624, October 2007.

CUDA-Based Acceleration of Post Deblocking Filter

Ting Liu
School of Communication
and Information Engineering,
Shanghai University,
Shanghai, China
Email:tingl1987@163.com

Chunchun chen
School of Communication
and Information Engineering,
Shanghai University,
Shanghai, China
Email:chunchun.chen2010@gmail.com

Eryan. Yang
School of Communication
and Information Engineering,
Shanghai University,
Shanghai, China
Email:eryan.yang@gmail.com

Abstract—**In this paper, an accelerated post deblocking filter algorithm based on CUDA (Computer unifiled device architecture) technology is proposed. The algorithm is used to remove block artifacts and improves the visual quality. It is computationally intensive and usually requires high speed processors to run in real time. A commonly efficient way can be obtained using the DSP (Digital Signal Processing) or other hardware resources. Here, we put forward a software solution method. Using NVIDIA's CUDA technology, we save more hardware resources and make up the low programmability of hardware. CUDA technology assists GPU (Graphics Processing Unit) to work for CPU for large computation. In the experiment, a frame was picked up from JM (Joint Model) decoder stream and filtered by the improved algorithm. Without changing subjective quality, the result shows that the processing time saved 5.69 times than CPU.**

I. INTRODUCTION

With the continuous development of high and new technology, no matter the monitoring management, video conference, digital TV, or streaming media, HD (High-Definition) and 3D technology will be hot spots in the future. They bring us true, rich visual experience. And at the same time, the video transmission burden and processing complexity will increase a lot. For higher compression ratio, some kind of methods are used to reduce video stream and improve real time application, like MPEG-4 [1] organized by MPEG (Moving Pictures Experts Group) and H.264/AVC standard [2], developed by ITU-T Video Coding Experts Group and ISO/IEC Moving Picture Experts Group. High compression based on blocks is their common features. And it takes a common defect, obvious block artifacts which reduce the video quality.

When decoding is finished, video/image appears the discrete or false boundary. We call this phenomenon is Block artifacts. The image is divided into many blocks and more details lost in the picture. It seriously affects the subjective feeling of the human eye, cause uncomfortable visual sense. So, joining deblocking filter in or after encoding/decoding process for good quality in the video is very necessary.

There are two ways of removing block artifacts. One is in-loop deblocking filter and the other is post deblocking filter. The former method operates data within the coding process where the filtered frames are used as reference frames for

motion compensation to subsequent coded frames or output for display [3]. It takes nearly 33% of the whole coding process time [4]. Post deblocking filter processes data in the display buffer and it is less affection of decoding parameters. The algorithms of post deblocking filter are flexible and can be used both MPEG4 and H.264/AVC. It is widely used and more valuable.

The key point of article is researching how to use the post deblocking filter removing block artifacts and improving processing efficiency. At present, many researchers put forward different processing methods from different angles. Andreas Rossholm proposed an adaptive deblocking post filter [5], Doina Petrescu proposed using the BOPS parallel architecture to implementation deblocking [6] and S.-C. Tai proposed an low complexity deblocking method [7]. Especially, the filtering effect of S.-C. Tai's algorithm is better but a little complexity and is used in MPEG4 standard. Here, drawing lessons from the algorithm of Tai, we propose a two mode post deblocking filter based on CUDA. It is more suitable the mainstream compression standard, H.264/AVC.

The experiment results shows real-time performance of algorithm improve 5.69 times on GPU than CPU. Under the premise of not reduce video subjective quality, we can realize more video processing algorithms based on CUDA for the real-time performance. It will be an exploratory direction of HD and 3D display.

In the following, we optimize the algorithm of post deblocking filter based on CUDA for H.264/AVC decoder. An introduction of CUDA technology is given in Section 2. And in Section 3, we optimize post deblocking filter algorithm with CUDA technology and present experiment results. Then, we draw the conclusion in Section 4.

II. CUDA TECHNOLOGY

CUDA is a general purpose parallel computing architecture, with a new parallel programming model and instruction set architecture that leverages the parallel compute engine in NVIDIA GPUs to solve many complex computational problems in a more efficient way than on a CPU.

Figure.1 is CUDA programming model. CUDA programming environment allows the GPU to be programmed through traditional CPU. It means you can use C++ language and compiler to realize operations on GPU. A fundamental building block of CUDA programs is the CUDA kernel function, which is a special C++ function. The kernel is downloaded to the GPU device that acts as a coprocessor to the CPU (host).

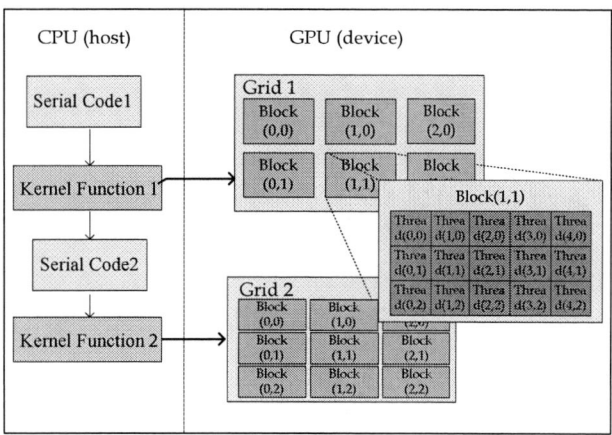

Fig. 1.   CUDA programming model

The kernel function is executed by threads which are organized in a block. There are maximally 512 threads in a block and the threads within a block can co-work with each other through the shared memory. Though the number of threads in a block is limited, we can co-operate multiple blocks or grid whose basic unit is a block to get enough threads for data parallel processing. Nevertheless, blocks within a grid cannot communicate with each other. The CUDA architecture provides access to three kinds of memory: Global Memory, Local Memory and Shared Memory. And memory instructions include any instruction that reads from or writes to shared, local or global memory. Global memory and local memory spaces are not cached and only the global memory can be both read and written by device. By contrast, share memory is on-chip, so it is much faster than any other two kinds of memories. So properly using the shared memory to reduce other kinds of memory access is a good way to reduce delay caused by memory access.

Nowadays, many developers, and researchers take advantage of GPU computing in many areas, like image and video processing, computational biology and chemistry, fluid dynamics simulation, CT image reconstruction, seismic analysis, ray tracing, and so on [8]. It has been proved that image/video processing with CUDA technology can get better efficiency [9] [10]. The following section is the implementation and optimization of post deblocking filter algorithm using CUDA technology.

III. CUDA-BASED POST DEBLOCKING ALGORITHM

Post deblocking filter is used to remove the block artifacts and improve the subjective quality. After post deblocking filter, data will be output directly. Here, we can consider HVS (human visual system) features and depend on former characteristics to realize post deblocking filter algorithm. For better filter affects and more widely implementation, we adopt filtering method related to decoder parameter QP (Quantization Parameter).

The whole filter process can be divided into four parts, including boundary judgment, mode selection, mode filter and realized filter on CUDA.

A. Boundary Judgment

Fig.2 shows the pixel of $4 * 4$ block boundaries. The difference between pixel V3 and V4 are used for judging real boundary. If

$$offset = V4 - V3 \geq 2.5 * QP \qquad (1)$$

It means the difference between V3 and V4 is big and the boundary is true. Otherwise, it is false boundary and the boundary should be filtered.

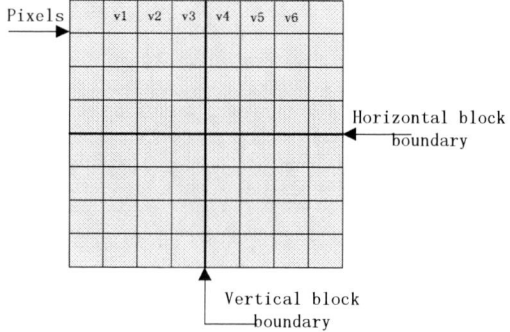

Fig. 2.   pixel of 4*4 block boundaries

B. Mode Selection

The HVS's reaction for flat area and complex area is different. We are more sensitive to block artifacts of the flat area but fuzzy detail of the complex area. So, the filter should be different in flat area and complex area. Below, draw a method from [11] to distinguish where is flat or complex.

It is based on the variable "var" which can reflect the change of sample's size in the area.

$$var = \sum_{i=1}^{5} \phi(V_i - V_{i+1}) \qquad (2)$$

Where

$$\phi(\triangle) = \begin{cases} 0 & \text{If } |\triangle| \leq S \\ 1 & \text{otherwise} \end{cases} \qquad (3)$$

If $var \geq T$, it is the smooth mode, otherwise it is the complex mode. Here, $T = 2$ and $S = 3$ are decided by the experiment value. After choosing the mode, each model has a corresponding filter for removing block artifacts.

978-1-4577-1608-9/11 $26.00 © 2011 IEEE

C. Mode Filter

1) Filter Of Smooth Mode: Under the smooth mode, block artifacts are very obvious. We should update four pixel around the border.

$$V_2' = V_2 + offset \gg 2 \qquad (4)$$

$$V_3' = V_3 + offset \gg 1 \qquad (5)$$

$$V_4' = V_4 - offset \gg 1 \qquad (6)$$

$$V_5' = V_5 - offset \gg 2 \qquad (7)$$

Then, we should guarantee the update pixel between 0 and 255.

2) Filter Of Complex Mode: Under the complex mode, we must avoid filtering too much to make detail fuzzy.QP is used for judging the strength of filter. If $|V2 - V3| < QP$,

$$V_3' = (V_2 + 2 \times V_3 + V_4) \gg 1 \qquad (8)$$

If $|V2 - V3| \geq QP$,

$$V_3' = V_3 + offset \gg 2 \qquad (9)$$

If $|V4 - V5| < QP$,

$$V_4' = (V_3 + 2 \times V_4 + V_5) \gg 2 \qquad (10)$$

If $|V4 - V5| \geq QP$,

$$V_4' = V_4 - offset \gg 2 \qquad (11)$$

Take the 26th frame of video "container" (QCIF QP=28) as an example. Figure.3 is original frame including block artifacts. And Figure.4 shows compare part of the original and processed frame.

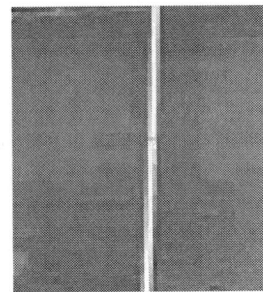

Fig. 3.　frame including block artifacts

Fig. 4.　compare effect of original and processed frame on CPU

D. Realized On CUDA

Based on CUDA technology, we accelerate algorithm for improving processing speed. The biggest characteristic of video is its great data. So, we first should consider choosing an appropriate filter for reducing the size of the memory in graphics cards and select more effective way of access in the GPU.

Post deblocking filter order, the same as in-loop deblocking filter, is based on MB (Macro Block). H.264/AVC standard defines filtering order that first does the horizontal filtering and then the vertical. When all the borders of each MB have been filtered, we will filter the next MB. Filter order in an image is determined by raster scanning way. But, YUV data for post deblocking filter are stored not as the order of MB. The whole image luminance value are stored from left to right, top to bottom, then U and V value. At the same time, because the display card memory is DRAM, the most efficient access way is series access. So, we propose filtering the image as the method of filtering a MB. In another way, we divide an image into $4 * 4$ blocks, and filtering the horizontal direction, then the vertical direction. Figure.5 (a) shows base filter order of Luminance of MB. Figure.5 (b) shows filter order of Luminance of image.

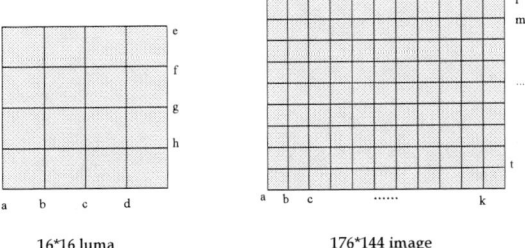

Fig. 5.　(a)Base filter order of luminance MB;(b)proposed filter order of image

We process the same picture to compare the efficiency under CPU and GPU. Based on CUDA, the picture load work is done in CPU and filtering for horizontal and vertical work are performed in GPU in a parallel fashion. How to allocate the threads is very important in the process. Our display card is NVIDIA GeForce GTX 260, its computing power is 1.3. It means every block has 512 threads and warp is consisting of 32 threads. For horizontal filter, in order to make thread's actual implementation results are continuous accessed, we use 36 blocks and each block has 176 threads been used. There are totally 176*36=6336 threads and each responsible for one boundary filtering. Figure.6 shows effect of original frame (left) and processed frame (right) on GPU.

Using HVS, we can see the filter effect is as good as the CPU processing method. Then,we will use clock function of C library to test filter time on CPU and GPU.Table.1 shows the results.

Here, we use proposed algorithm to filter a frame of "foreman" (QP=28, QCIF). Fig.7 shows the effects. The three frame image in this picture from left to right are: original

Fig. 6.   compare effect of original and processed frame on GPU

TABLE I

EXECUTION TIME OF POST DEBLOCKING ALGORITHM(TIME UNIT:S)

Execution Num	Times(CPU)	Times(GPU)	GPU clock cycle	speed up
1	0.0160	0.0027	104216	5.92
2	0.0150	0.0030	104184	5
3	0.0160	0.0026	104308	6.15

frame, filtered frame on CPU, filtered frame on GPU. Because the same size frame has same data volume to processing. So, its accelerate multiples has not changed a lot.

Fig. 7.   filtering effect contrast

From the experiment results,the purposed algorithm based on CUDA can more efficiently remove block artifacts than the algorithm realized on CPU. It also has been proved that this solution can be generally and widely applied.

IV. CONCLUSION

In this paper, we use CUDA technology to speed up the H.264/AVC post de-blocking filter. The results show it is an effective approach to deal with this highly processing algorithm. H.264/AVC is the most widely used standard and its optimization work is very necessary and important. For future work, we will continue to optimize other modules and reduce complexity of the whole decode process based on CUDA technology.

ACKNOWLEDGMENT

This project is Supported by the Key Project of Chinese Ministry of Education.(NO210069), STCSM NO. 09220502600, and STCSM NO. 09PJ1405100.

REFERENCES

[1] "Generic coding of audio-visual objects," *Part 2Video*, 1998.
[2] H.-T. R. 11496-10, "Advanced video coding for generic audiovisual services," *IEEE International Conference .Multimedia and Expo.*, March 2005.
[3] P. List, A. Joch, J. Lainema, G. Bjontegaard, and M. Karczewicz, "Adaptive deblocking filter," *Circuits and Systems for Video Technology, IEEE Transactions on*, vol. 13, no. 7, pp. 614 –619, july 2003.
[4] M. N. Bojnordi, M. R. Hashemi, and O. Fatemi, "A fast two dimensional deblocking filter for h.264/avc video coding," in *Electrical and Computer Engineering, 2006. CCECE '06. Canadian Conference on*, may 2006, pp. 2017 –2020.
[5] A. Rossholm and K. Andersson, "Adaptive de-blocking de-ringing post filter," in *Image Processing, 2005. ICIP 2005. IEEE International Conference on*, vol. 2, sept. 2005, pp. II – 1042–5.
[6] D. Petrescu, "Efficient implementation of video post-processing algorithms on the bops parallel architecture," in *Acoustics, Speech, and Signal Processing, 2001. Proceedings. (ICASSP '01). 2001 IEEE International Conference on*, vol. 2, 2001, pp. 945 –948 vol.2.
[7] S.-C. Tai, Y.-R. Chen, C.-Y. Chen, and Y.-H. Chen, "Low complexity deblocking method for dct coded video signals," *Vision, Image and Signal Processing, IEE Proceedings -*, vol. 153, no. 1, pp. 46 – 56, feb. 2006.
[8] C. Application, "http://www.nvidia.com/object/cuda."
[9] J. Cao, M.-C. Che, X. Wu, and J. Liang, "Gpu-aided directional image/video interpolation for real time resolution upconversion," in *Multimedia Signal Processing, 2009. MMSP '09. IEEE International Workshop on*, oct. 2009, pp. 1 –6.
[10] S. Yang, R. Cheng, and L. Zou, "Case study of programmable video post processing: Cuda-based novel edge directed video scaling," in *Multimedia and Expo (ICME), 2010 IEEE International Conference on*, july 2010, pp. 884 –889.
[11] W. Guo, J. Xie, and Z. Zhang, "Low complexity deblocking algorithm and implementation with a configurable processor," in *Advanced Computer Theory and Engineering, 2008. ICACTE '08. International Conference on*, dec. 2008, pp. 581 –584.

High Performance JPEG Decoder Based on FPGA

Junming Shan
School of Communication and
Information Engineering
Shanghai University
Shanghai, China
Email: junming.shan@gmail.com

Duyao Wang
School of Communication and
Information Engineering
Shanghai University
Shanghai, China
Email: wangduyao@yahoo.com.cn

Eryan Yang
School of Communication and
Information Engineering
Shanghai University
Shanghai, China
Email: eryan.yang@gmail.com

Abstract— Abstract: This paper describes designing a real-time Joint Photographic Experts Group (JPEG) decoder which is capable of high definition images decoding and realized in a Xilinx Vertex5 Field Programmable Gate Array (FPGA). We propose a highly efficient pipelining FPGA implementation of the two-dimensional inverse discrete cosine transformation. Ping-pong buffer is introduced in order to improve decoding performance. This JPEG decoder IP performs decompression of 1920 x 1080 pixels images with a speed of 30 frames per second at a required operating frequency as low as 100 MHz.

I. INTRODUCTION

Recently communication and storage cost are reduced by doing data compression, especially in the digital storage media area which is more reliable. As a result, a high performance decoder is required necessarily. Nowadays, there are many standards for image compression. Joint Photographic Experts Group (JPEG)[1] is one of the most poplular standars for photographic images decoding.

For real-time implementation of JPEG decoder, there are many different methods. However, only some implementations of JPEG decoder are efficient. Zulkalnanin mohdYousof, *et al* presented an FPGA based baseline JPEG decoder[2], which only decodes small-size images in real time. Mohammed Elbadri *et al* presented a hardware support of JPEG[3], which also has low frequency. Yoshihiro Noguchit proposed a method of extracting embedded binary data from jpeg bitstream using standard JPEG decoder[4].

This papers presents a soft intellectual property (IP) for high performance JPEG decoder based on FPGA, which can decode HD 1920 x 1080 pixels images with efficient results. This IP is also a reusable hardware block that can be used in many different designs.

In the following, Section II introduces the architecture of the proposed IP. Section III presents the verilog simulation and synthesis results. Finally, the conclusion of this work is presented in Section IV.

II. ARCHITECTURE

The proposed IP includes four modules: JPEG data, JPEG decoder, Post processing, VGA, as show in Fig.1.

A. JPEG data module

JPEG data module includes two parts: ROM and FIFO. The ROM stores the image data, and the FIFO sends the data into

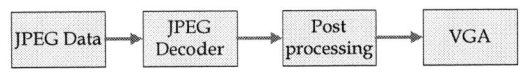

Fig. 1. Architecture of IP

the JPEG decoder with 100MHZ, as the JPEG decoder's data frequency is 100MHZ.

B. The architecture of JPEG decoder

The core of this soft IP is the JPEG decoder. The JPEG decoder includes Huffman decoder module, de-quantizer and inverse zig-zag module, Inverse Discrete Cosine Transformation (IDCT) module, color space conversion module, as show in Fig.2. Each block stage is designed to contain several pipeline stages.

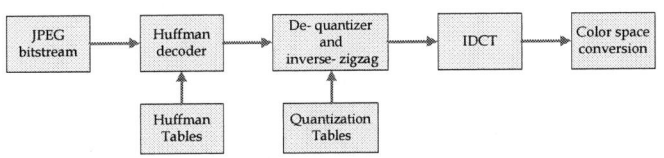

Fig. 2. block diagram of JPEG decoder

1) Huffman decoder module: Huffman encoding is a widely used compression algorithm. Because of the varying length of the coded codes, it increases the difficulty of decoding. There are many approaches for Huffman decoder, such as, binary tree search[5], and Canonical Huffman Table(CHT) [6]. In this paper we use the CHT algorithm because it produces smaller codes table and is faster in searching for the cord word and its corresponding symbol in the symbol list[7]. Further CHT algorithm greatly reduces the memory consumption, and the search time. And its block diagram is show in Fig.3.

The CHT algorithm includes three steps: judging the code length, calculation of memory address and reading the stored data for ROM. Because the Huffman tables include four tables: color DC Table, color AC table, luminance DC table, luminance AC table, as a result this architecture uses four ROM memory blocks to restore these Huffman tables, these tables are used to decoder bit streams.

First, the decoder gets decoding data and sends the data into the code length detector for interpretating the code length. Then according to the code length memory address

978-1-4577-1608-9/11 $26.00 © 2011 IEEE   57

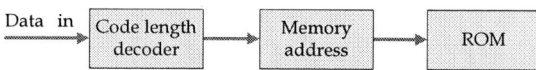

Fig. 3. block diagram of CHT

is calculated. Finally, it can get the code words. According to code words, it achieves the amplitude of the code from the input data, and restore the original coded data. In this design the Huffman decoding process is divided into seven steps as show in Fig.4.

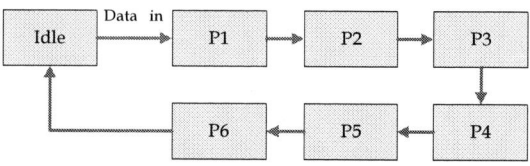

Fig. 4. Seven steps of Huffman decode

For this state machine, in the Idle state, there is no data coming in. Once there is data coming in, it will get in the P1 state. In the P1 state, it gets a table data and table number, then migrating to the P2 state. In the P2 state, it compares the calculated table data with the saved table and gets the code length. In the P3 state, the algorithm calculates the storage address. In the P4 state, it gets the symbol from the ROM. In the P5 state, according to the symbol, it gets the restored data and asserts how many data are decoded, and the address of the decoded data. In the P6 state, it asserts whether a data unit is finished. After finish these steps, the Huffman decoding has been finished.

2) De-quantizer and inverse- zigzag module: De-quantizer and inverse-zigzag is done by one module. De-quantizer is accomplished by multiplying the value with the quantization table in order to generate the reconstruction of the IDCT coefficients. Inverse-zigzag is implemented by using lookup table. In this block, the architecture designed for the inverse-zigzag buffer is consisting of two RAM memories that are time-interleaved for the reading and writing operations, which means using the ping-pong buffer and improves the efficiency of the inverse-zigzag.

3) IDCT module: The 2D-IDCT is important in the JPEG decoder. In order to restore the image information, the 2D-IDCT must be implemented. As 2-D IDCT calculation is a highly complex computation, which restricts its use in several applications. So a fast and efficient IDCT algorithm is important for image decoding. A 8*8 2-D IDCT is defined as:

$$f(x,y) = \frac{1}{4} \sum_{u=0}^{7} \sum_{v=0}^{7} C(u)C(v)F(u,v)$$
$$\cdot \cos \frac{(2x+1)u\pi}{16} \cos \frac{(2y+1)v\pi}{16} \qquad (1)$$

if u and v are zeroes, $C(u) = \frac{1}{\sqrt{2}}$, $C(v) = \frac{1}{\sqrt{2}}$, other else $C(u)$ and $C(v)$ are equal to 1.

TABLE I
1-D IDCT

1-D IDCT algorithm steps	
First step:	
	$A0 = (P0 + P4) * C4\_16$
	$A1 = (P0 - P4) * C4\_16$
	$A3 = (P2 * C2\_16) + (P6 * C6\_16)$
	$A2 = (P2 * C6\_16) - (P6 * C2\_16)$
	$A7 = (P1 * C1\_16) + (P7 * C7\_16)$
	$A4 = (P1 * C7\_16) - (P7 * C1\_16)$
	$A6 = (P5 * C5\_16) + (P3 * C3\_16)$
	$A5 = (P5 * C3\_16) - (P3 * C5\_16)$
Second step:	
	$B0 = A0 + A3 \qquad B3 = A0 - A3$
	$B1 = A1 + A2 \qquad B2 = A1 - A2$
	$B4 = A4 + A5 \qquad B5 = A4 - A5$
	$B7 = A7 + A6 \qquad B6 = A7 - A6$
Third step:	
	$A6 = (B5 + B6) * 181/256$
	$A5 = (B6 - B5) * 181/256$
Fourth step:	
	$D0 = B0 + B7 \qquad D1 = B1 + A6$
	$D2 = B2 + A5 \qquad D3 = B3 + B4$
	$D4 = B3 - B4 \qquad D5 = B2 - A5$
	$D6 = B1 - A6 \qquad D7 = B0 - B7$
in which Ci_16 means $\cos \frac{\pi i}{16}$	

The fast and efficient IDCT mainly has two approaches, i.e., row-column decomposition and direct decomposition. The speed of direct decomposition is fast, however it costs huge resource. The row-column decomposition's structure is simple, and more regular for hardware implementations. In this paper, we used the row-column decomposition. Particularly chen et al.'s algorithm[8] is employed. In this algorithm, as 2-D IDCT is a separable transform, it can be expressed in matrix notation as two 1-D IDCT[9]. The 8 data's 1-D IDCT is defined as:

$$f(x) = \frac{1}{2} \sum_{u=0}^{7} C(u)F(u) \cos \frac{(2x+1)u\pi}{16} \qquad (2)$$

if u is 0, $C(u) = \frac{1}{\sqrt{2}}$, else $C(u)=1$.

1-D IDCT algorithm has four completely independent steps, thus making it straightforward to pipeline these steps as show in TABLE I.

The Fig.5 shows the architecture for the 2-D IDCT, two 1-D IDCT processing and a Transpose RAM is required to connect these two 1-D IDCT modules.

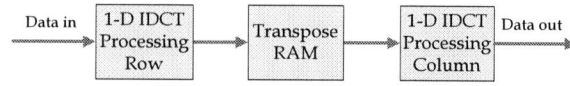

Fig. 5. block diagram of 2-D IDCT

With the data going through the four algorithm steps, then the data should be stored in Transpose RAM. The Transpose RAM is used to connect the two 1-D IDCT. And the results from the first 1-D IDCT should be stored line by line. After a basic unit data are finished by the first 1-D IDCT, the data in the Transpose RAM will be read out column by column by

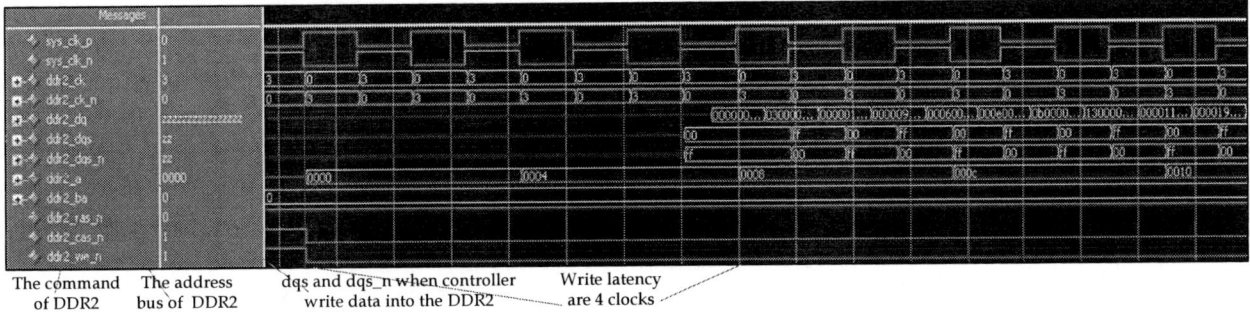

The command The address dqs and dqs_n when controller Write latency
of DDR2 bus of DDR2 write data into the DDR2 are 4 clocks

Fig. 6. The simulation of writing data into DDR2 SDRAM

the second 1-D IDCT. After the second 1-D IDCT is finished, the data will be send into the next module.

4) Color space conversion model: In this module, the JPEG decoding is almost completed. The next is to convert the color space from YCbCr to RGB for display purpose, and the conversion formula are shown as follows. Finally, the converted data is displayed on the display device.

$$R = Y + 1.402Cr \qquad (3)$$
$$G = Y - 0.344Cb - 0.714Cr \qquad (4)$$
$$B = Y + 1.772Cb \qquad (5)$$

C. Post processing module

This IP is designed for display purpose. A DDR2 SDRAM is designed in this module because the DDR2 SDRAM can store large data. This module includes two asynchronous FIFO, 24bit convert 128bit, two Liner buffers, 128bits convert 28bits and DDR2 as show in Fig.7.

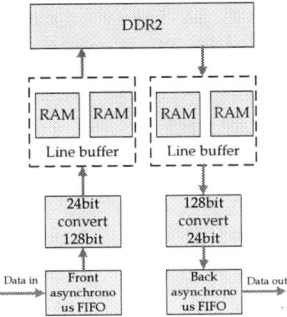

Fig. 7. block diagram of post processing IP

As data's clock frequency from the JPEG decoder is 100MHZ and DDR2 SDRAM receives data using clock frequency as 200MHZ, the front asynchronous FIFO is required. The data is written into the FIFO with 100MHZ and the data is read out from the FIFO with 200MHZ, which meet the DDR2's request. Since the VGA display module receive data using frequency as 65 MHZ, so the back asynchronous FIFO is also required.

The data from the JPEG decoder are 24 bits, however, DDR2 SDRAM receives data in terms of 128-bit bitwidth.

As a result the 24bit to 128bit converter module is designed. However when the data is send into the VGA, the VGA receiving data bit wide is 24 bits. The model 128bit converting to 24 bit is also designed.

Line buffer in the post processing module is used to improve the data transfer speed. This liner buffer uess the ping-pong buffer mechanism, which increases the efficiency of the buffer.

The DDR2 SDRAM can store large data, which can meet different requirements. Its frequency is 200 MHZ and its bit wide is 128 bits. The simulation of writing data into DDR2 SDRAM is show in Fig.6. The simulation of reading data out from DDR2 SDRAM is show in Fig.8, which show correctness of operating this DDR2 controller. The IP is used in the video decoder since a picture must be stored for the video decoder. In case it is the HD images, the data are so large and a DDR2 SDRAM must be used for store these data. Finally, when the post processing is finished, the output is send to the VGA for display.

III. VERILOG SIMULATION AND SYNTHSIS RESULT

This high-performance JPEG decoder IP is written in verilog, we use the modelsim SE 6.5 to simulate the IP's functionality. By comparing the simulating result with the results of Matlab JPEG decoder, it shows the result is correct. The simulate result is show in Fig.9. Fig.10 is the original image, and Fig.11 is decoded by the data from the JPEG decoder. From two pictures bit-by-bit comparison, we can also see the result is correct. As the original image has 1920*1080 pixels, so the JPEG Decoder can decompress larger image than the JPEG Decoder proposed in [2],and that JPEG Decoder only can decompress 320*240 image size.

Our platform is virtex 5 LX110T FPGA, The synthesis results are presented TABLE II.

TABLE II
DEVICE UTILIZATION OF VIRTEX 5 LX110T FPGA

Device utilization	used	total	utilization
Slices registers	1833	69120	2 %
Slice LUTs	1551	69120	2 %
Fully used LUT -FF pairs	1169	2215	52 %
Bonded IOBS	122	640	19%
Block RAM/FIFO	5	148	3%
BUFG /BUFGCTRCS	10	32	31%
Dcm_advs	3	12	25%

978-1-4577-1608-9/11 $26.00 © 2011 IEEE

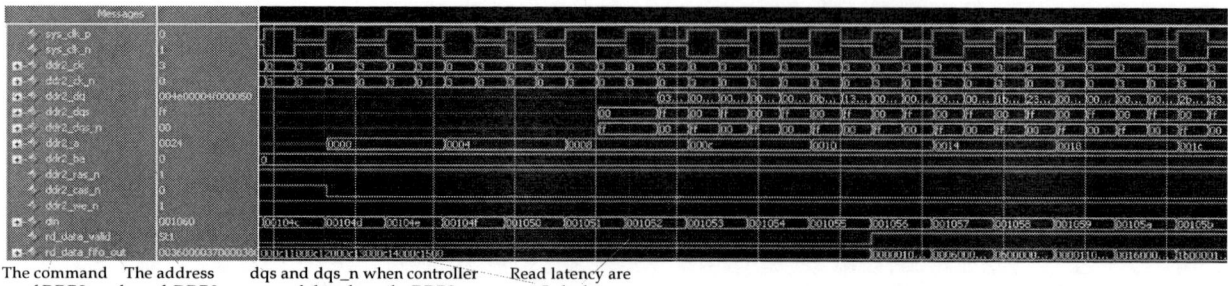

The command of DDR2 | The address bus of DDR2 | dqs and dqs_n when controller read data from the DDR2 | Read latency are 5 clocks

Fig. 8. simulation results of reading data out from DDR2 SDRAM

Fig. 9. simulation of JPEG Decoder

Fig. 10. original image

Fig. 11. decoded image

IV. CONCLUSION

In this paper, the IP design of high performance JPEG decoder based on FPGA is presented. The architecture of the IP was detailed in this paper, and the synthesis results after mapping to Xilinx FPGA were presented. The number of arithmetic operations of 2-D IDCT is reduced by utilizing row-column approach and the DDR2 SDRAM is used for the HD video decode. The IP performs decompression of 1920 x 1080 pixels images with the speed of 30 frames per second at a required operating frequency as low as 100 MHz.

ACKNOWLEDGMENT

This project is supported by the Key Project of Chinese Ministry of Education.(NO210069), STCSM NO. 09220502600, and STCSM NO.09PJ1405100.

REFERENCES

[1] K.-B. Lee and C.-C. Ju, "A memory-efficient progressive jpeg decoder," in *VLSI Design, Automation and Test, 2007. VLSI-DAT 2007. International Symposium on*, april 2007, pp. 1 –4.

[2] Z. Yusof, Z. Aspar, and I. Suleiman, "Field programmable gate array (fpga) based baseline jpeg decoder," in *TENCON 2000. Proceedings*, vol. 3, 2000, pp. 218 –220 vol.3.

[3] M. Elbadri, R. Peterkin, V. Groza, D. Ionescu, and A. El Saddik, "Hardware support of jpeg," in *Electrical and Computer Engineering, 2005. Canadian Conference on*, may 2005, pp. 812 –815.

[4] Y. Noguchi, H. Kobayashi, and H. Kiya, "A method of extracting embedded binary data from jpeg bitstreams using standard jpeg decoder," in *Image Processing, 2000. Proceedings. 2000 International Conference on*, vol. 1, 2000, pp. 577 –580 vol.1.

[5] K.-L. Tsai, P. Lan, S.-J. Ruan, and M.-C. Shie, "A low power high performance design for jpeg Huffman decoder," in *Electronics, Circuits and Systems, 2008. ICECS 2008. 15th IEEE International Conference on*, 31 2008-sept. 3 2008, pp. 1151 –1154.

[6] Y.-J. He, D.-L. Zhang, B. Shen, and L.-F. Geng, "Implementation of fast Huffman decoding algorithm," in *ASIC, 2007. ASICON '07. 7th International Conference on*, oct. 2007, pp. 770 –773.

[7] R. Hashemian, "Condensed Huffman coding, a new efficient decoding technique," in *Circuits and Systems, 2002. MWSCAS-2002. The 2002 45th Midwest Symposium on*, vol. 1, aug. 2002, pp. I – 228–31 vol.1.

[8] W.-H. Chen, C. Smith, and S. Fralick, "A fast computational algorithm for the discrete cosine transform," *Communications, IEEE Transactions on*, vol. 25, no. 9, pp. 1004 – 1009, sep 1977.

[9] X. Ma, J. Gao, and J. Chen, "A high-performance low-power 2d 8 times;8 idct processor with asynchronous pipeline," in *ASIC, 2005. ASICON 2005. 6th International Conference On*, vol. 1, oct. 2005, pp. 341 – 344.

978-1-4577-1608-9/11 $26.00 © 2011 IEEE

A fast noise variance estimation algorithm

Wenjiang Liu, Tao Liu, Mengtian Rong , Ruolin Wang, Hao Zhang

Department of Electronic Engineering, School of Electronic
Information and Electrical Engineering, Shanghai Jiao Tong University
Shanghai, China
liuwenjiang@sjtu.edu.cn, ttlyz@sjtu.edu.cn, rongmt@sjtu.edu.cn, wrlhrl@sjtu.edu.cn, zhanghao0953@yahoo.cn

Abstract—Noise estimation is an important part of video and image processing. In real time denoising of CMOS sensor environment, it is especially required to achieve fast and accurate noise estimation. This paper presents a fast and efficient noise estimation algorithm. The algorithm is block-based, in which noise of image or video is assumed to be additive zero-mean Gaussian noise. The algorithm requires $N1 \times N1$ window and the corresponding operation of computing variance. In order to achieve fast and accurate noise estimation, the method we use can avoid the sorting process making the complexity order of the algorithm down to $O(n)$ from either $O(n^2)$ or $O(n \log_2 n)$. In order to improve the accuracy of this algorithm, the main parameters in this paper are set to be adjusted adaptively according to the content of images or videos. By conducting experiments we find that the algorithm is fast and accurate.

I. INTRODUCTION

The inevitable existence of noise in videos and images will not only have an impact on people's subjective feelings, but also play a crucial role on some image processing algorithms. These algorithms include image restoration, edge detection, and image segmentation and so on. Noise can be added to the images or videos in a variety of ways, such as acquisition, recording, and transmission.

Generally, videos or images are assumed to be contaminated by the additive zero mean white Gaussian noise [2-5]:

$$I(i,j) = S(i,j) + N(i,j) \qquad (1)$$

where $S(i,j)$ represents the original image (noise-free), $I(i,j)$ denotes the observed noisy image, and $N(i,j)$ signifies the signal-independent noise.

Existing noise estimation algorithms can be divided into two categories: intra-frame and inter-frame approaches. Intra-frame approach can also be classified as block-based method, smoothing-based method and wavelet-based method. Olsen proposed a smoothing-based method [2], in which some other methods are also evaluated. John Immerkaer proposed a fast noise estimation algorithm [3]. The method uses a zero mean operator which is almost insensitive to image structure to evaluate noise. Dong-Hyuk Shin *et al.* proposed a block-based method [4], in which an adaptive Gaussian filter is used. Aishy Amer *et al.* proposed a fast block-based intra-frame method [5]. In this method, image structure is taken into account and a efficient estimate of big noise can be given. Donoho *et al.* proposed a wavelet-based noise estimation algorithm [6], but these wavelet-based algorithms are not suitable for fast implementation.

Some other noise models have also been considered. Olsen evaluated an algorithm in which multiplicative and additive noise are both taken into account [2]. Li Jin-Chao *et al.* also considered the quantization noise and estimated it [8]. Ce Liu *et al.* used the inverse function of Camera Response Function (CRF) to synthesize CCD noise [10].

The proposed algorithm which is a block-based intra-frame method, only takes zero-mean additive white Gaussian noise into account. However, the proposed algorithm also uses the information between frames to avoid sorting, thus speeding up the rate of noise estimation and being conductive to rapid implementation.

In general, existing denoising algorithms require accurate noise estimation [1][9]. Priyam Chatterjee *et al.* have compared some excellent denoising algorithms (BM3D, K-SVD, SKR, K-LLD) [7]. However, there are few papers discussing the effect of noise estimation on denoising algorithms. In this paper we will test how noise estimation affect denoising algorithms. In [9], Kostadin Dabov *et al.* proposed an excellent denoising algorithm named BM3D. The performance of this denoising algorithm is pretty good, we take it as the research object.

The rest of the paper is structured as follows. Section II tests the effect of noise estimation on BM3D denoising algorithm. Section III describes a block-based noise estimation algorithm. The improved algorithm is presented in Section IV. The calculation of the parameters is in Section V. Experiment results and discussion are shown in Section VI and conclusions are given in Section VII.

This work supported by Science and Technology Commission of Shanghai Municipality. project No:10706201300 and

Applied materials shanghai research development fund. project No:09700713900

978-1-4577-1608-9/11 $26.00 © 2011 IEEE

II. THE EFFECT OF NOISE ESTIMATION ON BM3D DENOISING ALGORITHM

Samples are two pictures, followed by 'barbara' and 'boat'. Test code is downloaded from the official website of the Department of Signal Processing, Tampere University of Technology (http://www.cs.tut.fi/~foi/GCF-BM3D/). Each image is contaminated by white Gaussian noise with same noise level sigma = 20. We artificially set the noise levels (sigma=15, 20, 25) which are supposed to be estimated by noise estimation algorithm. And we give the PSNR analysis of the images before and after denoising process. The results of BM3D denoising process which are based on three different noise estimates are shown in Figure 1. We give one part of the two noisy images (sigma=20) respectively, then demonstrate the same part of each image which is denoised with three different noise estimates.

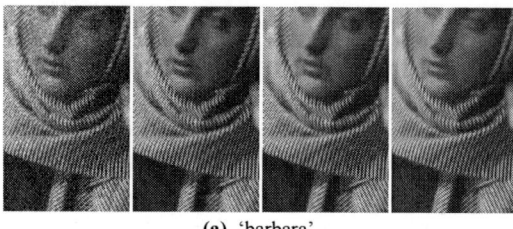

(a) 'barbara'

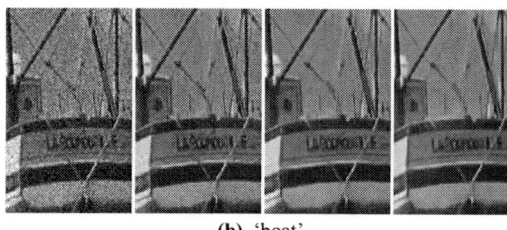

(b) 'boat'

Figure 1.   Fragments of noisy (sigma=20) grayscale images and the corresponding BM3D results. (a) sigma=20, PSNR=22.115dB. Esti-sigma=15, PSNR=29.538dB. Esti-sigma=20, PSNR=31.778dB. Esti-sigma=25, PSNR=31.435dB. (b) sigma=20, PSNR=22.115dB. Esti-sigma=15, PSNR=29.392dB. Esti-sigma=20, PSNR=30.882dB. Esti-sigma=25, PSNR=30.388dB.

It can be seen from Figure 1 that when the estimated noise differs from real noise, BM3D denoising results are affected and the increase of PSNR become smaller. When the noise estimate is 15, it is clear that two images are still noisy. When the noise estimate becomes 25, although we can't see noise clearly, the two images are all over-smoothing and the image details are covered up. So accurate noise estimation is the guarantee for successful denoising. Next, we will introduce a noise estimation algorithm.

III. BLOCK-BASED NOISE ESTIMATION ALGORITHM

A block-based noise estimation algorithm is described in [1]. Suppose the noise is additive zero mean white Gaussian noise, σ_N^2 is the variance of noise in the current frame. To estimate σ_N^2, the current frame is divided into blocks of equal size, and the blocks with small pixel variances are considered to be caused by noise. According to the experience of Liwei Guo *et al.* [1] and our experiment results, we choose 8×8

block size for videos with Common Intermediate Format (CIF) or lower resolution and 16×16 for larger videos. Suppose there are TBN blocks in the current frame. We select the minimum 3% out of the total TBN block variances in the current frame, and calculate their average $\bar{\sigma}_v^2$. We estimate the noise variance σ_N^2 from $\bar{\sigma}_v^2$ using the following equation.

$$\sigma_N^2 = \alpha \cdot \bar{\sigma}_v^2 \qquad (2)$$

where α is a correction factor which is chosen by experiments [1],

$$\alpha = 1 + 0.001(\bar{\sigma}_v^2 - 40) \qquad (3)$$

When selecting the minimum 3% out of the total TBN block variances, we inevitably use some sorting method. However, the complexity of currently used sorting method is $O(n^2)$ or $O(n \log_2 n)$, that is why fast noise estimation cannot be achieved. Therefore, this algorithm is not conductive to the rapid implementation either in software or hardware. Generally, video data is processed by software or hardware, the input order of data is from left to right and then from top to bottom. Based on this feature, we propose an improved fast and accurate noise estimation algorithm in this paper.

IV. IMPROVED FAST NOISE ESTIMATION ALGORITHM

Based on the noise estimation algorithm in [1], we propose an improved algorithm. For each video sequence, the following steps are needed to calculate the corresponding parameters:

A. Assuming a frame of size $M \times N$, the current frame is divided into blocks of equal size $N1 \times N1$. The value of N1 is determined by the size of the frame: when the frame is larger than CIF, N1 = 16; when the frame is less than or equal to CIF, N1 = 8.

B. Let the sliding window of size $N1 \times N1$ slide in the first row of current frame from left to right, calculate the variance of each block in the sliding window. And then calculate the mean variance ($\overline{n_1}$) of the blocks in this row.

C. Calculate the variance of the blocks in the remaining rows in turn according to step B, and get the mean variance of each row ($\overline{n_k}$, $2 \le k \le Y$). Y is the number of blocks per column, obtained by dividing M by N1 and then rounding up the result. Similarly, the number of blocks per row is obtained with the similar process.

D. Averaging the mean variance of each row, then get the mean value of all the blocks in the current frame ($\bar{v}$).

$$\bar{v} = \frac{\sum_{j=1}^{Y} \overline{n_j}}{Y}, 1 \le j \le Y \qquad (4)$$

The algorithm flow is as follows:

Initial: Determine the mean value of block variances of noise-free video frame ($\overline{vc}$). Above all, filter the first frame of video sequence with median filter. Then calculate the mean value of all block variances in the filtered frame ($\overline{vc}$) according to step A B C D described above.

978-1-4577-1608-9/11 $26.00 © 2011 IEEE

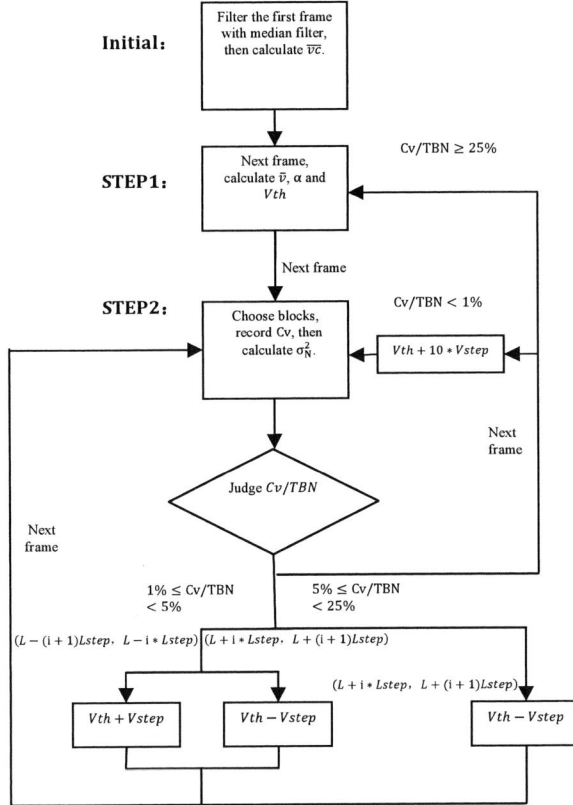

Figure 2. Flow chat of the proposed algorithm.

STEP1: Determine the initial reference threshold value (Vth). Choose the next frame as the current frame, the mean value of all block variances $\bar{v}$ and the obtained $\overline{vc}$ in **Initial** are needed to calculate the correction factor α:

$$\alpha = \frac{\sqrt{\bar{v} - \overline{vc}}}{50} \qquad (5)$$

where 50 is obtained by experiments. Then calculate the reference threshold value (Vth) with α and $\bar{v}$. The purpose is that the number of blocks that we choose by Vth is 3% of TBN.

$$Vth = \alpha\bar{v} \qquad (6)$$

STEP2: Choose the next frame as the current frame, divide it into blocks of equal size N1 × N1. As mentioned above, calculate the variance of each block using the sliding window. Take out the block of which the variance is smaller than Vth, at the same time record the number of chosen blocks (Cv). Averaging the variances of the obtained blocks and get $\bar{\sigma}_v^2$. Then calculate σ_N^2 according to equation (2).

Next, adjust Vth:

When $1\% \le Cv/TBN < 5\%$, Vth needs precise adjustment. Since the frame will meet this range under normal circumstances. Then select the corresponding adjustment step

$Lstep = \text{ceil}(TBN \times 0.005)$ and the adjustment center $L = \text{ceil}(TBN \times 0.03)$.

When $5\% \le Cv/TBN < 25\%$, Vth needs rough adjustment for fast implementation. Then select the corresponding adjustment step $Lstep = \text{ceil}(TBN \times 0.01)$ and the adjustment starting point $L = \text{ceil}(TBN \times 0.01)$. TBN is the number of N1 × N1 blocks in the current frame.

When $Cv \in (L - (i+1)Lstep, L - i*Lstep)$, $Vth = Vth + Vstep$. When $Cv \in (L + i*Lstep, L + (i+1)Lstep)$, $Vth = Vth - Vstep$. $Vstep$ will be given in next section.

Repeat **STEP2**. By adjusting several frames, Vth will be appropriate. The number of blocks that we choose by Vth will converge to 3% of TBN. At this time, the obtained σ_N^2 is a stable noise variance estimate. If $Cv/TBN > 25\%$, back to **STEP1**. The algorithm flow chart is shown in Figure 2.

V. PARAMETER CALCULATION

By experiments, we obtain $Vstep = i \times 0.01\alpha \times \bar{v}$. Since the noise of input video changes little under common circumstances, Vth of each frame will not change much and $1\% \le Cv/TBN < 25\%$. Vth needs fine-tuning according to Cv of each frame. When Cv increases, which means noise is decreasing, the corresponding Vth should decrease. When Cv decreases, which means noise is increasing, the corresponding Vth should increase.

When noise changes sharply, Cv will be out of the range $1\% \le Cv/TBN < 25\%$. At this time, the adjustment of Vth will be difficult to grasp. Therefore, the algorithm select: When $Cv/TBN \ge 25\%$, re-calculate the mean value of block variances ($\bar{v}$) in the next frame. Then we will get a new threshold value. For the current frame, we use the estimate of previous frame as a substitute. This is acceptable in continuous video sequences. When $Cv/TBN < 1\%$, $Vth = Vth + 10 * Vstep$, and then repeat **STEP2**.

This algorithm does not require too much storage space. It needs $L + \text{ceil}(TBN \times 0.22)$ storage units at most. And the operation process does not need to re-process the pixel data of N1 × N1 block. So the complexity order can be computed with only O(n). Since the block of size N1 × N1 is the basic block of various coding and denoising algorithms in which video data is processed from left to right and then from top to bottom. Therefore, this algorithm can perform noise estimation during the coding and denoising process.

The accuracy of noise estimation depends on the number of blocks. We use 3% of TBN [1] in this paper. If the blocks are too much, image structure is involved in estimation process and the noise will be over-estimated. If the blocks are too few, the estimate will also be inaccurate.

VI. EXPERIMENT RESULTS AND DISCUSSION

In this section, we present the experiment results of the two test sequences (Due to page limit, we can only present two sequences of our eight test sequences.). The two test sequences are 'city' and 'soccer', of which the frame size is 704×576. For 'city' and 'soccer', TBN = 6336, $L = 191(1\% \le Cv/TBN < 5\%)$, $L = 64(5\% \le Cv/TBN < 25\%$.

978-1-4577-1608-9/11 $26.00 © 2011 IEEE

$Lstep = 32(1\% \leq Cv/TBN < 5\%)$, $Lstep = 64(5\% \leq Cv/TBN < 25\%)$.

The screenshot of single frame for each sequence is shown in Figure 3. The convergence situation of noise estimation for 'city' is presented in Figure 4 in which noise changes sharply. Figure 4 shows that when noise changes significantly (Significant changes occur in frame 8 and frame 15.), the algorithm can achieve stable estimate within four frames. It is completely acceptable in a real video surveillance system. The performance comparison of the proposed algorithm and the original algorithm is shown in Figure 5, of which (a) and (b) illustrate the results of 'city' and 'soccer' respectively. Obviously, the two algorithms have high accuracy and consistency. It is because the nature of the two algorithms is same; the proposed algorithm is merely faster than the original algorithm which is shown in Table I. The proposed algorithm can obviously increase the speed of noise estimation. And frame of larger size involves more sorting times, especially for high definition video which is the current trend. So the proposed algorithm can reduce the complexity of algorithms which contain sorting process. Specifically, the complexity drops from $O(n^2)$ or $O(n \log_2 n)$ to $O(n)$. Furthermore, the speed of noise estimation is improved significantly.

TABLE I.   TIME RATIO COMPARISON BETWEEN THE PROPOSED AND ORIGINAL METHOD

Estimation methods	Proposed	Liwei Guo[1]
Time Ratio(city)	1	1.4240
Time Ratio (soccer)	1	1.4364

(a) 'city'(704 × 576)   **(b) 'soccer'(704 × 576)**

Figure 3.   Test video sequences in experiments.

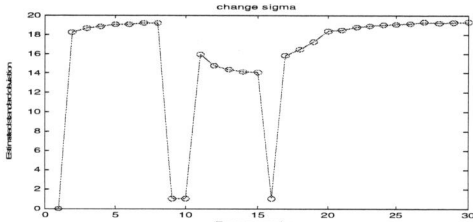

Figure 4.   Convergence situation when sigma changes.

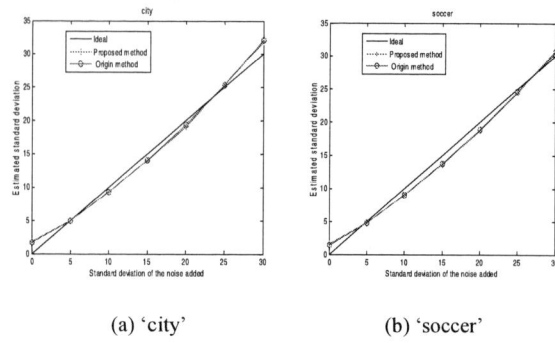

(a) 'city'   (b) 'soccer'

Figure 5.   Performance comparison of the two methods.

VII.   CONCLUSION

We proposed a fast and efficient noise estimation algorithm which is block-based and focused on video noise in this paper. The main parameters are set to be adjusted adaptively according to the information of video frames. So it can provide efficient estimates. At the same time, the algorithm uses a threshold comparison method to avoid the sorting process. Thus the complexity of algorithms which contain sorting process can be reduced significantly and real-time estimate of noise can be achieved.

REFERENCES

[1] Liwei Guo, Oscar C. Au, Mengyao Ma, and Zhiqin Liang, "Fast Multi-Hypothesis Motion Compensated Filter for Video Denoising," Journal of Signal Processing Systems, vol. 60, pp. 273-290, 2010

[2] S. I. Olsen, "Estimation of Noise in images: An Evalution," Graphical models and image processing, vol. 55, pp. 319-323, 1993

[3] John Immerkaer, "Fast Noise Variance Estimation," Computer Vision and Image Understanding, vol. 64, pp. 300-302, 1996

[4] Dong-Hyuk Shin, Rae-Hong Park, Seungjoon Yang, and Jae-Han Jung, "Block-Based Noise Estimation Using Adaptive Gaussian Filtering," IEEE Trans. Consumer Electronics., vol. 51, pp. 218-226, 2005

[5] Aishy Amer, and Eric Dubois, "Fast and Reliable Structure-Oriented Video Noise Estimation," IEEE Trans. Circuits and Systems for Video Technology, vol. 15, pp. 113-118, 2005

[6] D. L. Donoho, and I. M. Johnstone, "Ideal Spatial Adaptation by Wavelet Shrinkage," Biometrika, vol. 81, pp. 425-455, 1994

[7] Priyam Chatterjee, and Peyman Milanfar, "Fundamental Limits of Image Denoising: Are We There Yet?" IEEE International Conference on Acoustics, Speech and Signal Processing, pp. 1358-1361, 2010

[8] Li Jin-Chao, Tang Hui-Ming, and Lu Chao, "Noise Estimation in Video Surveillance Systems," World Congress on Computer Science and Information Engineering, vol.6, pp. 578-582, 2009

[9] Kostadin Dabov, Alessandro Foi, Vladimir Katkovnik, and Karen Egiazarian, "Image denoising by sparse 3D transform-domain collaborative filtering," IEEE Trans. Image Process., vol. 16, pp. 2080-2095, 2007

[10] Ce Liu, William T. Freeman, Richard Szeliski, and Sing Bing Kang, "Noise Estimation from a Single Image," IEEE Computer Society Conference on Computer Vision and Pattern Recognition, vol. 1, pp. 901-908, 2006

A Real-Time Heart Beat Detector and Quantitative Investigation based on FPGA

Cheng Dong, Chio In Ieong, Mang I Vai, Peng Un Mak, Pui In Mak, Feng Wan

Biomedical Engineering Laboratory, FST and State-Key Laboratory of Analog and Mixed-Signal VLSI,
University of Macau, Macao, China

Abstract—A Field Programmable Gate Array (FPGA) based system for single-lead electrocardiogram signal QRS complex detection is presented in this paper. The system consists of Quadratic Spline wavelet transform, moving average filter, signed squaring and a Modulus Maxima Pair Recognition module. The parallel and pipelined architecture of the system allows a maximum throughput equals 46MS/s. The QRS Complex detection accuracy is validated using MIT/BIH arrhythmia database, in which, sensitivity of 99.35% and predictivity of 99.70% are achieved. Less than 2000 logic elements are utilized to implement this algorithm in an Altera cycloneII FPGA. The performance and resource consumption show that the design also suits for digital ASIC, which benefits in processing high volume of ECG data.

I. INTRODUCTION

AUTOMATIC detection of the QRS complex is necessary for efficient extraction of beat-to-beat intervals (RR) from long electrocardiogram (ECG) recordings such as nighttime data or 24-hour Holter monitoring, which is useful for heart rate variability analysis, ECG classification and compression. QRS Complex is the most significant part of the ECG waveform and the detection of its position is helpful for the determination of other ECG characteristic points. In most cases, the temporal location of the *R*-wave is taken as the location of the QRS complex.

In the recent decades, many QRS complex detection approaches have been proposed; for example, algorithm base on band-pass filter and nonlinear transform [1], algorithm from the field of artificial neural networks [2], filter banks [3], etc. The performance of applying wavelet transform method to the task of QRS complex detection was reported in [4, 5]. This kind of method benefits from the time-frequency analysis property of wavelet transform. By employing several detection rules, the overall detection accuracy can exceed 99.8% [4]. However, some of these rules are too complex for hardware real-time implementation.

On the other hand, detection errors can be reduced by the application of computationally more expensive algorithms. However, particularly in the case of ASIC design, the computational complexity means larger chip area and larger power consumption. Hence, a tradeoff between complexity and detection performance needs to be carefully balanced.

The following sections will illustrate the whole profile of the QRS complex detection system, including feature extraction by linear and nonlinear transform, decision making.

The work is supported by the Research Committee of the University of Macau and the Science and Technology Development Fund (FDCT).

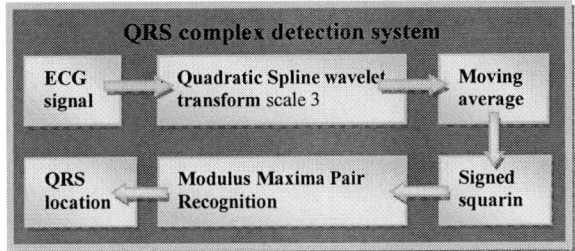

Fig. 1. System architecture of the proposed system.

Lastly, evaluation result on standard database is reported. Despite the performance of an algorithm on a database is not the ultimate answer as to its utility in a clinical environment, it provides a standardized means of comparing the basic performance of one algorithm to another [1].

II. ALGORITHM AND IMPLEMENTATION

The proposed beat detection algorithm consists of Quadratic Spline wavelet transform, moving average, signed squaring and Modulus Maxima Pair Recognition. The arrangement of these functional blocks is illustrated in fig.1.

A. Quadratic Spline wavelet transform

In the first step, Quadratic Spline Wavelet Transform (QSWT) is chosen. Theoretically, the discrete and inflexion points of a signal can show different obvious characteristics in multi-resolution after quadratic spline wavelet transform. Making use of this advantage, the high pointed QRS complex (especially R peak) in the ECG signal, after wavelet transform, will be transformed into pairs of positive maximum and negative minimum.

The Fourier transform of Quadratic Spline Wavelet $\psi(x)$ is[6,7]

$$\Psi(\omega) = i\omega(\frac{\sin(\omega/4)}{\omega/4})^4 \qquad (1)$$

The high pass and low pass filters H(ω) and G(ω) are Eq. 2 and Eq. 3 respectively.

$$H(\omega) = e^{i\omega/2}(\cos\frac{\omega}{2})^3 = \frac{1}{8}(e^{2i\omega} + 3e^{i\omega} + 3 + e^{-i\omega}) \qquad (2)$$

$$G(\omega) = 4ie^{i\omega/2}(\sin\frac{\omega}{2}) = 2(e^{i\omega} - 1) \qquad (3)$$

Let $Q_j(\omega)$ be the transfer function of the equivalent filter. Then, we have

978-1-4577-1608-9/11 $26.00 © 2011 IEEE

$$Q^{j=1}(\omega) = G(\omega) \tag{4}$$

$$Q^{j=2}(\omega) = G(2\omega)H(\omega) \tag{5}$$

$$Q^{j>2}(\omega) = G(2^{j-1}\omega)H(2^{j-2}\omega)...H(\omega) \tag{6}$$

From the equations above, we can derive that the equivalent filter coefficients are the integral multiple of 2 power an integer depend on the scale of wavelet j, that is,

$$FilterCoefficients = m \times 2^{-3(j-1)+1} \tag{7}$$

$$m = ...-1, 0, 1, 2...$$

This characteristic of Quadratic Spline wavelet makes it suitable for FPGA implementation, since all the equivalent filter coefficients can be represented by fixed point binary decimal without losing any information. The width of the binary decimal is equal to the absolute value of 4-3j. If the data being processed can be also fully express in fixed point binary form, the whole hardware filter can have the same result as the floating point version in software. This is just the case for MITBIH (Massachusetts Institute of Technology and Beth Israel Hospital) arrhythmia database, which we use to validate the proposed algorithm, because all the data collected in this database is obtained by an 11-bit ADC after an analog amplifier.

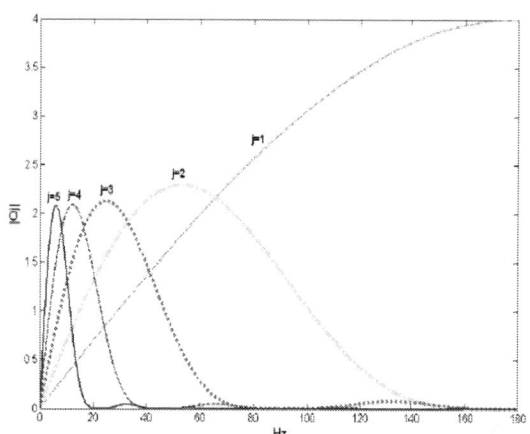

Fig. 2. The amplitude-frequency responses of equivalent filter Qj(ω) at different scales corresponding to 360 Hz sampling frequency; f=180ω/π.

TABLE I
THE BANDWIDTH OF $Q^1(\omega) \sim Q^5(\omega)$

	$Q^1(\omega)$	$Q^2(\omega)$	$Q^3(\omega)$	$Q^4(\omega)$	$Q^5(\omega)$
3dB Bandwidth	90.00~1 80.00	29.92~8 4.24	11.52~3 8.88	5.76~19 .44	2.88~9. 36

Taking the advantage of the fixed point nature of the filter and the data, the error of the hardware filter is easy to be estimated since it only consist of truncation error. We employ QSWT scale 3 in our system, the reason is not only its filter coefficients is shorter and also have the best performance in simulation while comparing the detection accuracy with other scale of QSWT. Thus only the filter in the path of wavelet

transform scale 3 is implemented.

From the equation above, the coefficient of QSWT scale 3 equivalent filter is the integral multiple of 1/32. As a result, 5 bit binary decimal is enough for the representation of coefficients of the wavelet filter. To reduce the area of the circuit, we truncated the binary fraction of the output, which will induce a white noise range from 0 to -1 with a -0.5 mean. This truncation error is negligible when comparing with the amplitude of the transformed signal.

B. Moving average filter

A 6-point moving average filter (MAF) is employed after the QSWT since the 60-Hz power line interference need to be eliminated. A FIR filter is designed to implement this step.

C. Signed squaring

The equivalent filter response of step A and step B enhance the spectra of QRS complex but also cover the spectra of P and T wave. In some records of MITBIH arrhythmia database, extraordinarily strong P wave or/and T wave exist. Absolutely they will attenuate by the QSWT and MAF, but sometimes they still have amplitude which is high enough to threaten the detection accuracy of the coming Modulus Maxima Pair Recognition module. Therefore, signed squaring as blow is used to enhance the QRS complex and suppress while maintain the sign of signal for the next step.

$$y[n] = sign(x[n]) \times x^2[n] \tag{8}$$

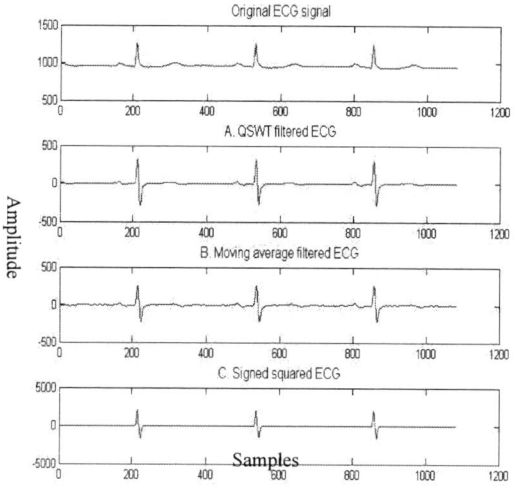

Fig. 3. Test with ECG signal from MIT/BIH Arrhythmia database record 102.

D. Modulus Maxima Pair Recognition

Compare to many other QRS complex detection algorithm with only one feature, that is the positive maximum amplitude of the transformed signal, we apply a two features recognition system, both the positive maximum and negative minimum are considered.

The detection of a QRS complex is accomplished by comparing the features against two threshold levels. In the proposed system, the threshold levels are computed to be signal dependent such that an adaption to changing signal characteristics is possible.

As a result, hardware implementation of the Modulus Maxima Pair Recognition (MMPR) module consists of two parts. The first part is threshold level computation and the second part making decision based on the comparison between transformed signal and the thresholds.

The guideline in selecting the threshold is given by empirical equations,

$$PT = APNL + 0.25\,APSL \qquad (9)$$
$$NT = ANNL + 0.25\,ANSL \qquad (10)$$

In (9) and (10), PT stands for positive threshold whereas NT for negative threshold. APNL is the average positive noise level and APSL is the average positive signal level; both these two levels are calculated by average of most recently 8 positive noise peaks or signal peaks. It is identical for the negative part which using average negative noise level (ANNL) and average negative signal level (ANSL). Thereby the two thresholds can be adjusted according to different signal amplitude and noise level to achieve higher detection accuracy.

The local extreme value of transformed signal has an amplitude large than this threshold is seen as Modulus pair candidate. It will be confirmed as a QRS complex if the subsequent zero crossing and opposite local extreme is found in a proper time interval. This job is accomplished by a state machine with several embedded rules. And the zero crossing between a maxima pair is considered as the location of the R peak.

The bit width of this part is minimized because it has linear correlation with the circuit area and power.

III. PERFORMANCE EVALUATION

A. Implement Result

The proposed system is implemented on an Altera FPGA chip. The data representation in the system is varying from module to module, since the minimum bit width requirement for enough accuracy of different module is not the same. A varying bit width can reduce the resource consumption while maintaining a low enough truncation error.

Table II describes the hardware usage, which consists of the proposed algorithm and also a UART data exchange circuit, of the proposed system. It uses about 3% of total resource for both logic elements and registers of target platform, which is a Cyclone II EP2C70F896C6N FPGA chip placed on an Altera DE2-70 board.

TABLE II
HARDWARE USAGE OF THE SYSTEM

	Total Logic Elements	Total Registers
Hardware Usage	1964	1244

B. Verification method

The verification data flow is MITBIH arrhythmia database stored in a PC and sends to the FPGA as fast as it can through UART. Although we already use the faster mode of UART, its speed is still much lower than the maximum processing speed of the proposed system. Thereby, the hardware can perform heart beat detection in real time. The result will be sent back to PC and compare to the standard annotations within the database to obtain detection accuracy of the proposed system. Our proposed system is a single lead detection system, thus the first channel of each record is used in the evaluation of the algorithm.

There are several modules of the system bring processing delay into the finally result, such as the QSWT, moving average filtering, MMPR modules and so on. This kind of delay will make the detected temporal location of QRS complex fall behind their real position. As a result, we move the location of detections ahead by a fixed number of samples, which depends on the total processing delay of the whole system, before it is compared with the reference annotations.

C. Detection accuracy

The entire hardware of the QRS detection has been tested correctly. The testing goes through all the 48 records (each lasts for 30 minutes) in MITBIH arrhythmia database to determine whether our detector has detected the beats correctly. The detected QRS complex temporal locations are considered as a true positive if they located around the corresponding reference annotation within 110 milliseconds. Thus it is not necessary to place annotations precisely on the major local extremum as in the MITBIH Data Base reference annotations [11].

$$Accuracy = 1 - \frac{FP + FN}{TB} \qquad (11)$$

There are totally 109,267 heart beats has been tested, with 741 missed beats (FN) and 330 extra detected points (FP), that is a sensitivity of 99.35% and a predictivity of 99.70%. According to Eq. (11), the accuracy of the system is 99.02%. This excludes episodes of ventricular flutter that occur on tape 207. Tap 105, 108, 203 and 210 contribute more or less half of the failed detections because of these taps have poor signal to noise ratio.

IV. DISCUSSION OF SYSTEM ARCHITECTURE

There are several possible architectures, such as different filtering method or different nonlinear transform, of the system can provide real time detection. We perform quantitative investigation on some of them to see which one is better for our application. The architecture of the proposed system is finally determined after we finish this kind of experiments.

A. Scale selection of QSWT

Typical frequency components of a QRS complex range from about 10 Hz to about 25 Hz [12]. Therefore, from table I, only QSWT scale 3 and 4 is possible for the extraction of QRS complex from ECG signal. And it seems that QSWT scale 4 befits more this task since the centre frequency of this scale best-fits the spectrum of QRS. In particular, we evaluate the detection accuracy of the system under the same conduction for both QSWT scale 3 and 4.

Compare to scale 4, scale 3 can provide higher detection accuracy under the same validation condition. The reason is that signal after QSWT scale 4 will contain more low frequency component. As a result, the transformed signal will suffer from the interference of high P and/or T wave and thus produce more false positives (FP).

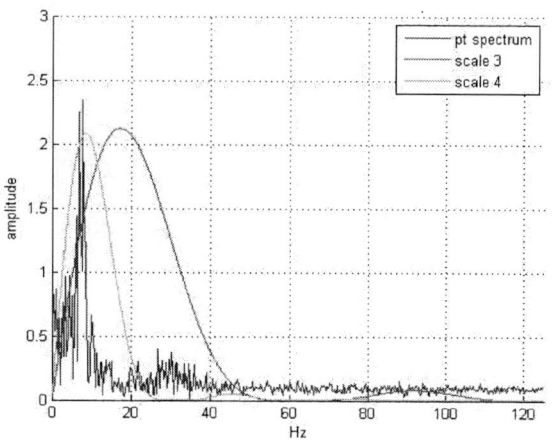

Fig. 3. The amplitude-frequency responses of equivalent filter Qj(ω) at scale 3 and 4 corresponding to 250 Hz sampling frequency; and spectrum of P and T wave of a typical ECG signal obtain from record "sel123" of MITBIH QT database.

B. Moving Average

The moving average stage is also necessary for the detection algorithm. By adding this filter, the high frequency noise can be attenuated and thus over 150 failed detections can be avoided.

C. Signed Squaring

Both QSWT and Moving Average filter are linear transform of the signal. The nonlinear transform is useful in this situation to improve the SNR of the feature signal.

Without using Signed Squaring, there are totally 596 FN and 1243 FP. Compare to the performance of system with signed squaring, the total failed detection decrease from 1884 to 1083, that is reduced by 801 failed detection or 42.52%.

The discussions above are summarized in fig.4.

V. CONCLUSION

A real-time QRS detection algorithm and its corresponding FPGA implementation are presented in this paper. The whole algorithm is written in verilog HDL and thus can operate in high speed. QRS complex can be detected reliable via this algorithm after the baseline wandering, power line noise, high P and T wave is removed by linear and nonlinear transform. The automatically threshold adjustment section enables the adaptability of decision making to diverse signal characteristics.

In the evaluation using MITBIH arrhythmia database, the algorithm failed to properly detect only 0.98 percent of the beats. The proposed algorithm is possible to transplant into ASIC because the low computational complexity and high detection accuracy of the circuit.

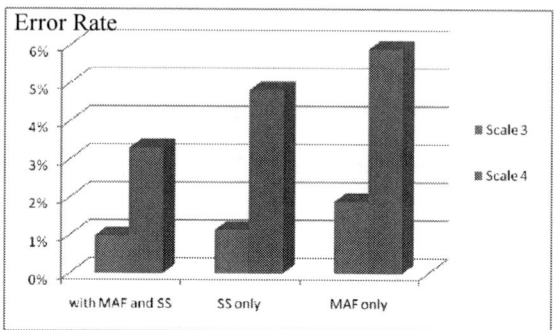

Fig. 4. Error rate of different system architecture evaluate by using MITBIH arrhythmia database; MAF stands for moving average filtering, SS stands for signed squaring.

REFERENCES

[1] J. Pan and W. J. Tompkins, "Quantitative Investigation of QRS Detection Algorithm," *IEEE Trans. Biomed. Eng.*, vol. 32, no. 3, pp. 203–236, 1985.

[2] Y.H. Hu, W.J. Tompkins, J.L. Urrusti, and V.X. Afonso, "Applications of artificial neural networks for ECG signal detection and classification," *J. Electrocardiology*, vol. 26 (Suppl.), pp. 66-73, 1993.

[3] V.X. Afonso, W.J. Tompkins, T.Q. Nguyen, and S. Luo, "ECG beat detection using filter banks," *IEEE Trans. Biomed. Eng.*, vol. 46, pp. 192-202, 1999.

[4] C. Li, C. Zheng, and C. Tai, "Detection of ECG characteristic points using wavelet transforms," *IEEE Trans. Biomed. Eng.*, vol. 32, pp. 21–28, 1995.

[5] J. P. Martínez, R. Almeida, et al, "A Wavelet-Based ECG Delineator: Evaluation on Standard Databases," *IEEE Trans. Biomed. Eng.*, vol. 51, no. 4, pp. 570-581, Apr. 2004.

[6] Weng Chi Chan, "ECG Parameter Extractor of Intelligent Home Healthcare Embedded System," *Master Thesis.*, pp. 35-72, pp. 117-118, University of Macau, Macao SAR., China, 2005.

[7] Chio In Ieong, Sio Hang Pun, Mang I Vai and Peng Un Mak, "Implementation of ECG Signal Processor in FPGA," in *Proc. China Biomedical Engineering Join Conference (CBME2007)*, Xi'an, Apr. 2007, pp. 192-195.

[8] P. S. Hamilton and W. J. Tompkins, "Quantitative investigation of QRS detection rules using the MIT/BIH arrhythmia database," *IEEE Trans. Biomed. Eng.*, vol. BME-33, pp. 1157–1165, 1986.

[9] MIT/BIH Database Distribution, Massachusetts Inst. Technol., Cambridge, MA

[10] Chio In Ieong, Mang I Vai and Peng Un Mak, "QRS recognition with programmable hardware," *the 2nd Int. Conf. Bioinformatics and Biomedical Eng. (iCBBE2008)*, Shanghai, China, 2008.

[11] Physionet – www.physionet.org

978-1-4577-1608-9/11 $26.00 © 2011 IEEE

[12] B.-U. Kohler, C. Hennig, R. Orglmeister, "The principles of software QRS detection," *IEEE Engineering in Medicine and Biology Magazine*, vol. 21, pp. 42-57, 2002.

The Design of a Universal and Configurable ASIC for Biological Stimulation

Weifeng Zhang[1], Yinan Dong[1], Jo-Yu Wu[2], Kea-Tiong Tang[2] and Guoxing Wang[1]
[1]School of Microelectronics, Shanghai Jiao Tong University, Shanghai, China
[2]Dept. of Elec. Eng., Tsing Hua Univ., Hsinchu, Taiwan
subrant@163.com

Abstract—**This paper presents the design of the digital part of an ASIC for biomedical stimulation. This design can generate different types of biphasic waveforms according to stimulation data and configuration data input from the outside. The width, amplitude level and polarity of the output pulse can be regulated by the data packet and the range of width and amplitude can be defined by the configuration packet. This enables the chip to adapt to different applications without the need of a large memory size and high input data rate. A two-stage architecture to distribute data received from the SPI interface and to reduce chip area has been adopted. The current design features an output of up to 64 channels and can be easily upgraded to support more channels. The chip is designed using Verilog HDL language, simulated in ISE stimulator, verified by Xilinx Virtex-2P FPGA board, synthesized in Synopsys Design Compiler environment, placed and routed in Synopsys Astro environment.**

I. INTRODUCTION

Nowadays the application of prostheses is booming in biomedical area, such as facial paralysis prosthesis [1], retinal prosthesis [2] etc. The core of the prosthesis is the stimulator which can generate voltage or current pulse to stimulate the body. Considering the size and power of the prostheses, it has the need to design a stimulation chip.

Considering the charge balance problem [3], this chip generates biphasic pulse [4]. The pulse waveform is showed in Fig. 1. Inside the figure, T1, T2, T3 and T4 describe the pulse width, A1 and A2 describe the pulse amplitude level. All of these parameters can be regulated by the user. The polarity of the pulse can also be set by the user through the data packet.

This chip consists of a digital part and an analog part. The most obvious features of this chip are flexibility and configurability. These features are mainly realized in the digital controller. This paper will focus on the architecture and design of the digital controller. This analog part, meanwhile, mainly serves as the DAC and current mirror [5].

This paper is organized as follows. In Section 2, the architecture and all the modules are described. The features and implementation are presented in Section 3. Simulation and verification results are presented in Section 4. In Section 5, it shows the layout after floorplanning, placement and routing. Finally, conclusion is given in Section 6.

The authors thank National Chip Implementation Center (CIC) for technical support.

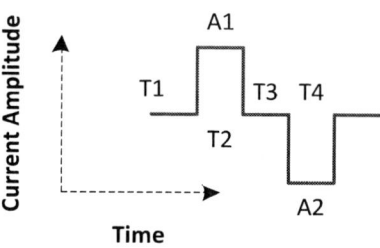

Fig. 1 Biphasic Current Pulse

II. ARCHITECTURE OF THE SYSTEM

The two-stage architecture [6] is chosen to distribute the stimulation data [7]. The advantage of choosing this architecture is that it only needs one header detector module and one CRC checker module. For two-stage architecture, there are two basic protocol types that can be chosen from-parallel bus topology and serial shift topology [8]. These topologies can be seen in Fig. 2. In parallel bus topology, every channel data has its own address, so it enables the stimulation arbitrarily. Its disadvantage is the increase in data rate. For higher number of channels in the design, this topology will need higher number of extra bits to execute every stimulation. However, the serial shift topology does not need these extra bits. Its disadvantage is that the flexibility is limited since the stimulation order is fixed.

In order to combine the advantages of the parallel bus topology and serial shift topology, the parallel bus topology is modified to adapt in this design. The output of the global controller is respectively connected to each channel. It can be seen in Fig. 2. After modification, the address redundancy is discarded. Another advantage of this design is that all the 64 channels' data can be delivered to the 64 timing generators in one clock cycle. It can guarantee all 64 channels to work together.

The global architecture of the chip and the external input unit are showed in Fig. 3. The external unit sends out the data from PC and these data are delivered through the SPI interface into the chip. The global controller serves as the SPI slave terminal. Three modules are included in the global controller. They are header detector, shift register and CRC checker. Header detector module is used to detect the header of the stimulation data packet or the configuration data packet. When the header is detected, the shift register is responsible for storing the stimulation information. The function of the CRC checker module is to verify the validity of the stimulation data packet or the configuration data packet. Only after the CRC check, the stimulation data can be decided to be delivered to the next stage or be discarded.

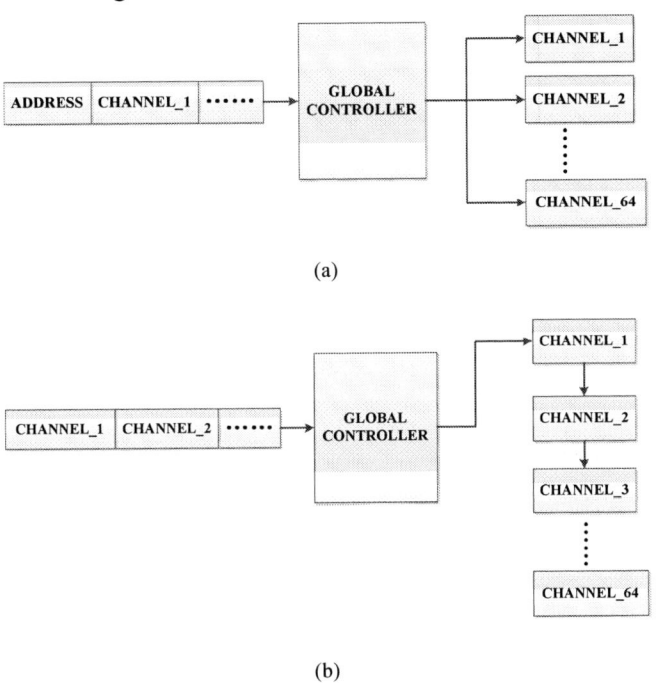

(a)

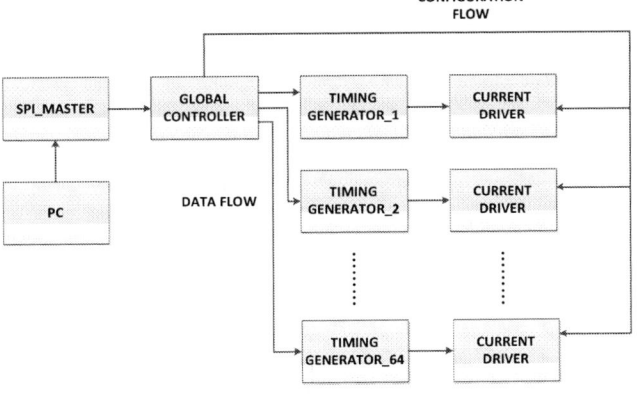

(b)

Fig. 2 (a) Parallel Bus Topology (b) Serial Shift Topology

Fig. 3 Global Architecture of the Stimulator ASIC

Three modules are included in the global controller. Fig. 4 illustrates the global digital controller.

The timing generator is used to control pulse width, amplitude level and polarity according to the stimulation data. The output of this module is connected to the input of the analog part. It will drive the electrode to stimulate the body. Safety is crucial in stimulation process. So for the safety

consideration, the analog part has charge cancellation module to guarantee the charge balanced.

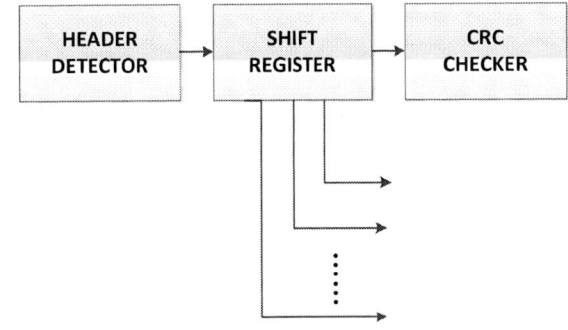

Fig. 4 Global Controller

III. FEATURES AND IMPLEMENTATION

A. Data Flow

According to the chosen topology, two types of packets used for control are shown in Table I, II and III.

As for the data packet showed in Table I, the first 16 bits are header. The next 3984 bits are 64 channels' stimulation data and the last 16 bits are CRC check codes. These data is serially stored into the shift register and simultaneously shifted into CRC checker module [9].

Each channel's 62 bits stimulation data is listed in Table II. The 1-bit charge cancel bit is used to keep charge balanced in the body and the 1-bit polarity bit is used to decide whether the leading pulse of the stimulation is cathodic or anodic.

TABLE I STIMULATION DATA PACKET

	Header	64 Channel Stimulation Data	CRC
Bit	16	3968(64*62)	16

TABLE II EACH CHANNEL'S INFORMATION

	A1	A2	T1	T2	T3	T4	Charge Cancel	Polarity
Bit	8	8	8	14	8	14	1	1

TABLE III CONFIGURATION DATA PACKET

	Header	Reference Current in DAC	Current Mirror	Not Used	CRC
Bit	16	2	10	3956	16

From Table I. and Table II, it is known the output current pulse can be flexibly changed in a fixed range. If the width and amplitude of the stimulation pulse is beyond the range, it needs the configuration packet to configure the current driver to generate the desired stimulation pulse. The configuration data packet is illustrated in Table III.

In the configuration data packet, the first 16 bits are the header. The last 16 bits are for CRC check. Only after the CRC check is verified to be correct, the current driver can be configured [10]. The 2 bits are used to configure the reference current level of DAC in the current driver. The 10 bits are used to select the number of copy current in current mirror. The 3956 bits are set to be zero because they are not used. These not-used bits exist here are used to keep with the same length as the stimulation data packet. The configuration data packet is not always transmitted. When it is transmitted, the range of the width and amplitude level of stimulation pulse will be changed. This kind of packet enables this chip to adapt to different applications.

B. Control Flow

As for the control flow, a state machine is designed to control all the modules to work harmoniously. The state transition diagram of this state machine is described in Fig. 5.

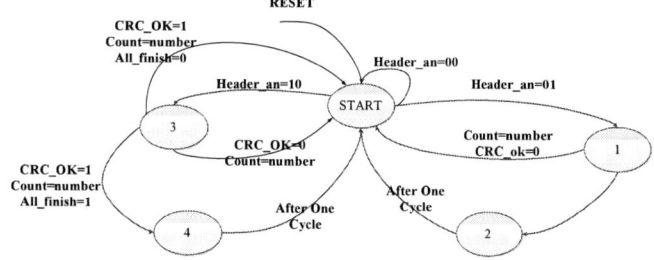

Fig. 5 State Transition Diagram

In the state transition diagram, there are five states included. The START state is reached when the falling edge of reset signal is detected. At that time all the modules are initialized. State 1 and State 2 are involved in the configuration process. State 3 and State 4 are involved in the stimulation process.

When the header detector module has detected the configuration data packet, the state machine will jump to state 1. At that time, the shift register and the CRC checker modules are receiving the configuration data. When the CRC check turns out to be valid, the state machine will jump to state 2. At this moment, the current driver is configured. If the CRC checker turns out to be invalid, the state machine will jump to state START to detect data packet header again.

When the header detector module has detected the stimulation data packet, the state machine will jump to state 3. At that time, the shift register and the CRC checker modules are receiving the stimulation data. When the CRC checker turns out to be valid and the timing generator is idle, the state machine will jump to state 4. At this moment, the current pulse is generated. If the CRC check turns out to be invalid or the timing generator is not idle, the state machine will jump to state START to detect data packet header again and the old data is discarded.

C. Flexilibity and Configurability

For various biomedical applications, different types of stimulation pulses are needed. These parameters include leading pulse polarity, pulse width and pulse amplitude level. In this design, the chip can generate different types of stimulation pulses according to different stimulation data packets and configuration data packets which have been received. If the data source is changed, the stimulation pulse will be simultaneously adjusted [11].

The pulses are flexible to change because one stimulation pulse generated by one data packet. So the stimulation pulses can be changed every time. This chip will be applied in different biomedical areas. Configurability is the feature to ensure its diversity.

This chip also has its own disadvantage to be overcomed when compared to dedicated ones [12]. In order to have the features of flexibility and configurability, the complexity has increased and the area will be large. It will be a challenge when the chip will be used to implantation. The solution to this problem is to optimize the architecture of the chip and share more resources together.

IV. SIMULATION AND VERIFICATION

The system is described by Verilog HDL language and simulated under Xilinx ISE simulator.

Fig. 6 shows the configuration data packet to configure the current driver. The configuration data packet is simplified here only for simulation. The configuration data is 101011011001. From the figure, we can see the function meets the requirement. Fig. 7 shows the stimulation data packet. It is also simplified here for the same reason. From the figure, we can see the output of the timing generator module. These data are connected to the current driver.

Furthermore, the circuit is verified on the Xilinx Virtex-2P FPGA board. The ChipScope analyzer serves as the logic oscillator to see the waveform. The verification result proves the validness of the circuit deeply from another side.

V. LAYOUT OF THE CHIP

After simulation, the system was synthesized in Synopsys Design Compiler Environment, floorplanned, placed and routed in Synopsys Astro Environment.

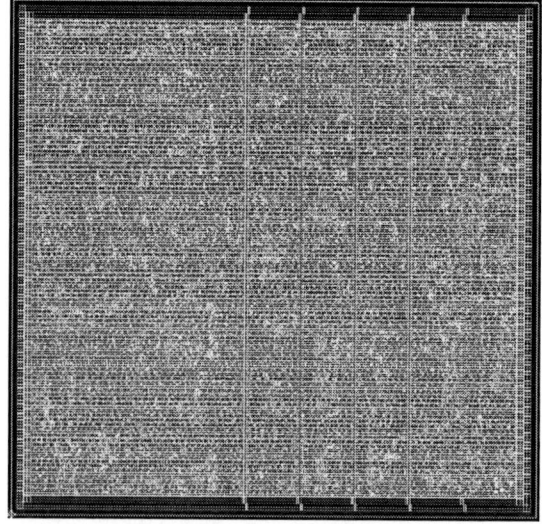

Fig. 8 Layout of the Chip

Fig. 8 is the layout of the design. The digital part of the chip is designed in TSMC 0.18 μm technology and occupies the die area of 1.618mm*1.618mm.

978-1-4577-1608-9/11 $26.00 © 2011 IEEE

VI. CONCLUSION

This paper presents the digital controller design of an ASIC for biological stimulation which can be used for different biomedical applications through its reconfigurability and universal structure. These features are fulfilled using two-stage architecture. Verilog HDL implementations are accomplished with this architecture. The circuit is simulated in ISE simulator and verified on FPGA board. From the results of simulation and verification, it can be seen that the system has met the design target. After that, the design is synthesized, placed and routed in Synopsys tools. The area can be optimized in future work. Finally it will be taped out with the analog part.

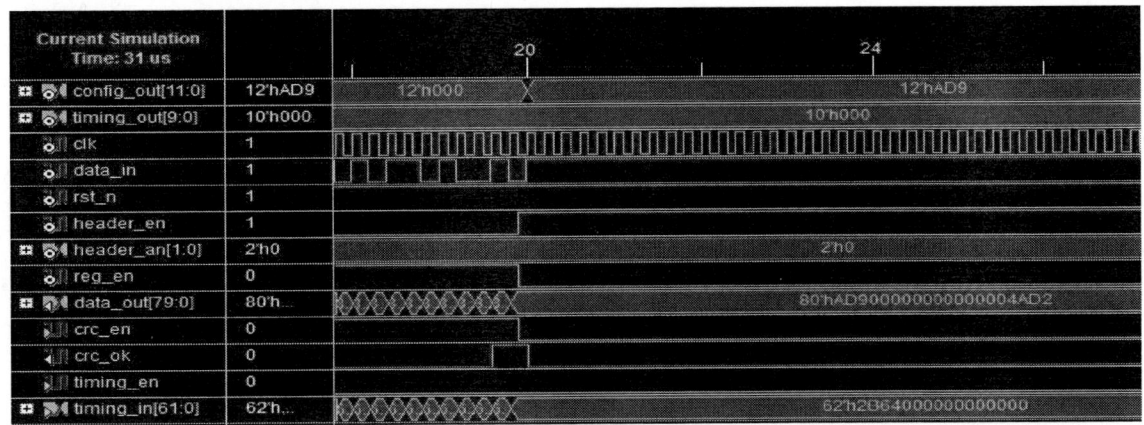

Fig. 6 Configuration Data

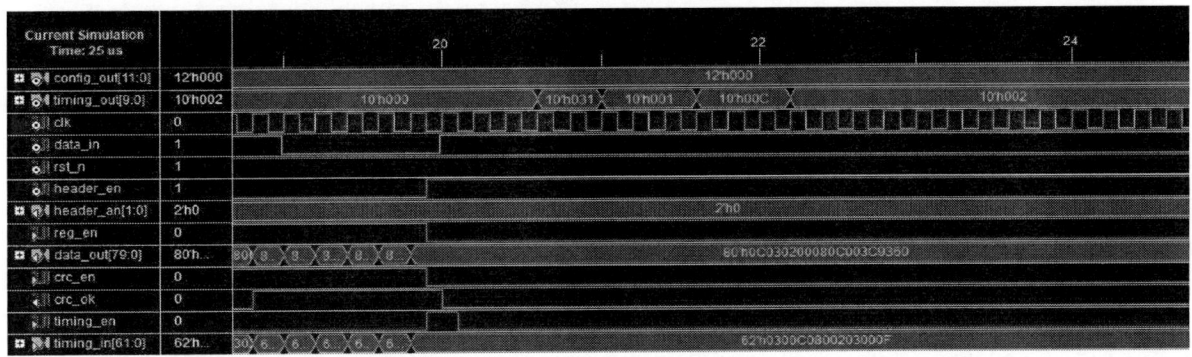

Fig. 7 Stimulation Data

REFERENCES

[1] Kuanfu Chen, Tung-Chien Chen, Cockerham, K. and Wentai Liu, "Closed-loop eyelid reanimation system with real-time blink detection and electrochemical stimulation for facial nerve paralysis", *Circuits and Systems*, pp. 549-552, 2009

[2] Sivaprakasam, M., Wentai Liu, Humayun, M.S. and Weiland, J.D., "A Variable Range Bi-Phasic Current Stimulus Driver Circuitry for an Implantable Retinal Prosthetic Device", *Solid-State Circuits*, pp. 763-771, 2005

[3] Kriangkrai Sooksood, Thomas Stieglitz and Maurits Ortmanns, "An Active Approach for Charge Balancing in Functional Electrical Stimulation", *Biomedical Circuits and Systems*, vol. 4, pp. 162, 2010

[4] Maurits Ortmanns, André Rocke, Marcus Gehrke, and Hans-Jüergen Tiedtke, "A 232-Channel Epiretinal Stimulator ASIC", *Solid-State Circuits, vol. 42*, pp. 2946-2959, 2009

[5] Kuanfu Chen, Tung-Chien Chen, Cockerham, K. and Wentai Liu, "Closed-loop eyelid reanimation system with real-time blink detection and electrochemical stimulation for facial nerve paralysis", *Circuits and Systems*, pp. 549-552, 2009

[6] Kuanfu Chen and Wentai Liu, "Highly programmable digital controller for high-density epi-retinal prosthesis", *Engineering in Medicine and Biology Society*, pp. 1592-1595, 2009

[7] Kuanfu Chen, Zhi Yang, Linh Hoang, Weiland, J., Humayun, M. and Wentai Liu, "An Integrated 256-Channel Epiretinal Prosthesis", *Solid-State Circuits*, pp. 1946-1956, 2010

[8] Sivaprakasam, M., Wentai Liu, Guoxing Wang, Mingcui Zhou, Weilana, J.D. and Humayun, M.S., "Architecture Tradeoffs in High-Density Microstimulators for Retinal Prosthesis", *Neural Engineering*, pp. 466-469, 2005

[9] Emilia Noorsal and Maurits Ortmanns, "An architecture for a universal neural stimulator with almost arbitrary current waveform", *Engineering in Medicine and Biology Society*, pp. 2931-2934, 2010

[10] Takashi Tokuda, Kohei Hiyama, Shigeki Sawamura, Kiyotaka Sasagawa, Yasuo Terasawa, Kentaro Nishida, Yoshiyuki Kitaguchi, Takashi Fujikado, Yasuo Tano and Jun Ohta, "CMOS-Based Multichip Networked Flexible Retinal Stimulator Designed for Image-Based Retinal Prosthesis", *Electron Devices*, vol. 56, pp. 2577-2585, 2009

[11] Pereira, M.C. and Kassab, F., "An Electrical Stimulator for Sensory Substitution", *Engineering in Medicine and Biology Society*, pp. 6016-6020, 2006

[12] Maoxin Wei, Mouine, J., Fontaine, R. and Duval, F., "A dedicated microprocessor for externally powered implantable pain controller", Engineering in Medicine and Biology Society, vol. 2, pp. 1133, 1995

A New ECG Signal Processing Scheme for Low-Power Wearable ECG Devices

Yibin Hong, Iniyal Rajendran, Yong Lian
Department of Electrical and Computer Engineering
National University of Singapore
Singapore 117576
hongyibin@nus.edu.sg

Abstract—A new ECG signal processing approach is proposed for use in ultra low-power wireless ECG recording devices. Significant energy saving is achievable with the adoption of level-crossing sampling, which adapts its sampling frequency according to signal activities. Linear Interpolation is then performed to convert the non-uniform samples into uniform ones to allow the use of existing signal processing tools. We show by an example that the proposed approach achieves 88.8% and 92.6% reductions in the number of sampling points and the order of post-processing FIR filter respectively.

I. INTRODUCTION

A conventional Nyquist-based digital signal processing (DSP) system takes samples from its analog input, digitizes and processes them periodically according to a clock. Such a system does not take advantage of the statistical properties of the analog input, but instead it samples the analog input at a constant rate that is at least twice the signal bandwidth, no matter how fast or how slow the signal changes. For biomedical signals, such as Electrocardiographic (ECG) signals, the fast changes only occur in a brief period while most of time the signal varies slowly. Sampling such sporadically varying signals using the Nyquist approach would give rise to a large number of samples that carry redundant information, wasting power not only in analog to digital conversions, but also in subsequent processing [1].

Unlike conventional DSP systems, which sample the input at a fixed clock frequency, level-crossing sampling takes the statistical properties of the inputs into consideration [2]. Only when a significant change occurs in the input will a new sample be generated. For low-frequency or inactive inputs, the constant-frequency sampling in conventional DSP systems simply wastes power. For level-crossing sampling however, slow inputs naturally result in sparse samples, which lead to lower dynamic power dissipation. During silent periods of the input, the system waits for a change in the signal while dissipating no dynamic power. This is illustrated in Fig. 1(b): significant power is consumed only during the two time intervals t_1-t_2 and t_3-t_4, when substantial changes occur in the analog input. The first interval also has a higher peak than the second interval since during t_1-t_2 the input changes more

drastically and therefore samples are being generated more frequently. During the rest of the time when the input remains more or less constant, the system consumes minimum static power mainly due to biasing current and transistor leakage. Such adaptive-rate sampling of level-crossing analog to digital conversion (ADC) makes it well suited for ECG signals that exhibit bust-like waveforms [3].

This however creates another problem since the samples obtained from a level-crossing ADC are not uniformly spaced

(a) Signal waveforms of a level-crossing ADC

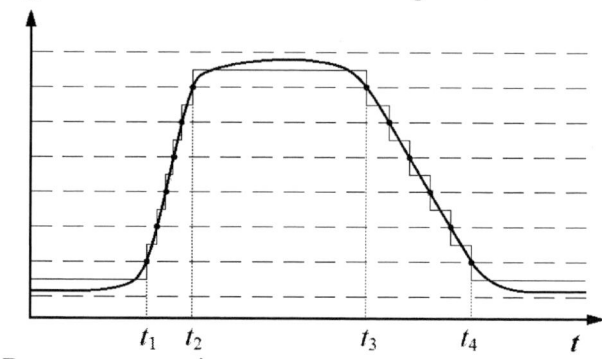

(b) Power consumption

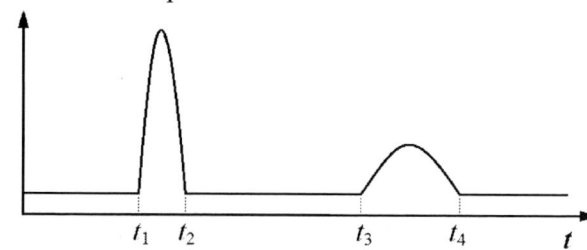

Fig. 1. (a) Level-crossing ADC: the thick solid curve represents the analog input, the thin staircase waveform represents the digital output. The set of quantization boundaries are represented by the dashed lines. (b) The corresponding power dissipation of the ADC and DSP combined.

in time, as is the case for a conventional Nyquist-based ADC. Most of the available digital signal processing theories and techniques were developed based on uniform sampling, which means they cannot be directly applied to the digital output of a level-crossing ADC. Although efforts have been made in developing the corresponding theories and techniques for non-uniformly sampled digital signals, no systematic approaches have been formed yet. Techniques like general discrete Fourier transform and Lomb's algorithm face problems such as noises in spectra [4]-[5].

Having these constraints in mind, a new system that combines level-crossing sampling and conventional synchronous processing with the aid of linear interpolation is proposed. The overall computation complexity and power consumption of this new processing scheme is expected to be significantly lower than those of the conventional Nyquist approach. Matlab simulation on real ECG signals showed that the signal quality is preserved with an average error of less than 2%, while the average sampling rate of such a new system is only 11.2% of the sampling rate of a Nyquist-based system.

The paper is organized as follows. Section II discusses in detail about level-crossing sampling, including the quantization scheme, the introduction of hysteresis and simulation using real ECG signals. Different interpolation techniques used to convert the obtained non-uniform samples into uniform ones are compared in Section III. In Section IV, an FIR high-pass filter is designed to filter out the unwanted baseline wandering noise in the original ECG signal. Section V concludes this paper.

II. LEVEL-CROSSING SAMPLING

A. Quantization scheme

A level-crossing ADC generates samples only when the input crosses a predefined set of regularly-spaced amplitude boundaries, as illustrated by the dashed lines in Fig. 1(a). Each solid line resting in the middle of two dashed lines is used to represent the quantized value for signals falling within the two boundaries. This finite number of solid lines can then be encoded using binary words, yielding the digital

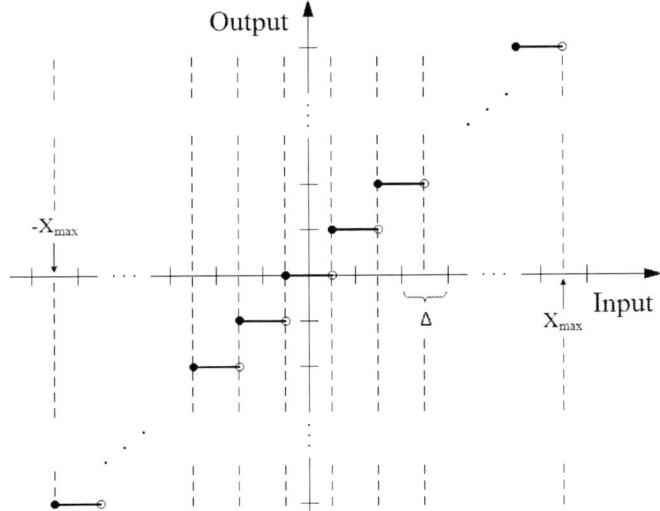

Fig. 2. Input-output transfer characteristic of a middle-tread quantizer. Δ denotes the quantizer resolution and X_{max} denotes the maximum allowable input range.

representation of a level-crossing ADC.

The input-output transfer characteristic of an N-bit middle-tread quantizer used for level-crossing sampling is shown in Fig. 2. The horizontal axis corresponds to the analog input, which is continuous, while the vertical axis corresponds to the digital output, which only takes discrete values. The regularly-spaced dashed lines again represent the quantization boundaries. Δ denotes the quantizer resolution and X_{max} is the maximum allowable input range. They can be calculated from:

$$\Delta = \frac{D}{2^N}, \tag{1}$$

$$X_{max} = D\left(1 - \frac{1}{2^N}\right), \tag{2}$$

where D denotes the full dynamic range. The reason why the maximum allowable input range is less than the full dynamic range is that one quantization level is purposely left unused, in order to preserve odd symmetry in the transfer characteristic.

B. Hysteresis

Biomedical signals such as ECG signals are usually very noisy. Such noises may result from patient body movement,

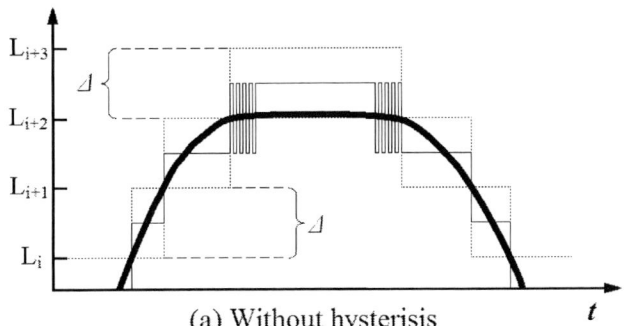

(a) Without hysterisis

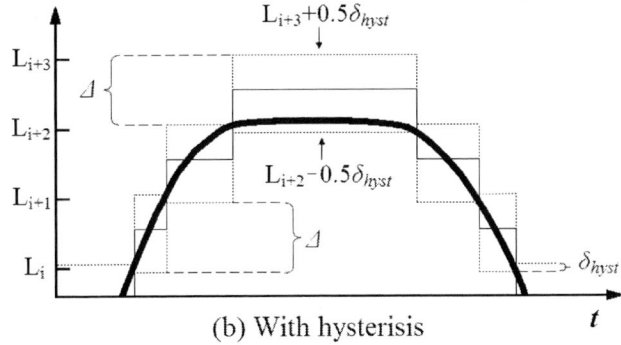

(b) With hysterisis

Fig. 3. Level-crossing ADC in the presence of noise for (a) quantizers without hysteresis and (b) quantizers with hysteresis. In both plots, the thick solid curve represents the analog input, the thin solid staircase waveform represents the digital output. The quantization boundaries are represented by the dotted lines. Δ denotes the quantizer resolution and δ_{hyst} denotes the amount of hysteresis introduced.

ambient electromagnetic interference as well as other various types of disturbances. This has a severe impact on level crossing sampling. Every time the analog input approaches a quantization boundary, the added noises make the signal crossing that boundary back and forth even though it actually is not, as shown in Fig. 3(a). This causes the level-crossing ADC to generate extra unnecessary samples. Such noise-induced samples provide no additional information about the

978-1-4577-1608-9/11 $26.00 © 2011 IEEE

ECG signal of interest, yet they consume significant power in both the ADC as well as the subsequent DSP.

To avoid such undesirable noise-induced toggling, hysteresis is purposely introduced in the quantizer, as illustrated in Fig. 3(b). Normally, when the analog input is situated between the two boundaries L_i and L_{i+1}, a new sample is generated once the following condition is violated:

$$L_i \leq x(t) < L_{i+1}. \tag{3}$$

To introduce hysteresis, the above condition is modified as

$$L_i - \frac{1}{2}\delta_{hyst} \leq x(t) < L_{i+1} + \frac{1}{2}\delta_{hyst}. \tag{4}$$

In other words, each quantization interval is now widened by an amount equal to δ_{hyst}. For any two consecutive intervals, the upper boundary of the lower interval is no longer aligned with the lower boundary of the higher interval, creating an overlapping between the two intervals. This way, when the analog input increases over a certain boundary causing the ADC to produce a new sample, it must decrease by at least δ_{hyst} before another sample to be generated. This prevents the input from going back and forth between two consecutive intervals in the presence of noise, as long as the amount of fluctuation is less than δ_{hyst}.

Even though such nonlinear operation inevitably causes a certain degree of distortion to the digitized output [6], the gain of noise immunity and better robustness through the introduction of hysteresis in the quantizer is considered more critical to the overall performance of the entire system.

C. Simulation of real ECG signals

Real ECG signals obtained from the MIT-BIH arrhythmia database are used to verify the performance of the proposed processing scheme. Fig. 4 shows a portion of the level-crossing sampling result of record 151 from the database. To achieve a reasonable degree of accuracy while keeping the average sampling rate low, the quantizer resolution is chosen to be N = 5. As represented by the series of bubbles, the level-crossing sampled output tracks the analog input (the solid

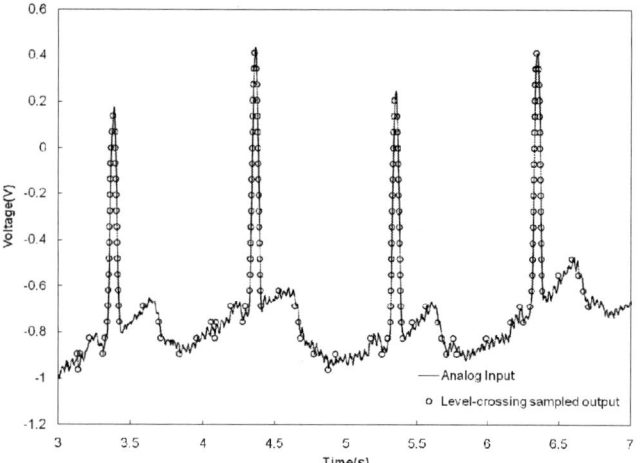

Fig. 4. Level-crossing sampling of ECG signal #151 retrieved from the MIT-BIH arrhythmia database. The solid curve represents the original analog input and the series of bubbles represent the level-crossing sampled output.

curve) very well. All P, Q, R, S, T waveforms in the original ECG signal are recorded by the level-crossing sampled output. The total number of samples obtained in a 10-second interval is 402, leading to an average sampling rate of only 40.2Hz. As compared to the 360Hz sampling frequency used in the original Nyquist-based system, a reduction of 88.8% is achieved.

The high frequency noise present in the original signal is also successfully suppressed by introducing a hysteresis of $\delta_{hyst} = 0.25\Delta$. This not only prevents unnecessary noise-induced samples from wasting system power, but also removes the high-frequency noise, which otherwise would have to be filtered out in the subsequent DSP.

III. INTERPOLATION

Due to the many reasons mentioned in Section I, the non-uniform samples obtained from the level-crossing ADC must first be converted to uniform format before any DSP can be carried out. Of course uniform oversampling would serve the purpose but that again makes the sampling rate high, canceling the benefit of reduced power consumption achieved by level-crossing sampling. Therefore, various interpolation techniques were investigated to achieve the uniform conversion while keeping the total number of samples the same.

Predicting signal values from neighboring samples may sound risky, but considering the event-driven nature of level-crossing sampling it is realized that the signal value between any two consecutive samples is in fact guaranteed to be bounded between those two samples, since any further crossing of quantization boundaries will trigger the ADC to produce additional samples. Therefore, it is fair to conclude that extracting information through interpolation is safe for level-crossing sampled data.

Higher order polynomial interpolations generally give lower interpolation errors, but they are computationally more expensive. This translates into higher dynamic power consumption in real circuit implementations. To achieve a reasonable degree of accuracy while keeping the power consumption low, classical interpolation techniques including nearest neighbor interpolation, linear interpolation and cubic interpolation are explored and compared.

Nearest neighbor interpolation is the simplest interpolation wherein the value at the interpolated point is assigned with the value of the closest point [7]. In terms of computational efficiency this is the best since it requires no multiplication at all. However, the signal quality of nearest neighbor interpolation is poor. A staircase-like waveform is obtained after the interpolation.

In linear interpolation, the weighted average of the two closest non-uniform samples at t_k and t_{k+1} is taken to be the value of the interpolated point at t [7], given by

$$x(t) = x(t_{k-1}) + (x(t_k) - x(t_{k-1}))\frac{t - t_{k-1}}{t_k - t_{k-1}}. \tag{5}$$

The computational complexity of linear interpolation is higher than that of the nearest neighbor interpolation, but the signal quality is significantly better. Fig. 5 shows the linear interpolated result from the non-uniform samples obtained in Fig. 4. The average error between the linear interpolated result and the original input is only 1.53%.

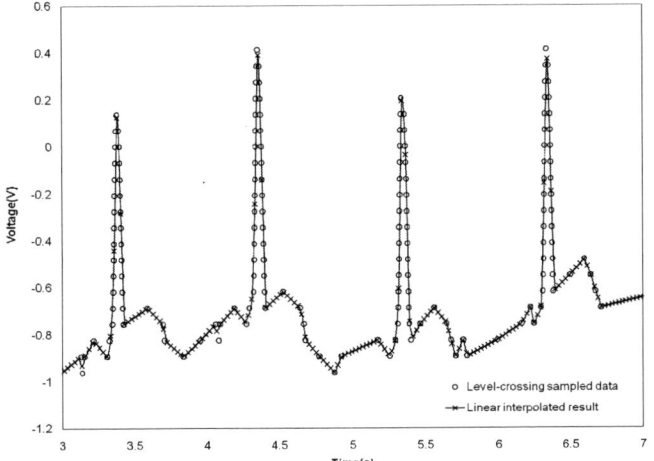

Fig. 5. Linear interpolation result of the level-crossing sampled data obtained in Fig. 4. The series of bubbles again represent the level-crossing sampled data, and the solid curve with crosses on it represents the linear interpolated result.

Cubic interpolation forms a third order polynomial that passes through the set of interpolated points between any two non-uniform samples [7]. The detailed algorithm is very complex and is not presented here. Such complexity translates into increased circuit area and power consumption, but the signal quality for this case is in fact not any better: The average error of the cubic interpolated signal is 1.59%, which is slightly higher than that of linear interpolation.

With the above comparison, linear interpolation is chosen for the proposed processing scheme due to its nice balance between signal quality and computational efficiency.

IV. DIGITAL FILTERING

Having the samples converted to uniform format, signal processing can be easily carried out using the available DSP tools. As an example, an FIR high-pass filter is designed to remove the low frequency baseline wandering noise in the original ECG signal.

The Remez exchange algorithm in Matlab is used to design a high-pass FIR filter with the normalized stop-band and pass-band edges at 0.0244 and 0.0733 respectively. The pass-band ripple is set at 0.01 and stop-band attenuation is 17dB. The interpolated ECG signal is filtered by the designed high-pass filter and the output is shown in Fig. 6. It is clear that the baseline wandering noise is successfully suppressed. The entire waveform also shifts upward due to the removal of the DC component, which carries no useful information for

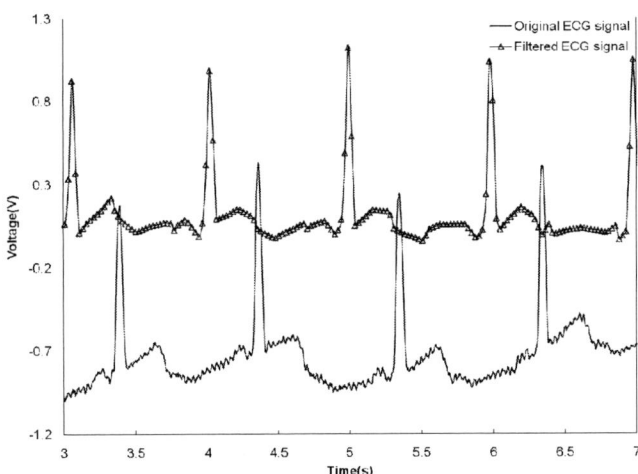

Fig. 6. High-pass filtered result of the linearly interpolated data. The solid curve represents the original ECG signal, while the curve with triangles represents the filtered output.

the analysis of ECG signals. The designed FIR filter has an order of 52. By comparison, filters designed for the Nyquist-based system has an order of 701 due to the higher sampling rate. This implies further power saving in the DSP by the proposed scheme.

V. CONCLUSION

A new ECG signal processing approach that combines level-crossing sampling with conventional uniform DSP through linear interpolation is presented. Significant power and area reduction is achievable from Matlab simulation, making it well-suited for wearable ECG devices that have stringent power requirements due to battery limitations. Although ECG signals are used for demonstration, it is believed that other burst-type signals could also derive the same benefits when using the proposed processing scheme.

REFERENCES

[1] F. Aeschlimann, E. Allier, L. Fesquet and M. Renaudin, "Asynchronous FIR filters: towards a new digital processing chain," *Proc. 2004 IEEE Int. Symp. Asynchronous Circuits and Systems (ASYNC '04)*, pp. 198-206.

[2] A. Baums, U. Grunde and M. Greitans, "Level-crossing sampling using micropocessor based system," *Proc. 2008 IEEE Int. Conf. Signals and Electronic Systems (ICSES '08)*, pp. 19-22.

[3] M. Kurchuk and Y. Tsividis, "Signal-dependent variable-resolution quantization for continuous-time digital signal processing," *Proc. 2009 IEEE Int. Symp. Circuits and Systems (ISCAS 2009)*, pp. 1109-1112.

[4] S. M. Qaisar, L. Fesquet and M. Renaudin, "Spectral analysis of a signal driven sampling scheme," *14th European Signal Processing Conference (EUSIPCO 2006)*, Florence, Italy, Sep. 2006.

[5] P. C. Bagshaw and M. Sarhadi, "Analysis of samples of wideband signals taken at irregular, sub-Nyquist, intervals," *Electronics Letters*, vol. 27, no. 14, pp. 1228-1230, 1991.

[6] B. Schell, "Continuous-time digital signal processors: Analysis and implementation," Ph.D. dissertation, Columbia Univ., New York, 2008

[7] E. Kreyszig, "Advanced engineering mathematics," Wiley Eastern, 2006.

A SiGe BiCMOS Class A Power Amplifier Targeting 5.5GHz Application

Yan Qiong[1], Hua Lin[1], Chen Lei[1], Ruan Ying[1], Su Jie[1], Zhang Shulin[1], Zhang Wei[1], Liu Shengfu[1] and Lai Zongsheng[1]

1.Institute of Microelectronics Circuit & System
East China Normal University
Shanghai, China
yanqiongfzu@163.com

Abstract—**This paper discusses a 5.5GHz fully integrated high-linearity Class A power amplifier based on 0.18μm SiGe BiCMOS technology. According to the post simulation results, the maximum output power can reach 24.18 dBm, and the PAE is 17.42% at P_{1dB}. The designed PA based on SiGe BiCMOS technology demonstrates a competive linearity performance compared with RF CMOS and GaAs when considering the factors of cost and process complexity. It can be used in 802.11a Wireless Local Area Network (WLAN) application.**

I. INTRODUCTION

SiGe BiCMOS technology is widely used in different application fields, such as consumer wireless (WiFi, WiMAX, CDMA transceiver, GPS receiver and Ultra wide band), wire line infrastructure, integrated TV, broadband wireless, precision analog, microwave applications and RF power.

High output power handheld device PA once has been monopolized by GaAs and RF LDMOS (Laterally Diffused Metal Oxide Semiconductor). This traditional design relies on multi-stage structure, its matching network design and power control are very difficult. Because of its low cost, high linearity, high power efficiency and high integration, SiGe BiCMOS PA began to gradually capture the market [1] [2].

SiGe technology has higher level of integration compared with compound semiconductors such as GaAs. SiGe BiCMOS has high-quality passive components integration by increasing the metal and dielectric layer stack to reduce parasitic capacitance and inductance. The germanium profile injected can be used to improve device behavior over temperature. SiGe BiCMOS is compatible with virtually all innovations in the silicon VLSI industry including SOI and trench isolation [3][4].

SiGe BiCMOS has lower cost than GaAs. Nowadays, a SiGe mainstream process on 200mm (8-inch wafers) is moving on towards the industry roadmap to 300mm. GaAs on 4- to 6-inch wafers, with the yield and processing cost disadvantages is associated with a smaller wafer size.

Table I shows the performance comparison of PA module based on different technologies, including GaAs, RF CMOS and SiGe BiCMOS. As can be seen, the actual design which the literature [5] proposed shows that the performance of SiGe HBT power devices at 28 dBm output power is comparable with GaAs technology.

TABLE I.   THE PERFORMANCE COMPARISON OF PA MODULE BASED ON DIFFERENT TECHNOLOGIES

Technology	Performance Comparison			
	PAE(%)	PGain (dB)	Pin (dBm)	PA Survive
RF CMOS	32.5	22.4	6.4	4.7
GaAs	41	27	1.4	10
SiGe	32.5	22.4	6.3	6.3

Technology	Performance Comparison		
	ACPR1 (dBc)	ACPR2 (dBc)	VCC at Failure
RF CMOS	-47	-56.8	4.8
GaAs	-47	-59.7	10.1
SiGe	-47	-57.5	6.5

II. CIRCUIT DESIGN

This paper uses SiGe HBT to design power amplifier since the device has the following characteristics: the compatibility is good between SiGe BiCMOS and the traditional RF CMOS process, so a power controlling circuit can be easily included. The RF performance of SiGe BiCMOS is much better than RF CMOS, i.e. 0.18μm SiGe BiCMOS technology is comparable to the 90nm RF CMOS [6] [7], so SiGe BiCMOS can be implemented by lower level technology, consequently, the cost at typical process node is reduced. SiGe HBT device can achieve better load matching, i.e. easier to implement the load matching to the common-used 50Ω impedance. The parasitics in SiGe HBT devices are less than CMOS devices. The doping

---

This paper is supported by National Key Project of China (2009ZX01034-002-002-001-02) and Application Specific Integrated Circuit (ASIC) and Systems State Key Laboratory under Grant 11KF008.

978-1-4577-1608-9/11 $26.00 © 2011 IEEE

concentration is lower on both sides of the emitter in SiGe HBT devices, so the E-B junction capacitance is small, which result in higher gain and better gain flatness at high frequencies when compared to CMOS devices.

This paper presents a 5.5GHz fully integrated power amplifier based on 0.18μm SiGe BiCMOS technology. The presented power amplifier schematic has a two-stage single-ended common-emitter structure with 3.3V voltage supply.

The PA is based on a 0.18μm SiGe BiCMOS process, which provides three kinds of HBT: High-speed, Standard and High-voltage. We utilize HV HBT for the following PA design. The device performance is demonstrated in Table II.

TABLE II.   THE HBT PARAMETERS PROVIDED BY PROCESS

Device	Performance Comparison				
	Peak f_T (GHz)	Ic @ Pk f_T (mA/μm^2)	β	Bvceo (V)	Ie @ Pk f_T (mA/μm^2)
Standard	45	0.7	150	4.1	1.25
High-speed	60	1.65	140	3.3	4.0
High-voltage	29	0.4	140	6.0	0.8

As demonstrated in Table II, High-voltage HBT has the highest breakdown voltage BVCEO (collector-emitter breakdown, base open), which is 6V, and its peak f_T is 29GHz, emitter current density is 0.8 mA/μm^2 at peak f_T. So we use High-voltage HBT in power stage to design the 5.5GHz PA.

As shown from Fig.1, The class A power amplifier is a linear amplifier including the pre-amplifier stage and the power stage. Note that the inductor L1 and the capacitor C1 work as a simple L-matching network, forming the input matching network. Similarly, the inductor L4 and the capacitor C5 have been taken to achieve good output matching. The RC series negative feedback circuit is adopted to improve the linearity of the circuit in both the pre-amplifier stage and the power stage. Note that the resistor R1 and the capacitor C2 work as the first stage of the negative feedback loop, while R2 and C4 form the second stage of the negative feedback loop. To ensure good regulative performance, the bias voltage is provided off-chip.

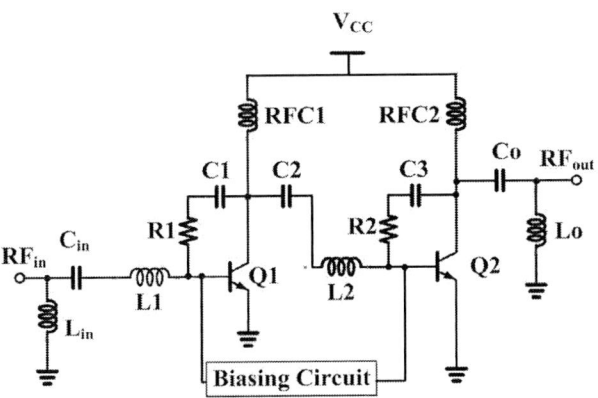

Figure 1.   The simplified schematic circuit of the power amplifier.

Assuming that the drain current is reasonably well approximated by:

$$i_D = I_{DC} + i_{rf} \sin \omega_0 t , \qquad (1)$$

Where I_{DC} is the bias current, i_{rf} is the amplitude of the signal component of the drain current, and ω_0 is frequency. The output voltage can be deduced from signal current and the load resistance. Since the RF choke inductor RFC1 and RFC2 force a substantially constant current through them, it can be easily got that the signal current is none other than the signal component of the drain current. Therefore,

$$v_o = -i_{rf} R \sin \omega_0 t , \qquad (2)$$

Finally, the drain voltage is the sum of the DC drain voltage and the signal voltage. RFC1 and RFC2 present a DC short, so the drain voltage has a symmetrically peak and bottom over V_{CC}.

Drain efficiency and Power added efficiency (PAE) is critical figure of merit in PA design, to get the PAE, firstly compute the efficiency by calculating the signal power delivered to the resistor R:

$$P_{rf} = \frac{i_{rf}^2 R}{2} , \qquad (3)$$

Next, compute the DC power supplied to the amplifier. Assuming the quiescent drain current I_{DC} is made just large enough to guarantee that the transistor does not ever cut off. That is,

$$I_{DC} = i_{rf} , \qquad (4)$$

so that the input DC power is:

$$P_{DC} = I_{DC} V_{CC} = i_{rf} V_{CC} , \qquad (5)$$

The ratio of RF output power to DC input power is a measure of efficiency (usually called the drain efficiency) and is given by:

$$\eta \equiv \frac{P_{rf}}{P_{DC}} = \frac{i_{rf}^2 (R/2)}{i_{rf} V_{CC}} = \frac{i_{rf} R}{2 V_{CC}} , \qquad (6)$$

Now, the largest value of $i_{rf}R$ can reach V_{CC}. Therefore, the maximum theoretical drain efficiency is just 50%. If one makes due allowance for nonzero minimum v_{DS}, variation in bias conditions, non-ideal drive amplitude, and inevitable losses in the filter and interconnect, values substantially smaller than 50% often result – particularly at lower supply voltages. Consequently, drain efficiencies of 30-35% is very normal in Class A amplifiers.

Besides efficiency, another important FOM is the stress on the output transistor. In a Class A amplifier, the maximum drain-to-source voltage is 2Vcc, while the peak drain current has a value of 2Vcc/R. So the device must be able to endure peak voltage breakdown and current melt of these magnitudes. Since scaling trends in IC process technology force reductions in breakdown voltage, the design of PAs becomes more and more difficult [8].

One common way to quantify the relative stress on the devices is to define another type of efficiency, called the normalized power output capability, which is simply the ratio of the actual output power to the product of the maximum device voltage and current. For this type of amplifier, the maximum value of this dimensionless figure of merit is:

$$P_N \equiv \frac{P_{rf}}{v_{DS,pk}\, i_{D,max}} = \frac{V_{CC}^2/(2R)}{(2V_{CC})(2V_{CC}/R)} = \frac{1}{8}. \quad (7)$$

The Class A amplifier thus provides linearity at the cost of low efficiency and relatively large device stresses.

III. DESIGN DISCUSSION AND POST SIMULATION

Since the frequency is quite high at 5.5GHz, when dealing with the layout design, the Binary Feed Line or the Fishbone structure is adopted in all the key current path structure design to transfer the current evenly to every single HBT.

To reduce the parasitic resistances and ensure that the parasitic capacitances do not increase too much, stack method is used in all the key current path interconnect lines design, and some intensive holes are added at special node, as shown in Fig.2.

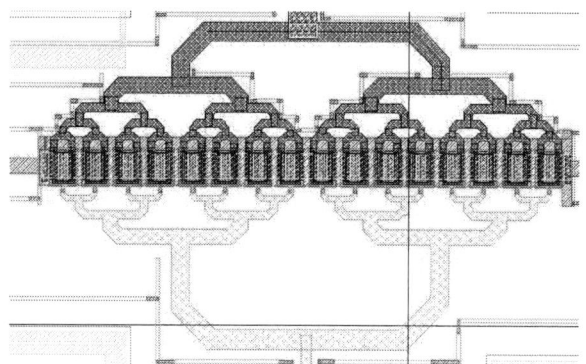

Figure 2. The layout optimization of the power amplifier current path.

On-chip inductors and capacitors are used as matching components, to form matching network. Completed layout of the presented power amplifier is shown in Fig.3.

The summary of post simulation results of the 5.5GHz PA is shown below. The S-parameter results are shown in Fig.4. The small gain S21 is 22.46 dB at 5.5GHz, S11 and S22 are less than -10dB at 5.5GHz, which shows good input matching.

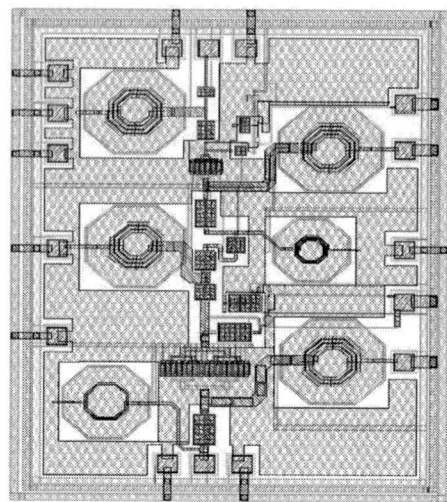

Figure 3. The layout of the 5.5GHz power amplifier.

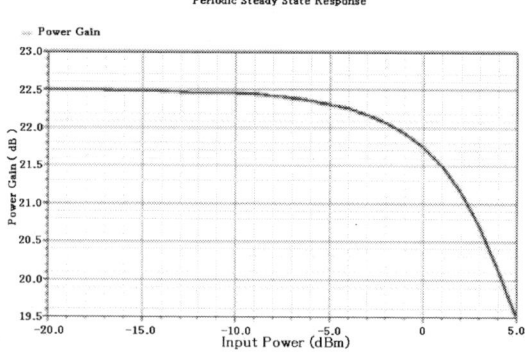

Figure 4. The S-parameter results of the 5.5GHz PA.

Fig.5 shows the simulation results of the power gain decreased when the input signal amplitude increased gradually.

Figure 5. The power gain of the 5.5GHz PA.

The maximum output power P_{MAX} is 24.18 dBm, which is more than 24dBm, shows good power output performance.

The post simulation results of the output 1dB compression point P_{1dB} is shown in Fig.6. The post simulation results of the

978-1-4577-1608-9/11 $26.00 © 2011 IEEE

efficiency PAE is shown in Fig.7. PAE reaches 17.42% at P_{1dB}.

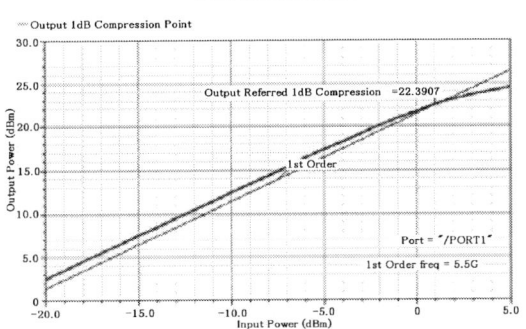

Figure 6.   The output 1dB compression point P1dB of the 5.5GHz PA.

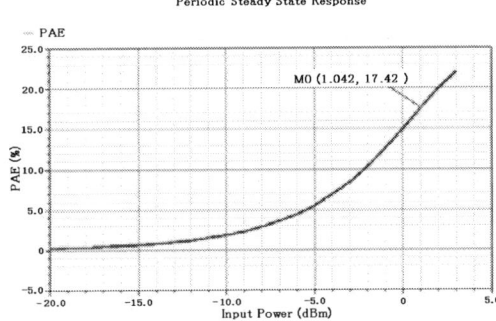

Figure 7.   The efficiency PAE of the 5.5GHz PA.

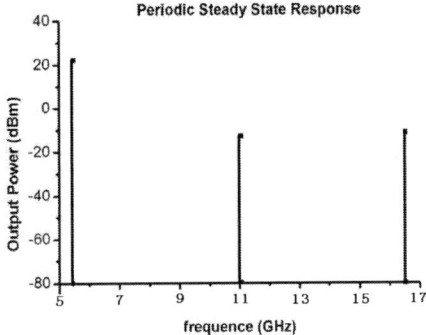

Figure 8.   The output hamonics of the 5.5GHz PA at the input power of 0dB.

The output harmonics at the input power of 0dB is shown in Fig.8. The second harmonics is -36dBc, and the third harmonics is below -33dBc. Consequently, fully integrated passive components for filtering and matching can exhibit good linearity performance.

IV.   SUMMARY

This paper discusses a 5.5GHz fully integrated and high-linearity Class A power amplifier based on 0.18μm SiGe Bi-CMOS technology. According to the post simulation results, the output power of this PA can reach 24.18 dBm, and the PAE is 17.42% at P1dB. The simulation results of the power gain decreased when the input signal amplitude increased gradually. It is suitable to be used in 802.11a Wireless Local Area Network (WLAN) frontend.

ACKNOWLEDGMENT

This paper is supported by National Key Project of China (2009ZX01034-002-002-001-02) and Application Specific Integrated Circuit (ASIC) and Systems State Key Laboratory under Grant 11KF008.

REFERENCES

[1]   H. H. Liao, J. King and A. Behzad,"A Fully Integrated 2*2 Power Amplifier for Dual Band MIMO 802.11n WLAN Application Using SiGe HBT Technology", IEEE Journal of Solid-state Circuits, VOL. 44, NO. 5, May 2009.

[2]   F. H. Raab, S. Cripps and N. O. Sokal, RF and Microwave Power Amplifier and Transmitter Technologies, High Frequency Electronics, May 2003.

[3]   Nickolas, Christina, "SiGe BiCMOS technology drives next-gen green, efficient analog ICs", Electronic Products (Garden City, New York), VOL. 50, NO. 8, July 2008.

[4]   R. Racanellli and P. Kempf, "SiGe BiCMOS Technology for RF circuit Applications", IEEE Trans on Electron Devices, VOL. 52, NO. 7, July 2005.

[5]   K. Nellis and P. Zampardi, A comparison of bipolar technologies for linear handset power amplifier applications, in Proc. BCTM, 2003. pp. 203-209.

[6]   J. S. Rieh, etc. and G. Freeman, SiGe HBT`s for millimeter-wave applications with simultaneously optimized f and f of 300GHz, in Proc. RFIC Symp., Jun. 2004, pp. 395-398.

[7]   D. Chowdhury, "A Fully Integrated Dual-Mode Highly Linear 2.4GHz CMOS Power Amplifier for 4G WiMax Applications", IEEE J. Solid-State Circuits, vol. 44, no 12, pp. 3393-3401, Dec. 2009.

[8]   Thomas H. Lee, The Design of CMOS Radio-Frequency Integrated Circuits, 2nd ed., Cambridge: The Cambridge University Press, 2004. pp. 495-497 .

NTF Zero Compensation Technique For Passive Sigma-Delta Modulator

Arshad Hussain, Sai-Weng Sin, Seng-Pan U, Rui. P. Martins[1]

State Key Laboratory of Analog and Mixed-Signal VLSI (http://www.fst.umac.mo/en/lab/ans_vlsi/website/index.html)
Faculty of Science and Technology, University of Macau, Macao, China
Email: arshad2105@hotmail.com, TerrySSW@umac.mo
(1-On leave from Instituto Superior Técnico /TU of Lisbon, Portugal)

Abstract— **A novel technique is proposed to compensate zeroes of Noise Transfer Function (NTF) for single loop low pass passive switched capacitor sigma-delta modulator (PSDM). PSDM uses switched capacitor (SC) integrator as a loop filter. Poles of loop filter are zeroes of NTF. The proposed technique for SC integrator shifts the poles of SC integrator from inside the unit circle to z=1 at DC. As a result zeroes of NTF for PSDM also get modified and shifts, which greatly improves the noise suppression and noise-shaping performance. Simulation results in simulink, shows that the technique of proposed SC integrator for 2nd Order PSDM with 4-bit quantizer having oversampling ratio (OSR) = 32 improves 18dB SNDR over traditional SC integrator.**

I. INTRODUCTION

Sigma-Delta modulator can attain medium to high resolution by noise-shaping and oversampling. Sigma-Delta circuitry can be implemented by DT (discrete-time) or CT (continuous-time). Discrete-time circuit implementation uses SC network with operational amplifier while CT (continuous-time) circuit is implemented by R and C network with operational amplifier for higher speed. Passive Sigma-Delta Modulator [1-3] is the promising area of low power and high dynamic range. Its loop filter can be defined as analog switches and capacitors only. Comparator is the only active component in the PSDM, so operational amplifier can be avoided. The SC integrator implementation is very simple and consumes no power as compared to operational amplifier. Switched capacitor integrator has charge leakage phenomena, which causes only a portion of charge from input capacitor is transfer to integrating capacitor. This results in NTF zero of PSDM to move inside the unit circle at DC in z-domain. It degrades the noise attenuation in signal band and destroys the noise-shaping property of sigma-delta modulator.

In this paper, NTF zeroes compensation technique is presented which can significantly improves the noise shaping of low-pass PSDM and results in higher SNDR. A 2nd order PSDM is simulated in the simulink to verify the NTF zeroes compensation technique. The organization of the paper is as follows. In section II basic understanding of real integrator and traditional switched capacitor integrator are presented. Section III presents, the proposed switched capacitor integrator. Section IV provides the modeling and simulation results and the section V conclude the paper.

II. REVIEW OF INTEGRATOR MODEL

A. Real Integrator Model

The commonly used delaying active integrator has transfer function H(z) as equation(1).

$$H(z) = \frac{z^{-1}}{1-z^{-1}} \qquad (1)$$

It has zero at origin and pole z=1 at DC of the unit circle in z-domain as shown in Figure 1.

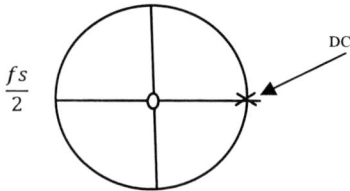

Figure 1. Pole-zero constellation plot for real integrator

B. Passive Switched Capacitor Integrator (Leaky Inegrator)

Switched capacitor integrator with non-overlapping clock can be realized by analog switches and capacitor only as shown in Figure 2.

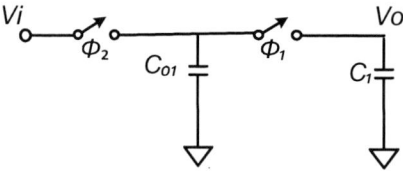

Figure 2. Switched Capacitor Integrator

The transfer function [4] of the switched capacitor integrator is

978-1-4577-1608-9/11 $26.00 © 2011 IEEE

$$H(z) = \frac{\alpha z^{-1}}{[1-(1-\alpha)z^{-1}]} \qquad (2)$$

Where α defines the ratio of capacitor as

$$\alpha = \frac{C_{01}}{C_1+C_{01}} \qquad (3)$$

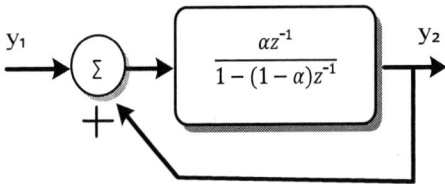

Figure 3. Pole-zero constellation plot for switched capacitor integrator

So this ratio of capacitors defines the pole position on unit circle and gain for the switched capacitor integrator. For stability, the pole position now still at DC but is shifted inside the unit circle as shown in Figure 3.

III. PROPOSED SWITCHED CAPACITOR INTEGRATOR

We propose a new SC integrator which removes the effect of leakage from the SC integrator as shown in Figure 4. So the transfer function of SC integrator behaves like a real integrator. The main leakage term in the denominator cancels and so it is a real integrator.

A model of delaying switched capacitor integrator is presented to prove the concept.

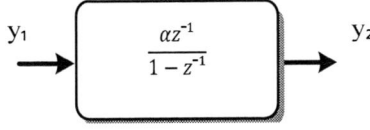

Figure 4. Proposed delaying switched capacitor integrator

Here is the mathematical prove of the proposed switched capacitor integrator

$$y_2 = (y_1 + y_2)\frac{\alpha z^{-1}}{[1-(1-\alpha)z^{-1}]} \qquad (4)$$

$$y_2\left[1 - \frac{\alpha z^{-1}}{[1-(1-\alpha)z^{-1}]}\right] = \frac{\alpha z^{-1}}{[1-(1-\alpha)z^{-1}]}y_1 \qquad (5)$$

$$y_2\left[\frac{1-z^{-1}+\alpha z^{-1}-\alpha z^{-1}}{[1-(1-\alpha)z^{-1}]}\right] = \frac{\alpha z^{-1}}{[1-(1-\alpha)z^{-1}]}y_1 \qquad (6)$$

$$y_2[1-z^{-1}] = \alpha z^{-1}y_1 \qquad (7)$$

$$y_2 = \frac{\alpha z^{-1}}{1-z^{-1}}y_1 \qquad (8)$$

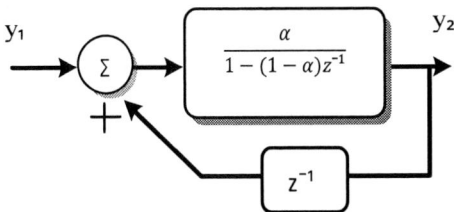

Figure 5. Proposed delaying switched capacitor integrator model

So the equation (8) proves that it become a real delaying integrator with numerator co-efficient as shown in Figure 5. The pole of the switched capacitor integrator is shifted from inside the unit circle to z=1 at DC.

Now considering non-delaying switched capacitor integrator, a delay is compensated in the feedback path so it needs at least one delay in the feedback as shown in Figure 6. So proposed model will be

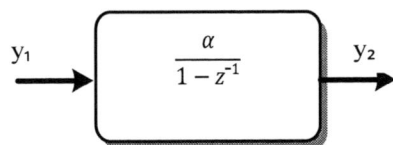

Figure 6. Proposed non-delaying switched capacitor integrator

This will result into non-delaying integrator and forward co-efficient in numerator as normal SC integrator.

$$y_2 = (y_1 + y_2 z^{-1})\frac{\alpha}{[1-(1-\alpha)z^{-1}]} \qquad (9)$$

$$y_2\left[1 - \frac{\alpha z^{-1}}{[1-(1-\alpha)z^{-1}]}\right] = \frac{\alpha}{[1-(1-\alpha)z^{-1}]}y_1 \qquad (10)$$

$$y_2\left[\frac{1-z^{-1}+\alpha z^{-1}-\alpha z^{-1}}{[1-(1-\alpha)z^{-1}]}\right] = \frac{\alpha}{[1-(1-\alpha)z^{-1}]}y_1 \qquad (11)$$

$$y_2[1-z^{-1}] = \alpha y_1 \qquad (12)$$

$$y_2 = \frac{\alpha}{1-z^{-1}}y_1 \qquad (13)$$

Figure 7. Proposed non-delaying switched capacitor integrator model

The equation (13) proves that assuming a delay z^{-1} in the feedback path, even this also verifies that pole of the integrator does not lies inside the unit circle at DC in z-plane. This means that the leaky term αz^{-1} of SC integrator automatically canceled, and a real integrator operation returns with numerator co-efficient. The Figure 7 shows a block of proposed non-delaying SC integrator.

IV. MODELING AND SIMULATION RESULTS

In this section, a design example of 2nd order PSDM is modeled for traditional SC integrator and proposed SC integrator. Matlab modeling and simulation results are presented.

A. DESIGN EXAMPLE

As proof of the concept, 2nd Order PSDM model is presented and simulated in simulink for traditional SC integrator and proposed SC integrator.

978-1-4577-1608-9/11 $26.00 © 2011 IEEE

First a 2nd Order PSDM with traditional SC integrator model is presented in Figure 8. The signal swing in front of quantizer is very small so large gain G is required and quantizer is the only active component. The poles of loop filter are zeroes of the NTF [5]. Traditional SC integrators causes zeroes of NTF to moves inside the unit circle at DC, which effects proper noise attenuation in signal band and degrade the noise-shaping of PSDM.

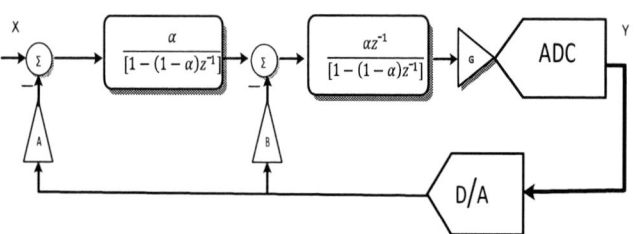

Figure 8. 2nd Order PSDM with traditional switched capacitor integrator model

The NTF(z) of Figure 8 for 2nd order PSDM is given by

$$\text{NTF (z)} = \frac{[1-(1-\alpha)z^{-1}]^2}{D(z)} \qquad (14)$$

$$D(z) = [1-(1-\alpha)z^{-1}]^2 + AG\alpha^2 z^{-1} + BG\alpha z^{-1}[1-(1-\alpha)z^{-1}]$$

Where A & B are the feedback co-efficient, G is the gain and $\alpha = 0.2$ defines the ratio of capacitors for SC integrator in the loop filter.

Now the proposed SC integrator with 2nd Order PSDM model is simulated as shown in Figure 9.

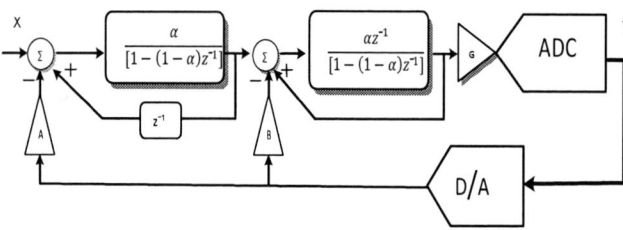

Figure 9. 2nd Order PSDM model with proposed integrator model

The NTF(z) of the proposed integrator model is

$$\text{NTF(z)} = \frac{[1-z^{-1}]^2}{D(z)} \qquad (15)$$

$$D(z) = [1-z^{-1}]^2 + AG\alpha^2 z^{-1} + BG\alpha z^{-1}[1-z^{-1}]$$

Equation (14) & (15) describes NTF(z) of traditional and proposed SC integrator. From equation (15), the NTF(z) zeroes shift to z=1 at DC for proposed SC integrator, which eliminates noise from signal band and improves the SNDR as compared to the traditional SC integrator. A comprehensive system level performance analysis for 2nd Order PSDM considering both types of SC integrators is performed as shown in Figure 10 and Figure 11. OSR versus SNDR for different no of quantizer bits for both types of SC integrators models are simulated. The OSR ranges from 5 to 150, while no of quantizer bit from 2 bit to 8-bit and SNDR is compared for both type of integrator models. As an example for a 4-bit quantizer traditional SC integrator having OSR=32 results in

SNDR=67dB while proposed SC integrator model with same no of quantizer bit and OSR results in SNDR=85dB. Its output spectrum for traditional and proposed SC integrators with input signal Vi = 0.94v is shown in Figure 12.

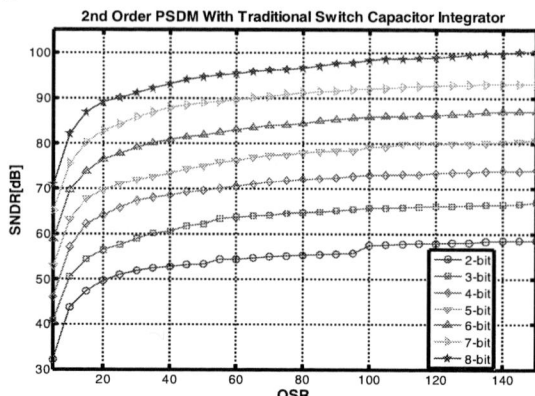

Figure 10. 2nd Order PSDM for traditional switched capacitor integrator

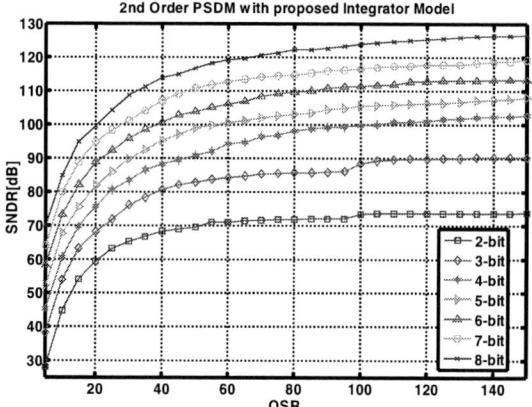

Figure 11. 2nd Order PSDM for proposed switched capacitor integrator

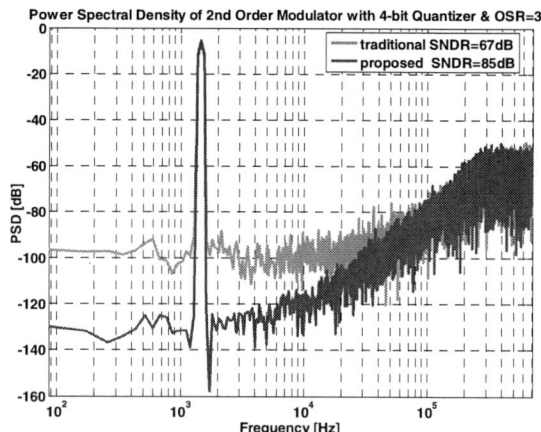

Figure 12. Matlab Simulated PSD for both integrator models

978-1-4577-1608-9/11 $26.00 © 2011 IEEE

Dynamic range comparison of 2nd Order PSDM having OSR=32 with 4-bit quantizer for both types of SC integrators is plotted in Figure 13. Proposed SC integrator clearly has more than 20dB higher dynamic range as compared to traditional SC integrator. Simulation results of 2nd Order PSDM for different no of quantizer bits for different OSR are summarized in Table I.

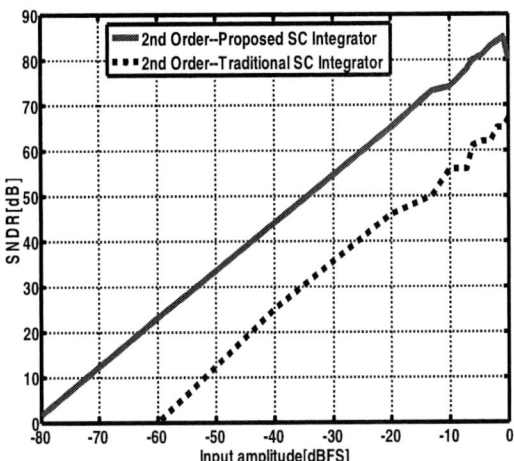

Figure 13. Dynamic range comparison for 2nd Order PSDM with 4-bit quantizer & OSR=32

TABLE 1: 2nd Order PSDM Results for proposed integrator

Specification	*Value*				
Peak SNDR	85dB	89dB	95dB	80dB	94dB
SNDR improvement	18dB	20dB	22dB	20dB	24dB
OSR	32	40	40	40	60
Quantizer	4-bit	4-bit	5-bit	3-bit	4-bit

B. INTEGRATOR WITH INCOMPELETE CHARGE TRANSFER

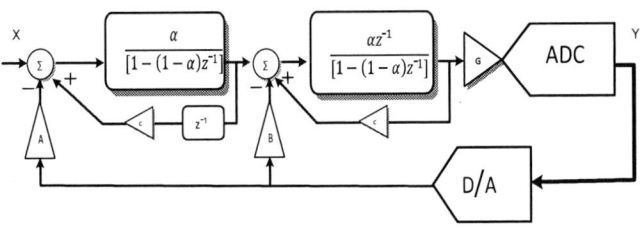

Figure 14. PSDM Model with non-ideal effects for 4-bit quantizer & OSR=32

The proposed integrator model for 2nd order PSDM having non-ideal effect, incomplete charge transfer effect C in the feedback path is modeled in Figure 14. The parameter C can be used to estimate the charge lost during the feedback path from output to the input of switched capacitor integrator. Performance degradation due to charge loss property is shown in Figure 15.

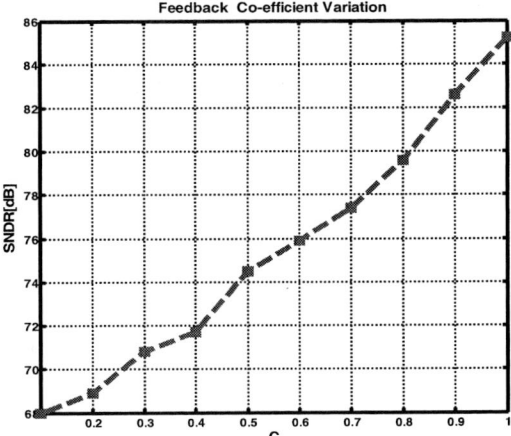

Figure 15. Feedback parameter C versus SNDR [dB]

V. CONCLUSION

A novel technique for passive switched capacitor integrator is proposed in this paper, which significantly improves the noise-shaping performance of PSDM. Moreover, it can remove the charge leakage phenomena proved mathematically and also simulink simulation is performed to verify this technique. Finally a design example of 2nd Order PSDM with 4-bit quantizer for OSR=32 is simulated and performance is compared between the traditional switched capacitor integrator and proposed switched capacitor integrator. The proposed model of switched capacitor integrator for 2nd order PSDM with 4-bit quantizer having OSR=32 achieves 18dB higher SNDR as compared to its traditional switched capacitor integrator.

ACKNOWLEDGMENT

THIS WORK WAS FINANCIALLY SUPPORTED BY RESEARCH GRANTS OF UNIVERSITY OF MACAU AND MACAU SCIENCE & TECHNOLOGY FUND (FDCT).

REFERENCES

[1] F. Chen, and B. Leung, " A 0.25-mW low-pass passive sigma-delta modulator with built-in Mixer for a 10-MHz IF Input", IEEE J. Solid-State Circuits, vol.32, no.6, Jun. 1997.

[2] F. Chen; Ramaswamy, S.; Bakkaloglu, B.; "A 1.5V 1mA 80dB passive ΣΔ ADC in 0.13μm digital CMOS process," ISSCC Dig. Tech. Papers, pp. 32-33, vol.1, 2003.

[3] F. Chen, Ramaswamy, S.; Bakkaloglu, B.; "Design and analysis of a CMOS passive ΣΔ ADC for low power RF transceivers" Analog Integr Circ Sig Process, pp. 129-141, 2009.

[4] Yousry, R.; Hegazi, E.; Ragai, H.F.; "A Third-Order 9-bit 10-MHz CMOS ΔΣ Modulator with one active stage", IEEE Tran on Circuit & Systems-I: Regular Papers, vol.55, no.9, pp.2469-2482, Oct. 2008.

[5] Richard Schreier and Gabor C. Temes, "Understanding Delta-Sigma Data Converter," IEEE Press, 2005.

[6] T. Song, and Shouli Yan, "A low power 1.1MHz CMOS continous-time delta-sigma modulator with active-passive loopfilters", IEEE-ISCAS, 2006.

[7] T. Song; Cao, Z.; Yan, S.; "A 2.7-mW 2-MHz continous-time ΣΔ modulator with a hybrid active-passive loop filter", IEEE J. Solid-State Circuits, vol.43, no.2, Feb. 2008.

A 0.7 V DTMOS-Based Class AB Current Mirror

Arnon Kanjanop and Varakorn Kasemsuwan
School of Electronics Engineering, Faculty of Engineering
King Mongkut's Institute of Technology Ladkrabang
Bangkok, Thailand
pf_sunday@hotmail.com, kkvarako@kmitl.ac.th

Abstract—**A 0.7 V DTMOS-Based class AB current mirror is presented. The circuit is developed based on a conventional class AB current mirror structure with a common-source output stage. The circuit is designed using a 0.13 μm CMOS technology and operates under a 0.7 V supply. SPICE with BSIM3V3 model parameters is used to verify the circuit performance. The maximum current transfer is found to be 7 times larger than the input bias current, while the DC current gain is -0.03 dB. The bandwidth and power dissipation are 540 MHz and 96 μW, respectively.**

I. INTRODUCTION

Current-mode circuits have been an active area of research for many years due to many intrinsic advantages over voltage-mode counterparts including low supply voltage requirement, wide bandwidth, tunable input impedances, high slew rates, and less susceptible to power and ground fluctuations [1], [2]. Current mirror (CM) is a versatile building block in current-mode circuits. CM is widely exploited in several active devices, e.g. current conveyor (CC), operational amplifiers, operational transconductance amplifiers (OTAs), and current amplifier (CA) [3]. CM is also used for realizing continuous and sampled-data analog filters [4, 5].

Class AB current mirror is well known to provide better dynamic range [6] compared to class A current mirror. In addition, class AB current mirror has reduced sensitivity to process tolerances [7] and increased quiescent current to signal amplitude ratio. These properties are essentially due to the complementary nature of the configuration which employs both n- and p-type current mirrors.

A conventional class AB current mirror is shown in Fig. 1 [8]. The circuit consists of a translinear loop (M1-M4) which sets both the bias voltage at the input terminal (that is equal to the analog ground) and the quiescent current in M3-M4. In the operation, the input signal current is mirrored to the output through the two complementary current mirrors M5-M7 and M6-M7. The circuit has been used in several applications, e.g. current amplifier [9], current differencing buffered amplifier (CDBA) [10], current comparator [11-12], and current conveyor [13]. The circuit has two main drawbacks; 1) its input impedance is rather high, and 2) the circuit requires a

supply voltage of at least equal to $2(V_T + 2V_{DS,SAT})$, thus preventing the circuit to operate under sub-1 volt supply. At present, operating the circuit under low voltage supply becomes essential mainly due to reliability and power consumption issues.

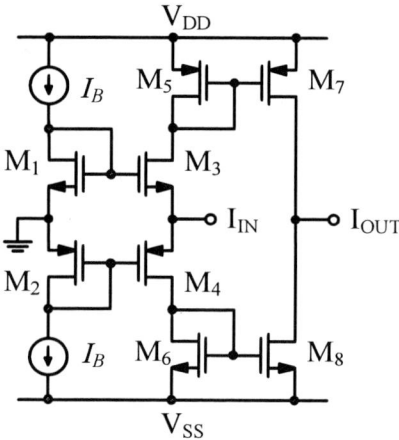

Figure 1. Conventional class AB current mirror.

Various attempts have been proposed to solve this problem. Self-cascode structure was proposed in [14]. This technique results in a circuit, which can operate under the supply voltage lower than the normal cascode structure. The circuit also achieved good current transfer characteristic. However, additional circuitry is required to implement class AB current mirror. This additional circuitry unfortunately demands quite high power consumption, high supply requirement, and causes input impedance to increase. [15] employed the dynamic biasing technique, enabling the circuit to operate under a low voltage supply. The circuit exhibited fast response with low settling time. However, a switched-capacitor network is required, making the circuit complex and prone to switching noise. [16] employed two auxiliary differential amplifiers and two poly resistors connecting in the negative feedback configuration to achieve low-voltage operation, good control of quiescent currents and input bias voltage. The resulting circuit can operate under a 1-V supply. Requirement of two poly resistor unfortunately affects the

978-1-4577-1608-9/11 $26.00 © 2011 IEEE

input impedance of the circuit. Quasi-floating gate was proposed to achieve low voltage class AB operation [17]. The circuit used differential amplifier to provide low input impedance, and to set bias voltage at the input. However, in order to bias the voltage at the middle of the supply, the circuit requires a supply voltage of at least $2V_T + 4V_{DS,SAT}$, hence making the circuit unsuitable for sub 1-V supply.

This paper presents a low voltage class AB current mirror. The circuit is developed using the dynamic threshold voltage MOS transistors (DTMOS) and two low voltage current mirror configured in the class AB operation. The circuit can operate under the supply voltage of 0.7 V, and demonstrates good performances. The input current amplitude can be 7 times larger than the input bias current. The bandwidth of the circuit is 540 MHz, while the DC current gain is found to be - 0.03 dB. The proposed structure offers input and output impedance of 264 Ω and 183 kΩ, respectively. The power dissipation under the quiescent condition is 96 μW.

II. PROPOSED CIRCUIT

The proposed current mirror is shown in Fig. 2. As seen, our mirror is developed based on a conventional class AB current mirror. Dynamic threshold voltage MOSFET (DTMOS) M_{ND1}-M_{ND2} and M_{PD1}-M_{PD2} form a translinear loop

which accurately sets the bias voltage at the input terminal (that is equal to the analog ground). DTMOS are employed, enabling the circuit to operate under the low voltage supply. It is known that DTMOS is suitable for low voltage operation due to its dynamic threshold voltage, larger transconductance, and lower noise [18]. Transistors $M_{N1(P1)}$-$M_{N4(P4)}$ are incorporated into the circuit to convert the difference between the currents through $M_{ND1(PD1)}$ and $M_{ND2(PD2)}$ ($i_{ND1(PD1)}$ and $i_{ND2(PD2)}$) into voltage, which is then fed to output DTMOS transistor M_{N8}-M_{N9} (M_{P8}-M_{P9}). This negative feedback configuration ensures low input impedance and equality between the input terminal and the source terminal of $M_{ND1(PD1)}$ and low input impedance. Composite transistors M_{N10}-M_{N11} (M_{P10}-M_{P11}) are employed, and operate as output stage with high output impedance.

The operation of the circuit can be explained as follows. When the input current I_{in} is forced into the input terminal, the gate-source voltage of M_{ND2} (M_{PD2}) will decrease (increase) resulting in current flowing into node N2 (P2). The difference of the current at nodes N1 (P1) and N2 (P2) will be converted to error signal at node N3 (P3), which connected to the gates of common source composite transistors M_{N8}-M_{N9} (M_{P8}-M_{P9}). This negative feedback configuration will force M_{N8}-M_{N9} (M_{P8}-M_{P9}) to sink all the input current. Since the gate

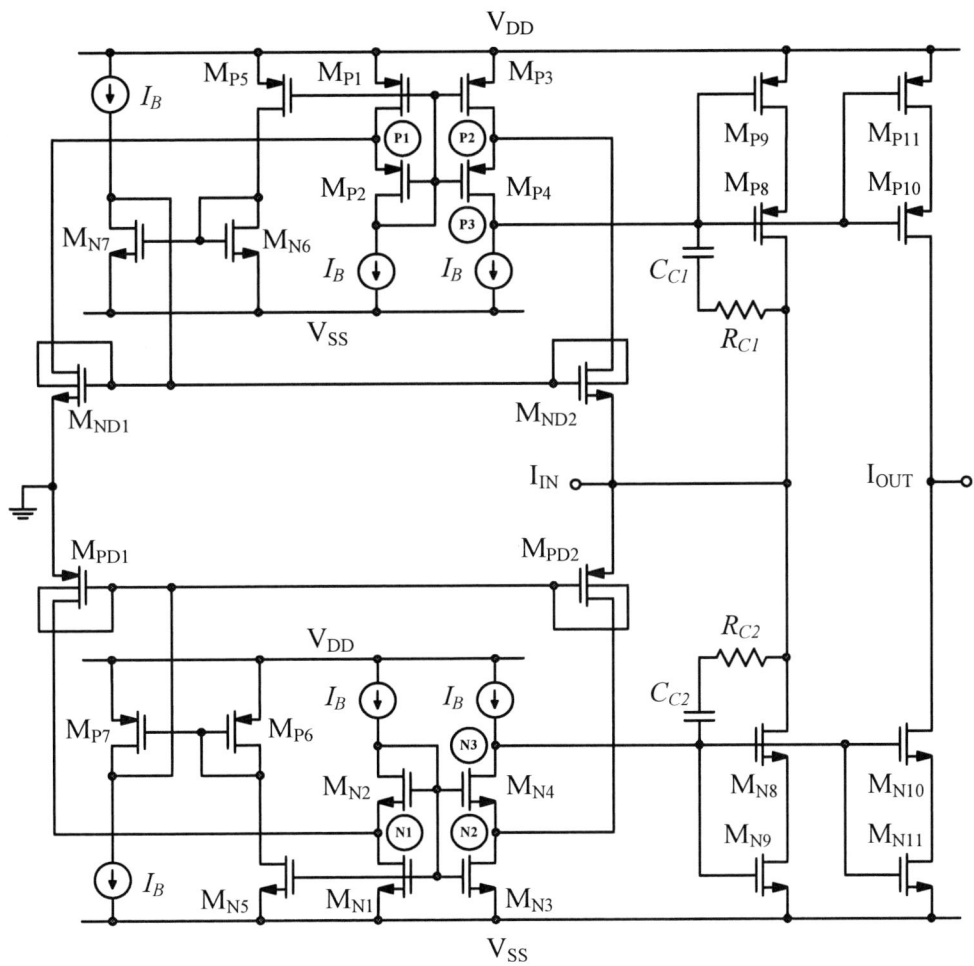

Figure 2. Proposed current mirror

terminals of the composite transistors M_{N8}-M_{N9} (M_{P8}-M_{P9}) are tied to the gates of M_{N10}-M_{N11} (M_{P10}-M_{P11}), the input current is therefore transfer to the output terminal.

On the contrary, when the input current I_{in} is forced out of the input node, the gate-source voltage of M_{ND2} (M_{PD2}) will increases (decreases) causing the current flowing out from node N2 (P2). The differential current between nodes N1 (P1) and N2 (P2) will be converted to error signal at node N3 (P3), which connected to the gates of common source composite transistors M_{N8}-M_{N9} (M_{P8}-M_{P9}). As a result, M_{N8}-M_{N9} (M_{P8}-M_{P9}) will source all the input current, which is also mirrored to the output terminal via M_{N10}-M_{N11} (M_{P10}-M_{P11}).

The input impedance of the proposed current mirror is quite small due to the negative feedback mechanism mentioned previously. To demonstrate this, let assume that the input current signal is forced into the input node causing the input node to increase. The current signal is then passed through $M_{ND2(PD2)}$, and converted to voltage at node N3 (and P3) via $M_{N4(P4)}$, which is configured as a common-gate amplifier. The signal at node N3 (and P3) is further amplified by the common-source composite transistors M_{N8}-M_{N9} (M_{P8}-M_{P9}). As a result, the voltage at the input terminal is forced to go low.

Straight forward small signal analysis shows that the input impedance of the proposed circuit is given by

$$R_{IN} = \frac{1}{\left(1 + g_{mP4} r_{oP4}\right) g_{mEffPCI} + \left(1 + g_{mN4} r_{oN4}\right) g_{mEffNCI}} \quad (1)$$

where r_o and g_m are the output impedance and transconductance of a transistor, respectively, $g_{mEffNCI}$ is the effective transconductance of composite transistor M_{N8-9} ($g_{mEffNCI} \cong g_{mN9}$) and $g_{mEffPCI}$ is the effective transconductance of composite transistor M_{P8-9} ($g_{mEffPCI} \cong g_{mP9}$).

Output impedance of circuit is large since composite transistors are employed at the output. This high output impedance is suitable for current signal processing circuit, which the output signal is current. The output impedance of the proposed circuit be expressed as [19]

$$R_{OUT} = r_{o,MN10,11} \parallel r_{o,MP10,11} \quad (2)$$

where $r_{o,MN10,11} \cong (g_{mN10} r_{oN11}\text{-}1)r_{oN10}$ and $r_{o,MP10,11} \cong (g_{mP10} r_{oP11}\text{-}1)r_{oP10}$.

The capacitor $C_{C1\text{-}2}$ and resistor $R_{C1\text{-}2}$ serve the purpose to compensate the circuit such that the closed-loop response of the proposed circuit is stable and demonstrate reasonable response times and bandwidth.

III. SIMULATION RESULTS

The proposed current mirror has been simulated using SPICE with BSIM3V3 model parameters. In the design, a 0.13 μm standard CMOS technology under the supply voltage of 0.7 V is employed. All the bias sinking and sourcing current sources (I_B) are implemented using a simple NMOS and PMOS, respectively, and set to 8 μA. The input and output impedance are found to be 264 Ω and 183 kΩ, respectively.

Fig. 3 shows the dc transfer characteristic between the input and the output currents. As seen, the input current is mirrored to the output over a wide range ($\pm$ 200 μA) with good linearity. Fig. 4 shows the step response of the proposed current mirror, when the square wave input current (10 μApp, 1 MHz) is applied. As seen, the proposed circuit shows stable characteristic, with settling time lower than 10 ns.

Fig. 5 shows the frequency response of the circuit. The DC current gain is found to be -0.03 dB and the bandwidth is 540 MHz. Fig. 6 shows the total harmonic distortion (THD) (at 1 MHz), when the output terminal is ground. The THD is found to be less than 0.5% for the input current less than 200 μA$_{p-p}$. Finally, the power dissipation under the quiescent condition is 96 μW.

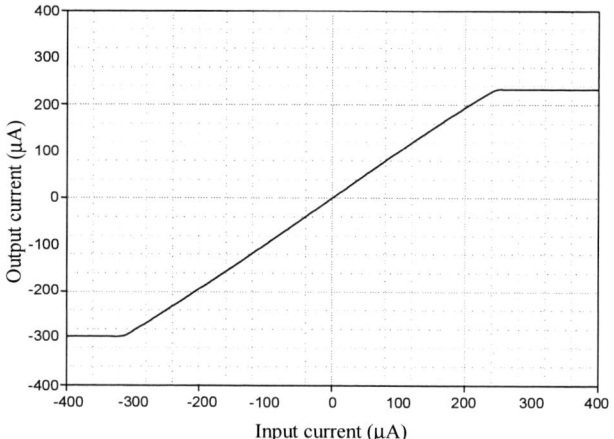

Figure 3.   DC transfer characteristic.

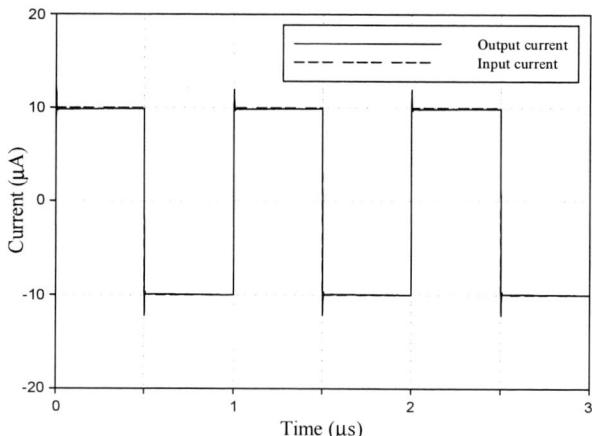

Figure 4.   Input and output waveform (Square wave : ±10 μA, 1 MHz).

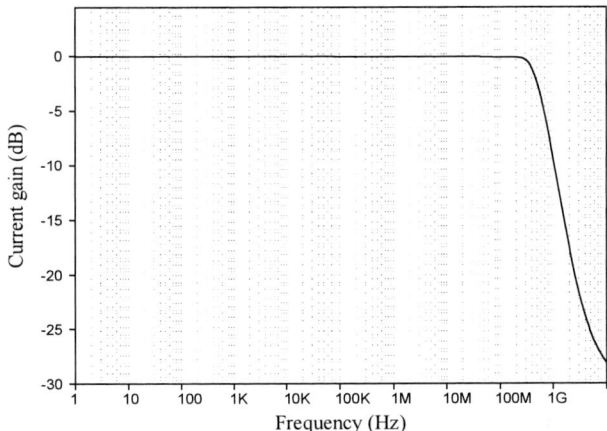

Figure 5. Frequency response.

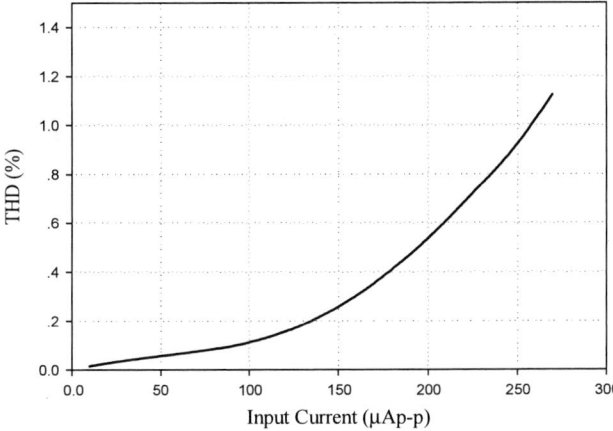

Figure 6. Total harmonic distortion (THD) at 1 MHz.

IV. CONCLUSION

A low voltage class AB current mirror is presented. The circuit is developed based on a complementary DTMOS structure and composite transistors with a common-source input and output stages. The circuit can operate under low voltage supply (0.7 V). The circuit performance is verified using SPICE with BSIM3V3 model parameters. The current mirror can support a wide range of the input current with good linearity. The DC current gain is -0.03 dB with the bandwidth of 540 MHz. The power dissipation under the quiescent condition is 96 μW.

REFERENCES

[1] G. Palmisano, G. Palumbo and S. Pennisi, "CMOS current amplifiers," Kluwer Academic Publishers, Dordrecht, Netherlands, 1999.

[2] K. Koli and A. I. Halonen, "CMOS current amplifiers: speed versus nonlinearity," The International Series in Engineering and Computer Science, Springer, 2002.

[3] C. Toumazou, F. Lidgey, and D. G. Haigh, "Analog IC design-the current-mode approach," Peter Peregrenus Ltd., United Kingdom, 1990.

[4] G. Souliotis and C. Psychalinos, "Current-mode linear transformation filters using current mirrors," IEEE Trans. on Circuits and Systems II, vol. 55, no. 6, pp. 541–545, 2008.

[5] S. S. Lee, R. H. Zele, D. J. Allstot, and G. Liang, "CMOS continuous-time current-mode filters for high frequency applications," IEEE J. Solid-State Circuits, vol. 28, no. 3, pp. 323–329, 1993.

[6] E. Bruun, "Worst case estimate of mismatch induced distortion in complementary CMOS current mirrors," Electronics Letters, vol. 34, pp. 1625–27, August 1998.

[7] S. Kawahito and Y. Tadokoro "CMOS class-AB current mirrors for precision current-mode analog-signal-processing," IEEE Trans. on Circuits and System II, vol.43, no. 12, pp. 843–845, December 1996.

[8] G. Palmisano and S. Pennisi, "Dynamic biasing for true low-voltage CMOS class AB current- mode circuits," IEEE Trans. on Circuits and Systems part II, vol. 47, no. 12, pp. 1569–1575, December 2000.

[9] G. Palmisano, G. Palumbo and S. Pennisi, "CMOS current amplifiers," Kluwer Academic Publishers,Boston, 1999.

[10] N. Tarim and H. Kuntman, "A high performance current differencing buffered amplifier," International Conference on Microelectronics, Morocco, pp. 153–156, 2001.

[11] D. Freitas and K. Current, "CMOS current comparator circuit," Electronics Letters, vol. 19, no. 17, pp. 695–697, August 1983.

[12] G. Palmisano and G. Palumbo, "Offset-compensated low power current comparators," Electronics Letters, vol. 30, no. 20, pp. 1637–1639, September 1994.

[13] A. Sedra, G. Roberts and F. Gohh, "The current conveyor: history, progress and new results," IEE Proc. Part G Circuits, Devices and Systems, vol.137, no. 2, pp.78– 87, April 1990.

[14] A. Zeki and H. Kuntman, "High-linearity low-voltage self cascode class AB CMOS current output stage," IEEE International Symposium on Circuits and Systems, pp. IV 257– 260, 2000.

[15] G. Palmisano and S. Pennisi, "A true low-voltage CMOS class AB current mirror," IEEE International Symposium on Circuits and Systems, pp. IV 249– 252, 2000.

[16] S. Pennisi, "1-V CMOS class AB current mirror," European Conference on Circuit Theory and Design, pp. II 93–96, 2001.

[17] A. J. Lopez-Martin, J. Ramirez-Angulo, R. G. Carvajal and J. M. Algueta, "Compact class AB CMOS current mirror," Electronics Letters, vol. 44, no. 23, 2008.

[18] F. Assaderaghi, S. Parke, D. Sinitsky, J. Bokor, P. K. Ko and C. Hu, "A dynamic threshold voltage MOSFET (DTMOS) for very low voltage operation," IEEE Electron Device Letters, vol. 15, no. 12, pp. 510– 512, 1994.

[19] E. Sanchez-Sinencio, "ELEN 607 Advanced analog circuit design techniques: low voltage circuit design techniques," Analog and Mixed-Signal Center, Texas A&M University.

978-1-4577-1608-9/11 $26.00 © 2011 IEEE

A High-Gain Fully-Differential Thermal Noise-Canceling CMOS Front-End Amplifier

Puttachai Chimpleekul and Varakorn Kasemsuwan
School of Electronics Engineering, Faculty of Engineering,
King Mongkut's Institute of Technology Ladkrabang (KMITL)
Bangkok 10520, THAILAND
E-mail: kkvarako@kmitl.ac.th

Abstract—**This paper presents a fully-differential CMOS front-end amplifier using g_m-boosting and noise-canceling techniques. The proposed front-end amplifier is designed based on a 0.18 μm standard CMOS process and 2 V supply. Simulation results show noise figure (*NF*) of 2.9 dB, while the voltage gain and bandwidth of the amplifier are 31.6 dB and 2 GHz, respectively. The power dissipation is 19 mW.**

I. INTRODUCTION

Nowadays, a wide spread of data communication systems is progressing rapidly especially in high bit rate for both short and medium ranges. The design of front-end amplifier (FA) is known to be the most challenging task, since it involves careful optimization of a number of trade-offs between bandwidth, gain, noise, stability and power dissipation. FA find various application in the hard-disk drive system [1], optical system [2], receiver [3-4] and tuner [5-6].

Noise is one among important issues in the design of FA, especially under a low voltage environment due to limited signal headroom. Noise represents a lower limit to the size of electrical signal that can be amplified by a circuit without significant deterioration in signal quality. Noise also results in an upper limit to the useful gain of an amplifier, because if the gain is increased without limit, the output stage of the circuit will eventually begin to limit on the amplified noise from the input stages. In the MOSFET, the channel can be treated as resistor. As a result, MOSFET exhibits thermal noise, which is usually a major source of noise in MOS transistor.

Recent FA using common-gate (CG) becomes a strong candidate since CG provides a wideband input matching over a wide frequency range. Unfortunately, the transconductance of the CG cannot arbitrarily be chosen but is set by the input impedance matching requirement, preventing improvements of the *NF*. Several approaches have been introduced to minimize the *NF* of the CG amplifier by incorporating noise-canceling circuitries into the system. [3-4] proposed a noise canceling broadband CMOS CG FA. The resulting circuits demonstrated low *NF*. However, single-ended structure makes the circuits sensitive to common-mode (CM) noises, which is

increasingly important especially in mixed-signal applications, where supply voltages could be very noisy. [5] employed noise-canceling technique based on a differential amplifier. Positive feedback was used to eliminate the correlation between the input impedance, and the transconductance of the CG input transistor. Although the transistor noise on one side of the differential pair can be eliminated, the feedback path allows the transistor noise from the other side of the pair to propagate to the output. A pseudo differential structure using noise-canceling circuitries was employed [7-10]. Their designs can successfully suppress noise, this structure is , however, susceptible to the CM noises. [11-13] proposed a fully balanced noise-canceling amplifier. The differential structure with tail current is employed to get rid of the CM noises. Unfortunately, the noise of the input transistor is partially suppressed.

In this paper, we present a 31.6 dB passband gain, 2 GHz bandwidth and 2.9 dB noise figure FA. The circuit is developed based on CG amplifier. The circuit employs differential amplifier with tail current to get rid of noise and to minimize CM noises. The circuit is realized using a 0.18 μm CMOS technology under 2 V supply. The paper is organized as follows. Section II explains the concepts of the g_m-boosting and noise-canceling techniques. Section III describes the operation of the proposed FA. Section IV presents the simulation results and, finally, the conclusion.

II. GM-BOOSTING AND THERMAL NOISE-CANCELING

A. GM-Boosting Technique

Fig. 1(a) illustrates a basic CG amplifier. Since the input impedance of the CG is $R_{in}=1/g_{m1}$. R_{in} can be matched to that of the transmission line (R_S) by setting $1/g_{m1}$ equal to R_S. The noise factor (*F*) of the CG can be shown as

$$F=1+\gamma/\alpha g_{m1}R_S \qquad (1)$$

where γ and α are bias-dependent noise parameters.

From Eq. (1), *F* of CG can be minimized by increasing g_{m1}. As seen, the noise performance of the CG is limited by

---

This work is supported by the College of Data Storage Innovation, King Mongkut's Institute of Tech. Ladkrabang, National Electronics and Computer Tech. Center and National Science and Tech. Development Agency.

978-1-4577-1608-9/11 $26.00 © 2011 IEEE

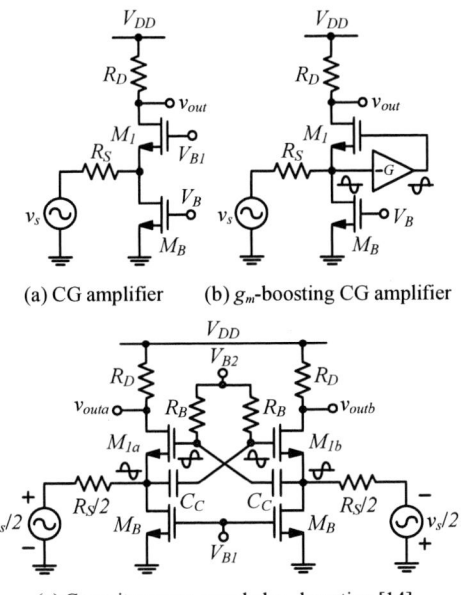

(a) CG amplifier   (b) g_m-boosting CG amplifier

(c) Capacitor cross-coupled g_m-boosting [14]

Figure 1. Schematic of CG amplifiers.

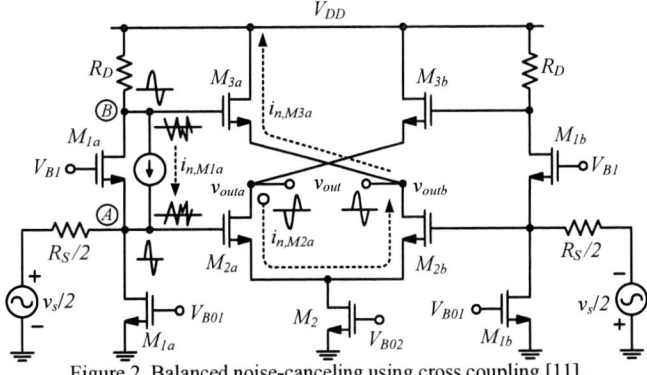

Figure 2. Balanced noise-canceling using cross coupling [11].

the input impedance matching requirement. One approach to decouple the noise performance from the input impedance matching requirement is to use g_m-boosting technique which its concept is shown in Fig. 1(b). As seen, the amplifier ($-G$) is inserted between the gate and source of M_1, which causes the gate to source voltage to increase, thus increasing the current conduction. This can be translated to an increase in the effective transconductance ($G_{m(eff)}$). One can easily see that ($G_{m(eff)}$) of the circuit in Fig. 1(b) is $G_{m(eff)}=(1+G)g_{m1}$.

One can also derive the input impedance R_{in} and F of the g_m-boosting CG amplifier in Fig. 1(b) as

$$R_{in}=1/(1+G)g_{m1}R_S \tag{2}$$

$$F=1+\gamma/\alpha(1+G)^2 g_{m1}R_S \tag{3}$$

From Eq. (2) input impedance matching can be obtained by setting $(1+G)g_{m1}$ to R_S, and, consequently, F becomes

$$F=1+\gamma/\alpha(1+G) \tag{4}$$

As seen, the F can be reduced by increasing $-G$. Since an active amplifier introduces additional noise itself, the idea of using the passive elements becomes attractive. Fig. 1(c) illustrates a differential topology, which allows the differential devices to be capacitor cross-coupled. One can see that the inverting gain ($-G$) is given by the capacitor division ratio

$$G=C_C/(C_C+c_{gs}) \tag{5}$$

In case that $C_C \gg c_{gs}$, $G\cong1$, and F is approximated as

$$F=1+\gamma/2\alpha \tag{6}$$

B. Noise-Canceling cross-coupling

Fig. 2 illustrates basic fully-differential noise-canceling exploits cross coupling [11]. The circuit consists of two CG ($M_{1a,b}$), and noise canceling circuitry ($M_{2a,b}–M_{3a,b}$). The circuit demonstrates low NF, low power and wide bandwidth. It is noticed that the noise canceling circuitry uses tail current, thus making the FA immune to CM noises.

The operation of the circuit can be explained as follows. The input signal $v_s/2$ (solid line) undergoes a feedforward amplification, i.e., the input signal is amplified by the CG ($M_{1a,b}$), and the noise-canceling circuitry (M_{2a-2b} and M_{3a-3b}). The differential output signal is then taken between nodes v_{outa} and v_{outb}. Notice that the signals v_{outa} and v_{outb} are 180 degree out of phase. As a result, the differential output signal v_{out} is larger than the signal at either node v_{outa} or v_{outb} alone. This translates to an increase in the voltage gain of the system.

In term of the channel thermal noise current (dotted signals) of M_{1a}, let us consider $i_{n,M1a}$, which flows between the drain and source terminals. This noise current flows out from node B and into node A, resulting in out of phase thermal noise voltages at nodes A and B. These voltages are converted to current via M_{2a} and M_{3a}. In the design, if $i_{n,M2a}$ is equal to $i_{n,M3a}$ the output thermal noise voltage associated with M_{1a} will be completely eliminated at node v_{outb}. However, it is noted that thermal noise voltage at v_{outa} still exists and could propagate to the next stage and further to the output of the system.

III. PROPOSED AMPLIFIER

Fig. 3 illustrates the proposed amplifier. As seen, the circuit is similar to Fig. 2. However, $M_{2a(b)}–M_{4a(b)}$ are incorporated to fully suppress the thermal noise of the input CG transistor and to enhance gain of the amplifier. $M_{3a(b)}$ in Fig. 3 is connected in the common-source (CS) rather than common-drain (CD) configuration as in Fig. 2, thus increasing the output impedance, and gain of the system. In addition, $M_{4a(b)}$, which is connected in the CG configuration, serves two proposes; 1) it is used to increase the output current by converting the amplified voltage signal at node B, and 2) it is used to fully suppress the thermal noise of $M_{1a(b)}$. C_C and M_B (in the cut off region) form the capacitor cross-coupled and passive resistor C_C and R_B, respectively, in Fig. 1(c).

The operation of the circuit can be explained as follows. The input signal is amplified by the CG ($M_{1a(b)}$). The signal is then further amplified by the noise-canceling circuitry ($M_{2a(b)}–M_{4a(b)}$), i.e., $M_{2a(b)}$ converts the input voltage to current, while $M_{3a(b)}$ and $M_{4a(b)}$ covert the amplified voltage at node B to current such that all these currents are constructively combined at the output. In the design, the bias current, size of $M_{1a(b)}$ and resistor R_D are chosen such that the input impedance matching is achieved. It is noted that the choice of R_D requires some considerations. Unlike Fig. 2, R_D does not need to be

978-1-4577-1608-9/11 $26.00 © 2011 IEEE

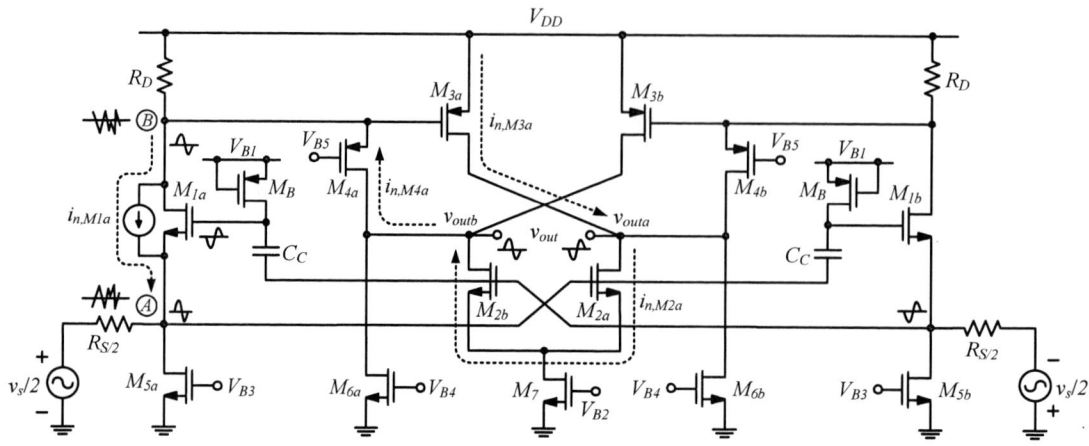

Figure 3. Proposed amplifier.

large to have high gain, since the noise-canceling circuitry ($M_{2a(b)}$–$M_{4a(b)}$), which acts as a second stage amplifier, helps increasing the overall gain. Moderate size of R_D relaxes the voltage headroom requirement and improves the output swing. Furthermore, the pole frequency associated with node B is improved due to reduced miller effect of the gate-drain capacitance of $M_{3a(b)}$. However, R_D should be set large enough, so that the noise of the CG is not too large. In our design, R_D is chosen to be 175 Ω.

In term of the channel thermal noise current (dotted signals), let us consider the noise current of M_{1a} ($i_{n,M1a}$), which flows into node A but out of node B. This results in two fully correlated noise voltages at nodes A and B with opposite phase. These two thermal noise voltages are further converted to currents $i_{n,M2a}$, $i_{n,M3a}$ and $i_{n,M4a}$ by M_{2a}, M_{3a} and M_{4a}, respectively. If the circuit is designed such that $i_{n,M2a}$ is equal to $i_{n,M3a}$ and $i_{n,M4a}$, the output thermal noise voltage associated with M_{1a} at v_{outa} and v_{outb} will be eliminated.

The passband gain of the proposed amplifier can be analyzed using small signal analysis and shown as

$$A_v \cong \frac{\left[g_{m2a(b)} + 2g_{m1a(b)}\left(g_{m3a(b)} + g_{m4a(b)} \right) R'_D \right] R_{out}}{\left(1 + s/\omega_{p(A)}\right)\left(1 + s/\omega_{p(B)}\right)\left(1 + s/\omega_{p(out)}\right)} \quad (7)$$

where R_{out} is $r_{o2a(b)} \| r_{o3a(b)} \| g_{m4a(b)} r_{o4a(b)} R_D$, $\omega_{p(A)}$, $\omega_{p(B)}$ and $\omega_{p(out)}$ are poles associated with nodes A, B and output, respectively, and given by $\omega_{p(A)}=2g_{m1a(b)}/[C_C+c_{gs1a(b)}+c_{sb1a(b)}+c_{gs2a(b)}+c_{gd2a(b)}(1+g_{m2a(b)}R_{out})+c_{gd5a(b)}+c_{db5a(b)}]$, $\omega_{p(B)}=(1+g_{m4a(b)}R_D)/R_D[c_{gd1a(b)}(1+g_{m1a(b)}R'_D)+c_{db1a(b)}+c_{gs3a(b)}+c_{gs4a(b)}+c_{sb4a(b)}+c_{gd3a(b)}(1+g_{m3a(b)}R_{out})]$ and $\omega_{p(out)}=1/(c_{gd2a(b)}+c_{db2a(b)}+c_{gd3a(b)}+c_{db3a(b)}+c_{gd4a(b)}+c_{db4a(b)}+c_{db6a(b)}+c_{gd6a(b)})$.

Since $\omega_{p(out)} \gg \omega_{p(A)}$, $\omega_{p(B)}$ the circuit has one dominant pole, and the bandwidth can be approximated as $\omega_{3dB} = \omega_{p(out)}$.

The noise factor of the proposed amplifier can be analyzed and shown as

$$F = 1 + 2\gamma g_{m1a(b)} \kappa^2 \left[\left(g_{m3a(b)} + g_{m4a(b)} \right) R'_D - g_{m2a(b)} R_S/2 \right]^2$$
$$+ \gamma g_{m2a(b)} \eta^2 + \gamma g_{m3a(b)} \eta^2 + \gamma g_{m4a(b)} \left(g_{m3a(b)} R'_D - 1 \right)^2 \delta^2 \eta^2$$
$$+ R_D \left(g_{m3a(b)} + g_{m4a(b)} \right)^2 \delta^2 \eta^2 \quad (8)$$

where $\kappa = (1-g_{m1a(b)})R_{out}/\sqrt{R_S} A_v$, $\eta = \sqrt{2}(1+g_{m1a(b)}R_S)R_{out}/\sqrt{R_S} A_v$ and $\delta = 1/(1+g_{m4a(b)}R_D)$.

From (8), the second term represents noise factor as a result of channel thermal noises associated with CG ($M_{1a(b)}$), while the third, the fourth, the fifth and the sixth term represents noise associated with $M_{2a(b)}$, $M_{3a(b)}$, $M_{4a(b)}$ and R_D, respectively. One can see that if $(g_{m4a(b)}+g_{m3a(b)})R_D$ is equal to $g_{m2a(b)}R_S/2$, the thermal noises from $M_{1a(b)}$ will be completely eliminated. It is instructive to note that, the noise from noise-canceling circuitry is also added into the system. As a result, its noise contribution to the amplifier has to be addressed. In fact, the noise contribution from noise canceling circuitry is relatively small, compared with the noises from $M_{1a(b)}$. This is because noise from noise-canceling circuitry is divided by gain of the CG, when referred to the input.

IV. RESULTS AND DISCUSSIONS

To verify the circuit performance, HSPICE is used to simulate the proposed circuit, using a 0.18 µm CMOS process with 2V supply voltage. The bias currents are chosen to optimize voltage gain, speed, NF, power dissipation and bandwidth. In addition, R_{in} is also designed to 50 Ω to ensure maximum power transfer and minimize any reflection.

Fig. 4 shows frequency response of the amplifier in Fig. 3. The passband gain of the proposed circuit is found to be 31.6 dB, while the bandwidth of the circuit is 2 GHz. Fig. 5 Show NF of the circuit in Fig. 3, which reads 2.9 dB. Fig. 6 show eye diagrams of the proposed amplifier at 2.8 Gb/s. The results are obtained by applying the pseudo random input to the amplifier, and subsequently superimpose the output data every 0.7 ns. The simulation shows 94.6 % and 89.7% horizontal and vertical eye opening, respectively. Table I summarizes and compares the performances of this design with other recently published CMOS FA. As seen, the proposed circuit exhibits large gain with small NF, while all other parameters are found to be quite comparable.

V. CONCLUSION

This paper uses g_m-boosting and noise-canceling techniques to improve the performance of the CG fully-differential FA. A noise-canceling circuitry can fully suppress noise from the input CG transistor. The differential structure with tail current makes the proposed circuit immune to CM noises. The simulation results show low *NF*, high gain and wide bandwidth. The proposed FA is suitable for a first stage amplifier such as the reading part of hard disk drive systems.

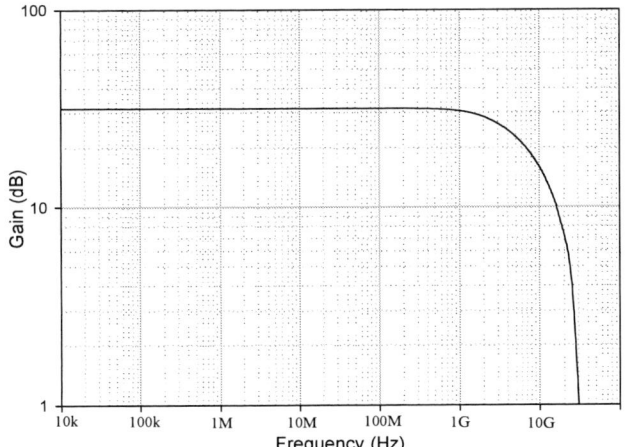

Figure 4. Frequency response.

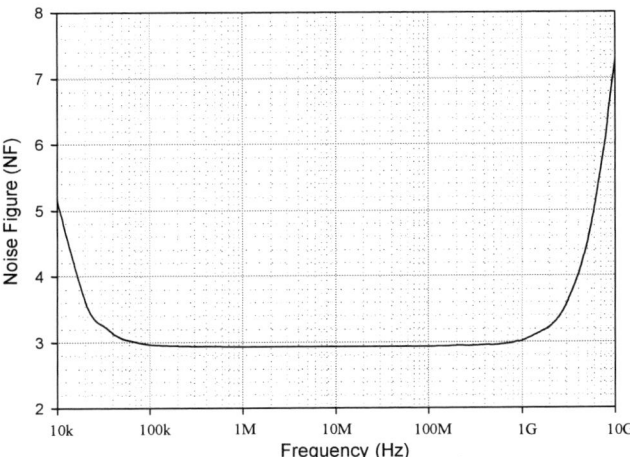

Figure 5. Noise Figure.

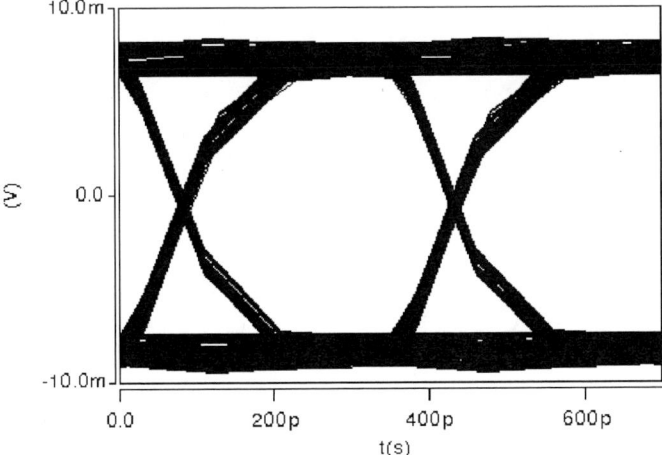

Figure 6. Eye diagram of the proposed amplifier.

TABLE I
PERFORMANCE COMPARISON WITH OTHER NOISE-CANCELING TECHNIQUE

Reference	[3]	[4]	[5]	[11]	This work
BW [GHz]	0.8-2.1	1.2-11.9	0.02-1.17	9	2
Gain	14.5dB	9.7dB	20.5dB	10dB	31.6dB
NF	2.6dB	4.5dB	3.3dB	3 dB	2.9dB
Differential	No	No	Yes	Yes	Yes
Technology	0.13um	0.18um	0.18um	0.18um	0.18um
Supply	1.5V	1.8V	1.8V	2V	2V
Power	17.4mW	20mW	18mW	19mW	19mW

REFERENCES

[1] I. Ranmuthu, et al., "A Design for High Noise Rejection in a Pseudodifferential Preamplifier for Hard Disk Drives," *IEEE J. Solid-State Circuits*, vol. 35, no. 6, pp. 911-914, June 2000.

[2] J. Martinez-Castillo, A.Diaz-Sanchez and J. L. Finol, "Characterization of Two opto-electronic Structures for High-Frequency Applications," *in IEEE Int. Midwest Symposium on Circuits and Systems*, pp.377-380, July 2004.

[3] W. H. Chen, G. Liu, B. Zdravko and A. M. Niknejad, "A Highly Linear Broadband CMOS LNA Employing Noise and Distortion Cancellation," *IEEE J. Solid-State Circuits*, vol. 43, no. 5 pp. 1164-1176, May 2008.

[4] C. F. Liao and S. I. Liu, "A Broadband Noise-Canceling CMOS LNA for 3.1-10.6-GHz UWB Receivers," *IEEE J. Solid-State Circuits*, vol. 42, no. 2 pp. 329-339, Feb. 2007.

[5] S.S. Song, D.G. Im, H.T. Kim, and K. Lee, "A Highly Linear Wideband CMOS Low-Noise Amplifier Based on Current Amplification for Digital TV Tuner Application," *IEEE. Micro. Wireless Compon. Lett.*, vol. 18, no. 2, pp. 118-120, Feb. 2008.

[6] D. Im, I. Nam, H.T. Kim, and K. Lee, " A Wideband CMOS Low Noise Amplifier Employing Noise and IM2 Distortion Cancellation for a Digital TV Tuner," *IEEE J. Solid-State Circuits*, vol. 44, no. 3, pp. 686-698, Mar. 2009.

[7] J. Jussila, and P. Sivonen, "A 1.2-V Highly Linear Balanced Noise-Celling LNA in 0.13-um CMOS," *IEEE J. Solid-State Circuits*, vol. 43, no. 3, pp. 579-587, Mar. 2008.

[8] J. Ollikainen, M. Kaltiokallio, K. Stadius, V. Saari, and J. Ryynanen, "A Wideband Interference Tolerant RF Receiver for Cognitive Radio Sensor Unit," *NORCHIP, The Nornic Microelectronic Conf*, vol. 43, no. 3, pp. 579-587, Mar. 2008.

[9] G. H. Zare Fatin, Z. D. Koozehkanani and H. Sjoland, "A Technique for Improving Gain and Noise Figure of Common-Gate Wideband LNAs" *Analog Interg. Circ Signal Process*, vol. 63, no.2, May 2010.

[10] H. Wang, L. Zhang, and Z. Yu, "A Wideband Inductorless LNA With Local Feedback and Noise Canceling for Low-Power Low-Voltage Applications," *IEEE Trans.Circuits Syst I, Reg. Paper*, vol. 57, no. 8 pp. 1993-2005, Aug. 2010.

[11] F. Bruccoleri, E. A.M. Klumperink, and B. Nauta, "Wideband Low Noise Amplifiers Exploiting Thermal Noise Cancellation," Dordrecht, The Netherlands: Springer, 2005.

[12] J. Liu, G. Chen, and R. Zhang, "Design of a Noise-canceling Differential CMOS LNA for 3.1-10.6 GHz UWB Receivers," *Proceeding of the 8th Annual Summer Interdisciplinary Conference*, pp. 1169-1172, Oct. 2009.

[13] D. Im, I. Nam and K. Lee, "A Low Power Broadband Differential Low Noise Amplifier Employing Noise and IM3 Distortion Cancellation for Mobile Broadcast Receivers," *IEEE Microw. Wireless Compon. Lett.*, vol. 20, no. 10, pp. 566-568, Oct. 2010.

[14] W. Zhuo and et.al. , "A Capacitor Cross-Coupled Common-Gate Low-Noise Amplifier," *IEEE Trans. Circuits Syst. II*, Vol. 52, no. 12, pp. 875-879, Dec. 2005.

Compact Distributed RLCG Interconnect Models—Signal Transient Response for Dispersionless Interconnect

Jing Xia[1], Chi Liu[1], Xinnan Lin[1,2]

[1]The Key Laboratory of Integrated Microsystems, Peking University Shenzhen Graduate School, Shenzhen, P. R. China
E-mail: xnlin@pkusz.edu.cn

Wei Zhao[2], Yu Han[2], Aixi Zhang[2], Jin He[1,2]

[2]Peking University Shenzhen SOC Key Laboratory, PKU-HKUST Shenzhen Institute, Hi-Tech Industrial Park South Shenzhen, P.R.China
E-mail: jinhe@socpku.net.cn

Abstract—New compact models that describe the transient response of dispersionless interconnect are rigorously derived with resistive and capacitive load terminations according to Heaviside's theory. The proposed compact model is proved to be highly accurate by HSPICE, and the physical insight of the developed compact models is also discussed.

I. INTRODUCTION

As the VLSI technology scaled into the deep submicron regime, interconnect parasitics such as capacitance, resistance and inductance have been playing a progressively more important role in integrated circuits [1-3]. Deep submicron physics indicates that it is interconnects rather than transistors that dominate power, performance and robustness [4].

In order to solve signal integrity problem, it is necessary to analyze the interconnect issues and inspect its potential impact to the actual circuits [5]. Therefore, a fast but accurate model for interconnect is desperately needed. In today's high-speed IC, inductive effect of interconnect must be considered by designers, and interconnect must be considered as a transmission line [6]. Sakurai solved a partial differential equation (PDE) and derived compact expressions for the transient response of a distributed resistance capacitance (RC) interconnect in [7]. Davis et al. developed the transient response and time delay models for the compact distributed RLC interconnect [8-11]. However, few works have been done on the situation of considering conductance, resistance, inductance and capacitance of interconnect transmission line.

In this paper, compact models for transient response of dispersionless interconnect with all kinds of load terminations are developed in the following parts.

II. COMPACT MODELS FOR TRANSIENT RESPONSE OF DISPERSIONLESS INTERCONNECTS AND VERIFICATION

With an internal resistor of R_s, and load Z_L finite distributed RLCG line is shown in Fig1. L is the interconnect length, and Rs is the resistance of the voltage source. $V_{in}(t)$ and $V(L,t)$ represent the input and output voltage respectively. The partial differential equation (PDE) that describes single distributed RLCG line is given by

$$\frac{\partial^2 V(x,t)}{\partial x^2} = lc\frac{\partial^2 V(x,t)}{\partial t^2} + (rc+gl)\frac{\partial V(x,t)}{\partial t} + rgV(x,t) \quad (1)$$

where r, g, l, c are conductance, resistance, inductance and capacitance per unit length between the line and ground respectively.

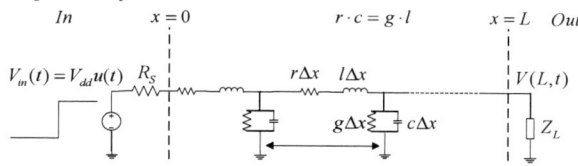

Fig1. The models for the finite distributed RLCG interconnect line driven by a resistive voltage and loaded with Z_L

According to Heaviside's theory, the signal propagation is dispersion-free on the condition of $r \cdot c = l \cdot g$ [12]. That means only the amplitude changes, but the signal keeps the same profile and the transmission exhibits distortion-free through the line. Hence the voltage expression of infinite dispersionless line is

$$V_\infty(L,t) = \lambda V_{dd} \cdot e^{-\alpha L} \cdot u(t - \frac{L}{\upsilon}) \quad (2)$$

where

Decay factor $\alpha = \sqrt{rg}$ (3)

Voltage wave transmission velocity $\upsilon = \dfrac{1}{\sqrt{lc}}$ (4)

Transmission time of the voltage wave $t_0 = \dfrac{L}{\upsilon}$ (5)

Characteristic impedance

$$Z_0 = \sqrt{\frac{r+sl}{g+sc}} = \sqrt{\frac{r}{g}} = \sqrt{\frac{l}{c}} = R_0 \quad (6)$$

Divide coefficient

$$\lambda = \frac{R_0}{R_S + R_0} \tag{7}$$

On the basic of telegraph equation [12] and the reflection diagram, new complete physical models for transient response of dispersionless interconnects are rigorously derived with resistive, inductive and capacitive load terminations. The voltage expression with a general load is given by

$$V(L,s) = V_\infty(L,s) \cdot (1+\Gamma_L) + V_\infty(3L,s) \cdot \Gamma_L \Gamma_S \cdot (1+\Gamma_L)$$
$$+ V_\infty(5L,s) \cdot \Gamma_L^2 \Gamma_S^2 \cdot (1+\Gamma_L) + \cdots$$
$$= \sum_{n=1}^{N} \left[V_\infty\big((2n-1)L,s\big) \cdot (\Gamma_L \Gamma_S)^{n-1} \cdot (1+\Gamma_L) \right] \tag{8}$$

The transient response for several load terminations is stated as below.

$$V_{R_L}(L,t) = \sum_{n=1}^{N} \left[\lambda V_{dd} \cdot e^{-\alpha(2n-1)L} \cdot (\Gamma_L \Gamma_S)^{n-1} \cdot (1+\Gamma_L) \cdot u\big(t-(2n-1)t_0\big) \right] \tag{9}$$

$$V_{R_L \| C_L}(L,t) = \sum_{n=1}^{N} \left\{ \begin{array}{l} \lambda V_{dd} \cdot e^{-\alpha(2n-1)L} \cdot \Gamma_S^{n-1} \cdot \Gamma_{LR}^{n-1} \cdot (1+\Gamma_{LR}) \cdot u\big(t-(2n-1)t_0\big) \\ \left[1 - \dfrac{\displaystyle\sum_{m=0}^{n-1}\left(\dfrac{1}{m!} \cdot \sum_{k=m}^{n-1}\binom{n-1}{k}(1+\Gamma_{LR}^{-1})^k\left(-\Gamma_{LR}^{-1}\right)^{n-1-k} \right)\cdot\left(\dfrac{t-(2n-1)t_0}{\tau_R}\right)^m}{e^{\frac{t-(2n-1)t_0}{\tau_R}}} \right] \end{array} \right\} \tag{10}$$

$$V_{R_L+C_L}(L,t) = \sum_{n=1}^{N} \left\{ \begin{array}{l} \lambda V_{dd} \cdot e^{-\alpha(2n-1)L} \cdot \Gamma_S^{n-1} \cdot \Gamma_{LR}^{n-1} \cdot (1+\Gamma_{LR}) \cdot u\big(t-(2n-1)t_0\big) \\ \cdot \dfrac{\displaystyle\sum_{m=0}^{n-1}\left(\dfrac{1}{m!}\binom{n-1}{m}\left(\Gamma_{LR}^{-1}-1\right)^m \right)\cdot\left(\dfrac{t-(2n-1)t_0}{\tau}\right)^m}{e^{\frac{t-(2n-1)t_0}{\tau}}} \end{array} \right\}$$
$$+ \sum_{n=1}^{N} \left[\begin{array}{l} \lambda V_{dd} \cdot e^{-\alpha(2n-1)L} \cdot \Gamma_S^{n-1} \cdot 2 \cdot u\big(t-(2n-1)t_0\big) \\ \left[1 - \dfrac{\displaystyle\sum_{m=0}^{n-1}\left(\dfrac{1}{m!}\cdot\sum_{k=m}^{n-1}\binom{n-1}{k}(1-\Gamma_{LR})^k\,\Gamma_{LR}^{n-1-k} \right)\cdot\left(\dfrac{t-(2n-1)t_0}{\tau}\right)^m}{e^{\frac{t-(2n-1)t_0}{\tau}}} \right] \end{array} \right] \tag{11}$$

where other definitions are:
Number of the reflections at the output (the' [ ]'in equation is integer operator. For example, [2.3] =2)

$$N = \left[\frac{1}{2}\left(\frac{t}{t_0}+1 \right) \right] \tag{12}$$

Source reflect factor

$$\Gamma_S = \frac{R_S - R_0}{R_S + R_0} \tag{13}$$

Output reflect factor

$$\Gamma_L = \frac{Z_L - R_0}{Z_L + R_0} \tag{14}$$

Output resistance reflect factor

$$\Gamma_{LR} = \frac{R_L - R_0}{R_L + R_0} \tag{15}$$

Time constant

for (10), $\tau = \dfrac{R_L R_0}{R_L + R_0} C_L$ (16); for (11), $\tau = (R_L + R_0) C_L$ (17)

According to our previous work [13-14], the model in (10~11) is compared to HSPICE simulation of an interconnect using 100 lumped r, l, c, g elements in Fig 2(a)-(b). This interconnect has a length of 1cm, and a driver resistance of 23.54Ω. The output transient response at

different output load is shown in Fig.2. The error between HSPICE simulations and the proposed model is negligible.

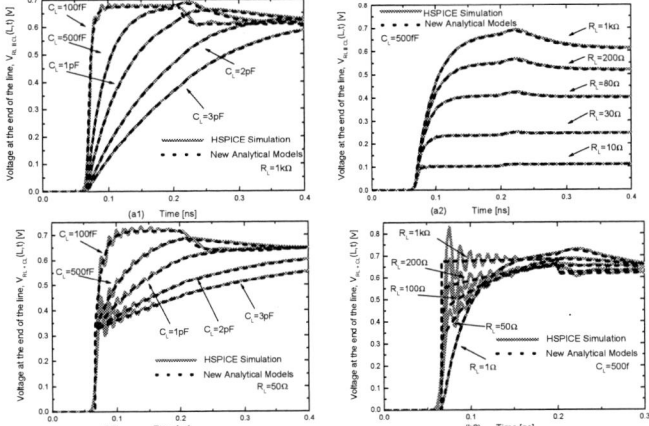

Fig2. Output transient response of a dispersionless interconnect to a step input waveform—analytical models versus HSPICE simulations (100 lumped r, l, c, g elements) with different output load

$R_s = 23.54\Omega, V_{dd} = 1V, L = 1cm, r = 42.5\Omega/cm, l = 4.05nH/cm,$
$c = 1.1pF/cm, g = 0.0116(\Omega \cdot cm)^{-1}, (a1)R_L = 1k\Omega, C_L = 100f, 500f, 1p, 2p, 3pF;$
$(a2)C_L = 500fF, R_L = 10,30,80,200,1k\Omega; (b1)R_L = 50\Omega, C_L = 100f, 500f, 1p, 2p, 3pF;$
$(b2)C_L = 500fF, R_L = 1,50,100,200,1k\Omega$

III. PHYSICAL INSIGHT

The physical meaning of the proposed models is quite clear. As the characteristic resistance of a dispersionless line is a constant, a voltage of $\lambda \cdot V_{dd}$ is passed on through the line, at a constant speed of υ. After a time of L/υ, the voltage has travelled a distance of L, and has decayed exponentially by a factor of e^{-aL}.

A. Increments

If there is a load at the end of the line, on the basis of reflection diagram, when time reaches $t = (2n-1)t_0$, the voltage wave arrives at the output termination, and adds a value to the total output voltage. It is then reflected back to the voltage source, and then again, reflected to the load, and so on.

For resistive load, each time the increment will be added to the total output voltage as followed:

$$\Delta V_{R_L}(L,t) = \lambda V_{dd} \cdot e^{-\alpha(2n-1)L} \cdot (\Gamma_L \Gamma_S)^{n-1} \cdot (1+\Gamma_L) \tag{18}$$

Since resistances absorb energy, each reflection cripples the voltage by a factor. Γ_L is the factor for the output end, and Γ_S is the factor for the input end in equation (18). In the item $1+\Gamma_L$, "1" represents the arriving voltage, while "Γ_L" is the reflected factor. If a dispersionless line opens at the end, no reduction happens when a voltage is reflected at the end, hence we have $\Gamma_L = 1$. That means the voltage will be doubled at the end. Equation (19) is used this special situation.

$$\Delta V_0(L,t) = \lambda V_{dd} \cdot e^{-\alpha(2n-1)L} \cdot \Gamma_S^{n-1} \cdot 2 \tag{19}$$

With a capacitive load, the output voltage increment is expressed as (20).

$$\Delta V_{C_L}(L,t)=\Delta V_0(L,t)\cdot\left\{1-\dfrac{\displaystyle\sum_{m=0}^{n-1}\left[\left(\dfrac{1}{m!}\cdot\sum_{k=m}^{n-1}\binom{n-1}{k}2^k(-1)^{n-1-k}\right)\cdot\left(\dfrac{t-(2n-1)t_0}{\tau}\right)^m\right]}{e^{\frac{t-(2n-1)t_0}{\tau}}}\right\} \qquad (20)$$

Since the capacitance doesn't absorb energy or cause any discontinuous change, the increment of capacitive load is as (20), which is the results of (19) multiplied by a factor which changing from 0 to 1, thus the increment changes from 0 to $\Delta V_0(L,t)$, and change rate is determined by $\tau=R_0C_L$.

If the load is an inductance, the increment is expressed as (21).

$$\Delta V_{L_L}(L,t)=\Delta V_0(L,t)\cdot\dfrac{\displaystyle\sum_{k=0}^{n-1}\left[\left(\dfrac{1}{k!}\cdot\binom{n-1}{k}(-2)^k\right)\cdot\left(\dfrac{t-(2n-1)t_0}{\tau}\right)^k\right]}{e^{\frac{t-(2n-1)t_0}{\tau}}} \qquad (21)$$

The difference between (20) and (21) is only the factor. The factor in (21) changes from 1 to 0, reversed to the situation of capacitive load. The increment decreases from $\Delta V_0(L,t)$ to 0, and reduction rate is determined by $\tau=L_L/R_0$.

B. Steady state

When $t\rightarrow\infty$, the voltage at the output termination goes to a steady state. Deduced from the expressions above, the steady value is expressed as

$$V_{R_L}(L,t\rightarrow\infty)=\lambda V_{dd}\cdot e^{-\alpha L}\cdot(1+\Gamma_L)\cdot\dfrac{1}{1-e^{-2\alpha L}\Gamma_L\Gamma_S} \qquad (22)$$

$$V_{C_L}(L,t\rightarrow\infty)=\lambda V_{dd}\cdot e^{-\alpha L}\cdot 2\cdot\dfrac{1}{1-e^{-2\alpha L}\Gamma_S} \qquad (23)$$

$$V_{L_L}(L,t\rightarrow\infty)=0 \qquad (24)$$

It is found that the value of the steady-state voltage is only determined by the resistance at the input and output terminations. When $t\rightarrow\infty$, each voltage wave arriving at the output termination will go to its steady value, but capacitance and inductance do not absorb any energy, so that the value will only be determined by the resistance at the input and output terminations, and so does the sum of all these voltage waves. Hence, equation (25) is a unified steady-state voltage expression for any kind of load terminations

$$V(L,t\rightarrow\infty)=\lambda V_{dd}\cdot e^{-\alpha L}\cdot(1+\Gamma_{LR})\cdot\dfrac{1}{1-e^{-2\alpha L}\Gamma_{LR}\Gamma_S} \qquad (25)$$

Output resistance reflect factor $\Gamma_{LR}=\dfrac{R_L-R_0}{R_L+R_0}$ (26)

R_S=0 or ∞ for a capacitance or inductance load respectively.

C. Delay

During $t_0<t<3t_0$, the time delay of dispersionless line with capacitive or inductive loads is found from equations (20), (21), and the factors affecting time delay of dispersionless line are clearly shown as below.

$$t_{C_L}=t_0+\tau\ln\ln\dfrac{1}{1-\dfrac{V_{C_L}(L,t)}{V_{dd}}\cdot\dfrac{1}{2\lambda e^{-\alpha L}}} \qquad (27)$$

$$t_{L_L}=t_0+\tau\ln\ln\dfrac{1}{\dfrac{V_{L_L}(L,t)}{V_{dd}}\cdot\dfrac{1}{2\lambda e^{-\alpha L}}} \qquad (28)$$

D. Crosstalk

Taking the situation of capacitive load as an example, the crosstalk between two coupled dispersionless lines is discussed in Fig 3. The two same interconnects interact through mutual capacitance c_m and inductance l_m (per unit length) are shown as below.

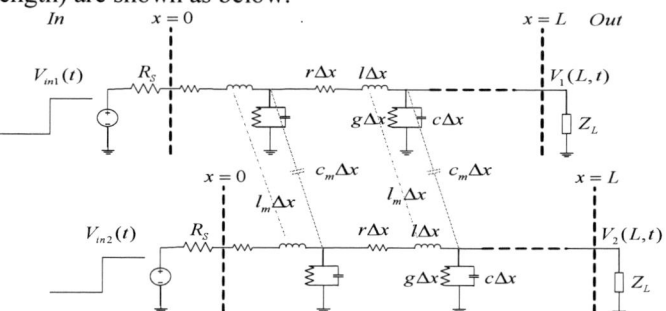

Fig 3. Diagram of two interconnects on a chip with mutual capacitance and inductance.

Considering the voltage and current on Line 1 gives

$$-\dfrac{\partial V_1(x,t)}{\partial x}=I_1(x,t)r+l\dfrac{\partial I_1(x,t)}{\partial t}+l_m\dfrac{\partial I_2(x,t)}{\partial t} \qquad (29)$$

$$-\dfrac{\partial I_1(x,t)}{\partial x}=V_1(x,t)g+c\dfrac{\partial V_1(x,t)}{\partial t}+c_m\dfrac{\partial[V_1(x,t)-V_2(x,t)]}{\partial t} \qquad (30)$$

The boundary conditions for Line 1 are
$$V_1(0,t)=V_{in1}(t)-I_1(0,t)R_s \qquad (31)$$

$$V_1(L,t)=I_1(L,t)R_L \qquad (32)$$

Similarly, for Line 2, the equations and boundary conditions are

$$-\dfrac{\partial V_2(x,t)}{\partial x}=I_2(x,t)r+l\dfrac{\partial I_2(x,t)}{\partial t}+l_m\dfrac{\partial I_1(x,t)}{\partial t} \qquad (33)$$

$$-\dfrac{\partial I_2(x,t)}{\partial x}=V_2(x,t)g+c\dfrac{\partial V_2(x,t)}{\partial t}+c_m\dfrac{\partial[V_2(x,t)-V_1(x,t)]}{\partial t} \qquad (34)$$

$$V_2(0,t)=V_{in2}(t)-I_2(0,t)R_s \qquad (35)$$

$$V_2(L,t)=I_2(L,t)R_L \qquad (36)$$

(29) + (33), (30) + (34), (31) + (35), (32) + (36) give

$$-\dfrac{\partial V_+(x,t)}{\partial x}=rI_+(x,t)+l_+\dfrac{\partial I_+(x,t)}{\partial t} \qquad (37)$$

$$-\dfrac{\partial I_+(x,t)}{\partial x}=gV_+(x,t)+c\dfrac{\partial V_+(x,t)}{\partial t} \qquad (38)$$

$$V_+(0,t)=V_{in+}(t)-I_+(0,t)R_s \qquad (39)$$

$$V_+(L,t)=I_+(L,t)R_L \qquad (40)$$

where $V_+ = V_1 + V_2$ $I_+ = I_1 + I_2$, $V_{in+} = V_{in1} + V_{in2}$ and $l_+ = l + l_m$.

From observation it is found that the equations and boundary conditions for the sum of the voltages on the two interconnects are the same as the voltage of a single line with parameters of r, c, l_+, g , and with input and output voltages to be V_{in+} and $V_+(L,t)$ respectively. Thus if $rc = l_+ g$, the two interconnects can be used as one distortionless interconnect. For special, if $V_{in1}(t) = V_{in2}(t)$, the two voltages on the two interconnects are all the same, thus $V_1(L,t) = V_2(L,t)$, and $V_1(L,t)$ or $V_2(L,t)$ is the distortionless output voltage for the input voltage $V_{in1}(t)$ or $V_{in2}(t)$, i.e. each line behaves as a distortionless line.

Similarly, (29) - (33), (30) - (34), (31) - (35), (32) - (36) give

$$-\frac{\partial V_-(x,t)}{\partial x} = rI_-(x,t) + l_-\frac{\partial I_-(x,t)}{\partial t} \tag{41}$$

$$-\frac{\partial I_-(x,t)}{\partial x} = gV_-(x,t) + c_-\frac{\partial V_-(x,t)}{\partial t} \tag{42}$$

$$V_-(0,t) = V_{in-}(t) - I_-(0,t)R_s \tag{43}$$

$$V_-(L,s) = I_-(L,s)R_L \tag{44}$$

where $V_- = V_1 - V_2$, $I_- = I_1 - I_2$, $V_{in-} = V_{in1} - V_{in2}$, and $l_- = l - l_m$, $c_- = c + 2c_m$.

The equations and boundary conditions for the difference of the voltages on the two interconnects are the same as the voltage of a single line with parameters of r, c_-, l_-, g and with input and output voltages to be $V_{in-}(t)$ and $V_-(L,t)$ respectively. Thus if $rc_- = l_- g$, the two interconnects also can be used as one distortionless interconnect. For special, if $V_{in1}(t)$ and $V_{in2}(t)$ are opposite, each line also behaves as a distortionless line.

IV.   CONCLUSION

In this paper, compact models for transients response of dispersionless interconnect are rigorously derived with resistive and capacitive load terminations. These models with clear physical insights enable accurate estimation of transient response of dispersionless interconnects, and the simulation speed of HSPICE will be also greatly improved by inserting the proposed models.

ACKNOLEDGMENT

This work is supported by the national natural sciense funds of china(61076036), the Guangdong science and technology project (2009A011604001) (2009B09060008), the Shenzhen science & technology foundation (JSA200903160146A) and the dean's foundation of the Shenzhen graduate school of Peking University (2009010).

PEFERENCES

[1] international technology roadmap for semiconductors (ITRS) 2007

[2] J. D. Meindl, J. A. Davis, P. Zarkesh-Ha, C. S. Patel, K. P. Martin, and P. A. Kohl, "Interconnect opportunities for gigascale integration", IBM J. Res. Develop., vol. 46, no. 2/3, Mar.–May 2002, pp.245-263

[3] H. B. Bakoglu. Circuits, Interconnections, and Packaging for VLSI. Addison–Wesley, 1990

[4] Jari Nurmi, Interconnect-Centric Design For Advanced SOC And NOC, Boston: Kluwer Academic Publishers , 2004

[5] S. H. Hall, G. W. Hall, and J. A. McCall. High–Speed Digital System Design: A Handbook of Interconnect Theory and Design Practices. John Wiley and Sons, 2000

[6] Howard W. Johnson and Martin Graham, High-speed digital design---a handbook of black magic, Prentice Hall, 2003

[7] Takayasu Sakurai, "Closed-Form Expressions For Interconnect Delay, Coupling ,and Crosstalk in VLSI's", IEEE Trans. On Electron. Devices, Vol. 40, No. 1, January 1993 , pp.118-124

[8] Jeffrey A. Davis and James D. Meindl, "Compact Distributed RLC Interconnect Models—Part I: Single Line Transient, Time Delay, and Overshoot Expressions", IEEE Trans. On Electron. Devices, Vol. 47, No. 11, Nov, 2000, pp.2068-2077

[9] Jeffrey A. Davis and James D. Meindl, Compact Distributed RLC Interconnect Models—Part II: Coupled Line Transient Expressions and Peak Crosstalk in Multilevel Networks, IEEE Trans. On Electron. Devices, Vol. 47, No. 11, Nov,2000, pp.2078-2087

[10] Raguraman Venkatesan, Jeffrey A. Davis, and James D. Meindl, Compact Distributed RLC Interconnect Models—Part III: Transients in Single and Coupled Lines with Capacitive Load Termination, IEEE Trans. On Electron. Devices, Vol. 50, No. 4, April 2003, pp.1081-1093

[11] Azad Naeemi, Jeffrey Alan Davis, and James D. Meindl, Compact Physical Models for Multilevel Interconnect Crosstalk in Gigascale Integration (GSI), IEEE Transactions on Electron Devices, vol. 51, no.11, november 2004, pp.1902-1912

[12] Thomas H. Lee, The Design of CMOS Radio-Frequency Integrated Circuits 2ed, Cambridge University Press, 2004

[13] Chi Liu, Zhize Zhou, Xinnan Lin, Xing Zhang, Jin He, "Compact Modeling of Signal Transients for Dispersionless Interconnects with Capacitive and Inductive Terminal Load Terminals", IWCM, 2010

[14] Chi Liu, Zhize Zhou, Xinnan Lin, Jing Xia, Xing Zhang, Jin He, "Compact Modeling of Signal Transients for Dispersionless Interconnects with Resistive, Capacitive and Inductive Terminal Loads", NSTI-Nanotech 2010, Vol.2, June ,2010, pp.809-812

Multiple Continuous Error Correct Code for High Performance Network-on-chip

Bin Wang, Jing Xie, Zhigang Mao, Qin Wang
School of Microelectronics, Shanghai Jiao Tong University
Shanghai, 200240, P.R.China
{wangbin2673, skingshere, maozhigang, qinqinwang}@sjtu.edu.cn

Abstract—**Network-on-chip (NoC) is considered as a reasonable solution for integrating plenty of IP blocks on a single chip in Nano-scale technology. But its proper function is suffered from crosstalk errors, which is becoming more and more serious with the rapid shrink of the technology. Meanwhile chip in hard environment where full filled with energetic particles such as nuclear power plant, eagerly calls for strong reliability. In this paper, we proposed a coding method for NoC channel, multiple-continuous-error-correct-coding (MCECC) which can strongly correct continuous error bits with 14-bit maximum fault tolerance. Besides, an adaptive error control scheme is proposed to work with this code to enhance the robust communication. Experiments show that the proposed design can reduce nearly 50% wire capacitance energy dissipation and stronger error correcting ability, especially for continuous errors.**

I. INTRODUCTION

Network-on-chip (NoC) has emerged as a methodology to achieve high integration with thousands of IPs [1].But with shrinking geometry, the ratio between the interconnection delay and the gate delay will increase to 9:1 according to ITRS prediction. And as the network becomes larger and more complex, the net area and wire power will become an important factor in designing NoC. It indicates on-chip interconnection architectures will dominate the performance of NoC.

With the shrinking geometry, NoC architecture is increasingly suffering with different sources of transient noise, which affects signal integrity and system reliability. Data-dependent crosstalk between adjacent wires is a major source of such transient noise. Two neighbor wires transit in opposite directions will cause the worst crosstalk case with respect to the victim wire. The decreasing of inter-wire spacing, reduction of operation power will deteriorate this negative effect [2]. Meanwhile cosmic ray is another great enemy towards large-scale integrated chips. It will cause single event upset (SEU) or single event latchup (SEL) [3] which makes the chip more vulnerable.

Many techniques were proposed to reduce the coupling capacitance effect and to avoid SEU/SEL. But most of solutions are impractical with largely increasing interconnect area, routing complex or high cost. Among them, bus encoding is a reasonable, economical and effective technique with less cost. The performance of different Crosstalk Avoidance Codes (CAC), Error Correct Codes (ECC) and joint CAC/ECC schemes in NoC architecture is discussed in [4-7]. In [6] the authors presented scheme called Crosstalk Avoiding Double Error Correction Code (CADEC), combining Hamming Code and Duplicate Add Parity (DAP). Hamming coding stage enables detect and correct error bits and DAP stage eliminates crosstalk effect. Authors of [7] proposed another coding scheme called self-corrected green code (sc_green), combining Low Power Coding (LPC), serialization and triplication. LPC stage encodes low power data. Serialization divides a flit into several pieces to reduce wire power consumption. And triplication eliminates crosstalk and provides error-correct ability. Both of the codes have strong ability in processing distributed errors, but cannot process continuous errors.

In this paper we proposed a configurable and powerful error control scheme to ensure the robust of NoC. Assuming the original data is 32 bits, our multiple-continuous-error-correction-code (MCECC) can fix up to 14-bits continuous errors. And combining with configurable transmitting scheme, NoC carries out a better reliability.

II. MULTIPLE CONTINUOUS ERROR CORRECT CODE

This section we will introduce our coding scheme (MCECC) which can process continuous errors meanwhile avoid crosstalk effectives. Main module contains (7, 4) Hamming, (5, 4) Forbidden Pattern Codes (FPC), interleaver, controller and evaluation.

Fig. 1 shows the whole encoding flow. Firstly the 32-bits data will be separated into 8 groups of 4 bits. Hamming stage encodes each 4-bit group into 7-bit hamming group hamming-1, …., hamming-8, and bits of hamming-i are i-1, …, i-7. Interleaver stage results in 7 groups of 8-bit inter. Inter-i contains 1-i, 2-i… 7-i, 8-i, here i-j means the bit j of hamming-i. Final stage is (5, 4) FPC. The 56-bit interleaved data is separated into 14 groups of 4-bit, the encoding method is shown in Eq. 1. "abcd" are the original bits, x0~x4 are encoded bits. And we duplicate the least significant bit (LSB) and the most significant bit (MSB) of each FPC group. LSB of FPC-1 and MSB of FPC-14 need not to be duplicated since has no adjacent group. And finally the MCECC codes is 32/4*7/4*7-2=96 bits long which is much shorter than sc_green. The decoding flow is a simple inverse of encoding.

$$
\begin{cases}
x0 = a \\
x1 = a == d\ ?\ b\ :\ (ad == 2'b01\ ?\ bc\ :\ b + c) \\
x2 = a! = d\ ?\ b\ :\ (ad == 2'b00\ ?\ b + c\ :\ bc) \\
x3 = a == d\ ?\ c\ :\ (ad == 2'b01\ ?\ b + c\ :\ bc) \\
x4 = d
\end{cases}
\tag{1}
$$

Now we will deeply analysis the error correct ability of MCECC. Since the only contribution of FPC is reducing bit error probability ε by eliminating crosstalk effective, we focus on the interleaved data. We assume ε is as Eq. 2. The noise model is Gaussian distributed noise which is widely accepted. Its voltage is V_N and variance is σ_N^2.

$$
\varepsilon = \frac{1}{\sqrt{2\pi}} \int_x^\infty e^{-\frac{y^2}{2}} dy,\ x = \frac{V_N}{2\sigma_N}
\tag{2}
$$

According to the character of interleaver, the error bits, no matter continuous or distributed, will be relocated into different groups of hamming after de-interleaved. And the character of hamming code can achieve error-free flit if no more than 1 error exists in each hamming group. The error in parity bit will not harm the data. So each hamming group's error-free probability Pi-correct is given by Eq. 3

$$
P_{i-correct} = (1 - \varepsilon)^4 + C_4^1 \varepsilon (1 - \varepsilon)^3
\tag{3}
$$

Because hamming groups are linear independent, we can get the word error-free probability Ptotal-correct.

$$
P_{total-correct} = \prod_1^8 P_{i-correct} = ((1 - \varepsilon)^4 + 4\varepsilon(1 - \varepsilon)^3)^8
\tag{4}
$$

For small probability of ε, (4) simplifies to

$$
P_{total-correct} \approx 1 - 48\varepsilon^2 + 64\varepsilon^3
\tag{5}
$$

The word error probability P_{error} is given by (6).

$$
P_{error} = 1 - P_{total-correct} = 48\varepsilon^2 - 64\varepsilon^3
\tag{6}
$$

By contrast, the word error probability of sc_green is $96\varepsilon^2 - 64\varepsilon^3$ [7] which is nearly double than our proposal.

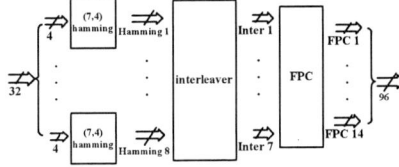

Figure 1.   Encoding data flow for 32-bit flit

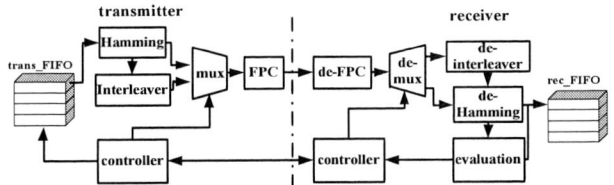

Figure 2.   Adaptive error control scheme

Besides our proposal can handle the continuous error bits which most codes cannot. When FPC added, the error probability can be further reduced. And the maximum number of continuous error bits we can process is 14 which other codes cannot.

III. ADAPTIVE ERROR CONTROL SCHEME

This section focuses the data link layer transfer regardless the routing issues such as deadlock and starvation. We proposed an error control scheme attached to the router for switch-to-switch transfer.

A. Function overview

Basically, this scheme is a kind of retransmitting mechanism. As shown in Fig. 2, data is first stored in the trans_FIFO. Then hamming codes and interleaved codes are both generated. Controller will decide which code to be selected according to the receiver's feedback. After FPC stage, the encoded data and control signals will be sent into channel. The receiver will carry out the decoding process according to the received control signals. The receiver evaluates the current transmitting and sends information back to transmitter as a feedback. If the evaluation is no good, the transmitter will adjust the encoding method or retransmit the former flit.

B. Transceiver design

The main controller module structure of transmitter is shown in Fig. 3. The transmitting type differs depending on the receiver's feedback. The feedback includes SUCCESS (S), SUC_BUT_CHANGE (SC), FAIL (F), WAIT_ONE_CYCLE (W). The corresponding control signals for trans_FIFO are SEND_NEW, PAUSE and RESEND. Signals for mux are REMAIN, CHANGE and INTERLEAVER. The controller module gives out the information of coding method ENCODING_TYPE to the receiver. Fig.4 shows the block diagram of receiver. According to the ENCODING_TYPE, controller sends appropriate signals to the mux. The evaluator will gives out the feedback to transmitter, and STORE or NO_STORE signal to rec_FIFO.

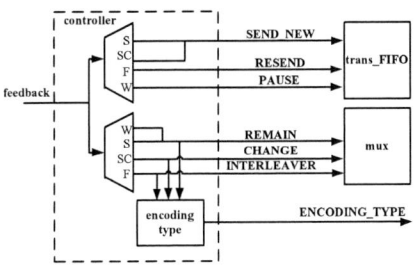

Figure 3.   Block diagram of the transmitter

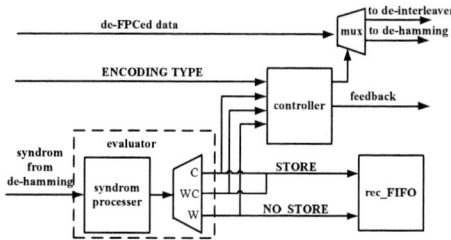

Figure 4. Block diagram of the receiver

IV. EXPERIMENTAL RESULT AND ANALYSIS

In this section, we present experimental results to demonstrate the improvement and the tradeoffs arising there by employing MCECC and adaptive error control scheme.

A. Error propability calculation

We took an experiment with Synopsys HSpice to get the error probability caused by crosstalk affective. The wire model is shown in Fig. 5. Our simulation parameter is as following. Source supply: v1=0v, v2=0.9v, TD=0, TR=TF=50p, PW=0.4n, PER=1n; Ci=0.3p; Cl=0.13p; R=26ohm; L=0.9n; K13=0.4, K12=K23=0.81. We set all three power supply's TD into 0ns, to simulate 111 to 000, and set the middle's TD into 0.5ns to simulate 101 to 010. The output voltage is shown in Fig. 6, Fig. 7.

Assume that 0.5V is the sample gate voltage, and we take 0.1ns as the maximum correct sample time. Case "111 to 000" can make a right transfer on these constrains. Case "101 to 010" has to take 0.15ns for a right sample, which costs 50% more delay time. This large delay can easily cause crosstalk error. So we take 0.5 as the error possibility for C_L+4C_i, C_L is load capacitance and C_i is coupling capacitance. By this method, we can get 0.2 for C_L+3C_i.

B. Simulation for the proposed error control scheme evaluation

We established a C++ model to estimate the power consumption by calculating the turnovers of every bit. When the bit line inverts, we consider an operation of charging/discharging C_L happened. When adjacent 2 bits turnover in different directions, we consider C_i is charged/discharged. And if the total effective capacitance is larger than C_L+4C_i, we consider an error may happen.

We randomly generated 256 *32-bit data, and encode them by different methods. The statistic results shows in Table I. Further processing the data in Table I, we focus on the power consumption per line. Compared with uncoded, power consumption on coupling capacitance reduced a half by employing MCECC.

We combine continuous error model and the crosstalk error model together. For simulation, we set the continuous error rate at a high value of 10e-5, and the number of error bits is from 1 to 14 randomly. We find out MCECC has the highest rate of error-corrected/total-error, shown in Fig. 8. When the number of error bits is 14, MCECC can correct 19% errors while the other codes can hardly recover.

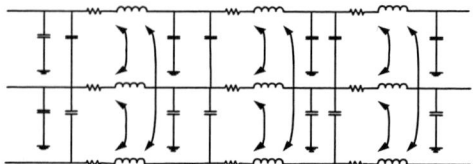

Figure 5. RCL model for the simulation

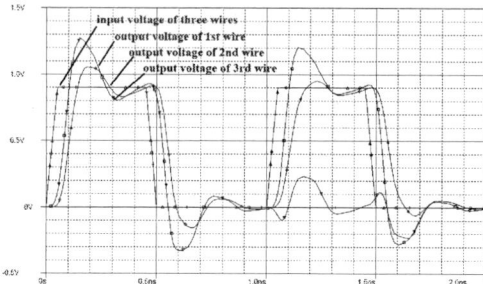

Figure 6. Data-dependence crosstalk case, transition from 111 to 000 with charging C_L

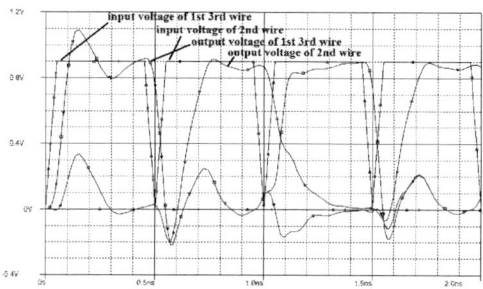

Figure 7. Data-dependence crosstalk case, transition from 101 to 010 with charging C_L+4C_i

Another further experiment is taken to combine MCECC and error control scheme as a whole, which is MCECC/ECS. When the 8 syndromes come out at least 4 non-zeros, we consider this transfer needs a retransmission. The maximum number of continuous error bits is 14. The other parameter is the same as former. We make a compare with other codes as shown in Table II. Cycles taken by MCECC/ECS is about 26% of that taken by SC_GREEN, 102% of that taken by CADEC. And the word error is obviously the lowest. This character is suitable for reliable high performance NoC.

On viewing of these data, we can find that MCECC can not only handle with crosstalk error but also with long continuous errors. When combined with error control scheme, a more reliable NoC is ensured.

TABLE I. CODE CHARACTERS

coding scheme	lines	C_L	C_i	error bits	C_L/line	C_i/line
uncoded	32	4109	5859	652	128.4	183.1
CADEC	77	9902	7223	0	128.6	93.8
SC_GREEN	120	14577	6960	0	121.475	58
DAP	65	8048	6060	0	123.8	93.2
MCECC	96	12042	8935	0	125.4	93.1

TABLE II. CYCLES TAKEN AND NUMBER OF INCORRECT WORD OF DIFFERENT CODING SCHEME

coding scheme	cycles	number of incorrect word
Uncoded	1024	34
SC_GREEN	4096	8
CADEC	1024	13
MCECC	1024	5
MCECC/ECS	1049	2

TABLE III. INCREMENT TRADEOFFS OF DIFFERENT CODECS

coding scheme	encoder		decoder	
	power/mw	area	power/mw	area
uncoded	0	0	0	0
SC_GREEN	0.1443	508	1.1145	1478.95
CADEC	2.287	4158.8	2.743	4128
MCECC	1.117	2277.7	1.787	2768.78

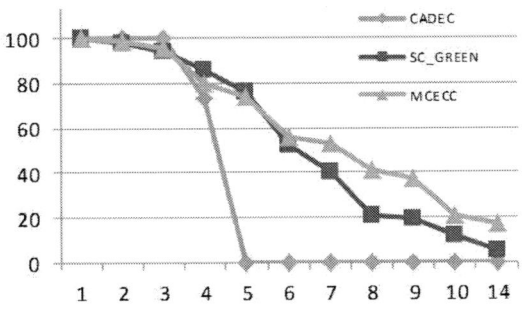

Figure 8. Correct ability of different codes without error control scheme. y-axis means percentage of corrected_error_bits/total_error_bits, x-axis is total_error_bits

C. Implementation analysis

We established RTL models for these codes to get more accurate data. For high frequency consideration, we pipelined the codec instead of combinational logic. The simulation is based on TSMC 90nm CMOS technology with Synopsys Design Compiler at 1 GHz. Assuming all codecs combined with error control scheme, we focusing on the area and power of en/decode modules. The result is a difference with uncoded data as shown in Table III. These data shows us that MCECC achieves high fault tolerance with reasonable tradeoffs.

V. CONCLUSION

As the scale of chip grows larger and larger, the on-chip interconnect becomes more and more complex. The power consumption, reliability, delay on interconnect is placed more importance on NoC design. In this paper, an error correction coding and an adaptive error control scheme is presented to construct reliable interconnection for NoC. The error correction coding (MCECC) is divided into three stages, hamming stage, interleaving stage and FPC stage. By employing groups of (7, 4) hamming code, MCECC provide distribute single error correction. MCECC also have the ability to handle with continuous errors by employing interleaver. And FPC can avoid crosstalk-induced error with the least line cost. The simulation results show MCECC can reduce 50% word error rate compared with SC_GREEN, and 49.2% coupling capacitance-induced energy saving.

REFERENCES

[1] L. Benini and G. De Micheli, "Networks on chips: A new SOC paradigm," IEEE Comput., vol. 35, no. 1, pp. 70–78, Jan. 2002.

[2] K. Aingaran, F. Klass, C.M. Kim, C. Amir, J.Mitra, E. You,J. Mohd, and S. K. Dong. "Coupling Noise Analysis for VLSI and ULSI Circuits". IEEE ISQED 2000, pages 485–489, March 2000.

[3] A. P. Frantz, M. Cassel, F. L. Kastensmidt, E. Cota and L. Carro, "Crosstalk and SEU-Aware Networks on Chips", IEEE Design & Test of Computers, 2007, pp.340-350.

[4] N. R. Shanbhag and M. Zhang, "Soft-error-rate-analysis (SERA) methodology," IEEE Trans. Comput.-Aided Design Circuits Syst., vol. 25, no. 10, pp. 2140–2155, Oct. 2006.

[5] Q. Yu, B. Zhang, Y. Li and P. Ampadu, "Error control integration scheme for reliable NoC," in Proc. 2010 IEEE Intl. Symp. on Circuit and Syst. (ISCAS'10), pp. 3893-3896, May 2010

[6] Amlan Ganguly, Partha Pande, Benjamin Belzer, "Crosstalk-Aware Channel Coding Schemes for Energy Efficient and Reliable NoC Interconnects", IEEE Transactions on VLSI (TVLSI) Vol. 17, No.11, pp. 1626-1639, November 2009.

[7] P-T. Huang , W.-L. Fang , Y.-L. Wang and W. Hwang "Low power and reliable interconnection with self-corrected green coding scheme for network-on-chip", Proc. IEEE NoC Symp., pp. 77 2008

Non-Linear Partitioning for Decimal Logarithm Approximation

Chetan Vudadha, Sreehari Veeramachaneni, M.B.Srinivas
Department of Electrical Engineering,
Birla Institute of Technology and Science (BITS),
Hyderabad Campus, Hyderabad, India

Abstract— **This work presents a systematic approach for the implementation of decimal logarithmic converter. In this approach the decimal logarithm is divided into different regions and linear approximation is applied to each of them. The novelty of this algorithm lies in the selection of regions for linear approximation. The regions are selected in such a way that only a minimum number of coefficients is to be stored. All the other coefficients for linear approximation can be generated from the stored coefficients. A 10-region approximation method is explained in detail. Simulation results show that this method achieves a maximum positive and negative error of 0.0044 and 0.0015 respectively. It can also be applied to a 20-region approximation, which has a maximum positive and negative error of 0.000811 and 0.000948 respectively.**

Keywords-Decimal; logarithm; linear approximation;

I. Introduction

The generation of logarithm of a number finds use in many areas of science and engineering. It also plays a very important role in the implementation of processors based on logarithmic number system. The use of logarithmic unit is preferred in many computationally intensive applications such as 3D graphics processing. This is due to the capability of logarithmic unit in reducing the computationally intensive multiplication and division operations to addition and subtraction respectively. A hybrid logarithmic unit, which uses logarithmic unit for multiplication/division and normal arithmetic unit for addition/subtraction, has earlier been implemented [1].

Though the binary arithmetic unit is widely used, the inclusion of decimal floating point (DFP) operation in latest IEEE754-2008 standard [2] resulted in more researchers devoting their effort in developing algorithms for decimal multiplication [12], division etc.

Many algorithms have been proposed for the hardware computation of base-2 logarithm of binary numbers [13]. These algorithms can be broadly classified into iterative and LUT based. Iterative (digit recurrence) algorithms are efficient from an area and accuracy perspective but have longer latencies [3] [13]. The LUT based methods are efficient in terms of latencies but result in larger area, while providing good accuracy [4-5].

Many decimal logarithmic converters have been proposed recently. The most general and intuitive algorithm is to convert the input decimal number in to a binary number and convert it into base-2 binary logarithm number. This number is further converted to base-10 binary number and is converted back to decimal number [6]. This algorithm consists of multiple conversions and is not efficient in terms of hardware implementation. A decimal to decimal logarithmic converter which uses linear approximation has been proposed in [6]. An iterative decimal logarithmic converter based on selection by rounding was proposed in [7]. This algorithm gives accurate results but has longer latency. Recently a 64-bit decimal logarithmic converted was proposed in [8]. This algorithm results in reduced number of clock cycles when compared to [7], but the implementation includes a power 10 block, which results in increased hardware complexity.

In this paper, a straight line approximation based approach for decimal logarithm is proposed. The novelty of the proposed approach lies in the region selection for linear approximation. The theory behind the region selection criteria and the advantages are explained in detail. A 10-region decimal logarithm approximation is presented, which can be extended to 20-region approximation. Finally a brief overview of the architecture of the proposed 10-region approximation is presented.

The rest of the paper is organized as follows: Section II describes existing straight line approximation based approaches for logarithm computation. Section III describes the proposed linear approximation technique. Section IV presents a brief overview of the hardware architecture while Section V draws the conclusions.

II. Related Work

One of the earliest approaches to approximate the binary logarithm of a number was provided by Mitchell [9]. This method was based on a straight line approximation to logarithm function and resulted in a very high error. There have been many error correction techniques proposed that can reduce the error in Mitchell's algorithm [10-11] without involving complex operations.

A linear approximation algorithm to calculate the decimal logarithm of a decimal number ($\log_{10}(dec)$) was proposed in [6]. This approach is summarized as below:

Assume N to be a decimal number in the interval $10^j \leq N \leq 10^{K+1}$, where $j = 0, \pm 1, \pm 2, ..., k = 0, \pm 1, \pm 2, ...,$ and $k \geq j$. $N = z_k...z_1z_0.z_{-1}z_{-2}....z_j$. Now, N can be represented by the equation

$$N = \sum_{i=j}^{k} 10^i z_i \qquad (1)$$

where, z_i = '0', '1',... '8','9'. We assume that z_k here is not zero since it is the MSD (Most Significant Digit). N can be expressed as:

$$N = 10^{k+1} \sum_{i=j}^{k} 10^{i-k-1} z_i$$

Factoring by 10^{k+1}, the above equation reduces to

$$N = 10^{k+1}(0 + \sum_{i=j}^{k} 10^{i-k-1} z_i) \qquad (2)$$

Letting the term

$$\sum_{i=j}^{k} 10^{i-k-1} z_i) = m \qquad (3)$$

Since $k \ge j$, m will be in the interval $0.1 \le m < 1$, and

$$N = 10^{k+1}(0 + m)$$

The logarithmic result of N is obtained as

$$log_{10} N = k + (1 + log_{10}(m)) \qquad (4)$$

Actually, m is the decimal fraction number. Since $0.1 \le m < 1$, the logarithm characteristic is k and the logarithm mantissa is only a function of m. A linear approximation of $1 + log_{10}(m)$ is described as

$$1 + log_{10}(m) \approx am + b \qquad (5)$$

Where 'a' is slope coefficient and 'b' is intercept coefficient The approximation error is shown to be

$$E = 1 + log_{10}(m) - am + b \qquad (6)$$

The coefficients 'a' and 'b' of the linear approximation equation in equation (5) are found in [6] by fine tuning to minimize the mean square error. This method also determines the bounds of linear approximation regions.

Though this approach results in error of $-0.000994 \le E_{absolute} \le 0.000962$, the bounds on the regions are not uniform. Also in this method of 16-region partitions, 32 4-digit values are to be stored in ROM.

III. PROPOSED LINEAR APPROXIMATION OF DECIMAL LOGARITHM

A. Novel Approach to Partition Region Selection

In this section we explain the proposed approach for region selection for decimal logarithm function given by equation (5), where $0.1 \le m < 1$. The graph of m v/s $(1 + log_{10}(m))$ is shown in figure 1. The function $(1 + log_{10}(m))$ varies such that, it is more linear as m tends to 1 and becomes more non-linear as m tends to 0.

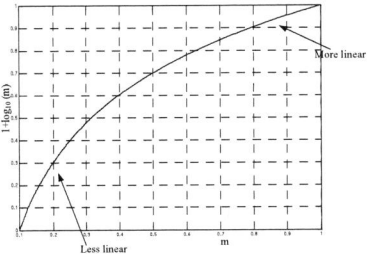

Figure 1. 1+log$_{10}$(m) Curve

Hence partitioning of regions with increased range (less number partitions) as m -> '1' and reduced range (more

number of partitions) as m -> '0', results in uniform error throughout. However, it also results in non-uniform ranges. The selection of the non-uniform ranges is not random and is guided by the observations explained below:

Let x_1 and x_2 be the two bounds of a partition region I such that $x_1 \le m < x_2$, as shown in figure 2(a), where x-axis varies with m and y-axis varies with $1 + log_{10}(m)$. Let the linear approximation coefficients of this region be a_1 (slope) and b_1 (intercept).

We define another partition region II with the bounds x_3 and x_4 such that $x_3 \le m < x_4$ as shown in figure 2(b), m for x-axis and $1 + log_{10}(m)$ for y-axis. Let the linear approximation coefficients of this region be a_2 (slope) and b_2 (intercept).

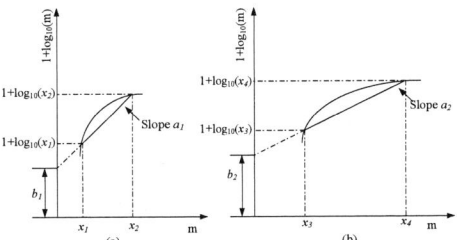

Figure 2. Partition Region Selection

Now if we change the range of partition region II such that

$$x3 = n.x_1 \text{ and}$$
$$x4 = n.x2 \qquad (8)$$

the slope coefficient a_2, which defines the slope of the linear approximation in partition region II, will be given by

$$a_2 = \frac{(1 + log_{10}(x_4)) - (1 + log_{10}(x_3))}{x_4 - x_3} \qquad (9)$$

Substituting equation (8) in (9), we have

$$a_2 = \frac{(1 + log_{10}(nx_2)) - (1 + log_{10}(nx_1))}{nx_2 - nx_1}$$

$$a_2 = \frac{(1 + log_{10}(x_2) + log_{10}(n)) - (1 + log_{10}(x_1) + log_{10}(n))}{n(x_2 - x_1)}$$

$$a_2 = \frac{(1 + log_{10}(x_2)) - (1 + log_{10}(x_1))}{n(x_2 - x_1)} \text{ which results in}$$

$$a_2 = (1/n) a_1 \qquad (10)$$

where, a_1 and a_2 are the slope coefficients of the partition regions I and II respectively.

Lemma 1: If the partition range I is multiplied by 'n' to result in a partition region II, the slope coefficient of partition region I 'a_1' is to be multiplied by '$1/n$' to get the slope coefficient of partition region II 'a_2'.

978-1-4577-1608-9/11 $26.00 © 2011 IEEE

Similar to the slope coefficient relation of equation (10), the intercept coefficient relation can also be derived. In partition region I at a point x_1 (starting bound after linear approximation)

$$\left(1 + log_{10}\left(x_1\right)\right) = a_1 x_1 + b_1$$
$$b_1 = \left(1 + log_{10}\left(x_1\right)\right) - a_1 x_1 \qquad (11)$$

In partition region II at a point x_3 (starting bound after linear approximation)

$$\left(1 + log_{10}\left(x_3\right)\right) = a_2 x_3 + b_2$$
$$b_2 = \left(1 + log_{10}\left(x_3\right)\right) - a_2 x_3 \qquad (12)$$

Assuming the condition in equation (8), i.e. $x_3 = n \cdot x_1$, and substituting values from equation (10), equation (12) reduces to

$$b_2 = \left(1 + log_{10}\left(n x_1\right)\right) - (1/n) \, a_1. \, n. \, x_1$$
$$= \left(1 + log_{10}\left(x_1\right)\right) - a_1 x_1 + log_{10}\left(n\right)$$
$$b_2 = b_1 + log_{10}\left(n\right) \qquad (13)$$

Lemma 2: If the partition range I is multiplied by 'n' to result in a partition region II, the intercept coefficient of partition region I 'b_1' is to be added by $log_{10}\left(n\right)$ to get the intercept coefficient of partition region II 'b_2'.

Thus the selection of the partition regions is made in such a way that some of the partition ranges are multiples of the other partition ranges. This results in a significantly reduced storage requirement. Also the value of 'n' can be chosen optimally to result in reduced complexity in generating other coefficients from stored coefficients.

B. A '10-region linear approximation' of Decimal Logarithm

Based on the above lemmas, a 10-region linear approximation of decimal logarithm is proposed in this work. The different regions selected for the proposed 10-region are shown in Table I. The regions for which the coefficients are to be stored are marked gray in Table I. The coefficients for other regions can be generated from the stored coefficients. In the Table I, a1 (b1), a2 (b2) and a3 (b3) represent slope (intercept) coefficients of regions 8, 9 and 10 respectively. These regions are selected as reference and the other partitions are selected such that, their ranges are 'n' times the range of regions 8, 9 or 10. The value of 'n' is selected as $(1/2^z)$, where z=1, 2, 3. The operations to be carried out on the stored coefficients of regions 8-10, to get the coefficients of region 1-7 are also shown in Table I.

For example the coefficients of the region 7, whose range is (1/2) times the range of region 8, are obtained as below

Slope coefficient of region - 7: $a7 = 2 * a3$, where $a3$ is the slope coefficient of region-10. Similarly,

Intercept coefficient of region - 7: $b7 = b3 - log_{10}\left(2\right)$, where b3 is the intercept coefficient of region 10.

The value of 'n' is selected to be 2, since 2 times any decimal number can be generated using a simple hardware as shown in [12].

TABLE I.   PARAMETERS OF LINEAR APPROXIMATION FOR 10-REGION DECIMAL LOGARITHM CONVERTER

Region	Range	a	b	Max. Abs +ve error (10^{-2})	Max. Abs −ve error (10^{-2})
1	[0.100, 0.125)	8*a3	b3-3*log₁₀(2)	0.445	0
2	[0.125,0.150)	4*a1	b1-2*log₁₀(2)	0.126	0.093
3	[0.150,0.200)	4*a2	b2-2*log₁₀(2)	0.366	0.151
4	[0.200,0.250)	4*a3	b3-2*log₁₀(2)	0.438	0
5	[0.250,0.300)	2*a1	b1-log₁₀(2)	0.121	0.083
6	[0.300,0.400)	2*a2	b2-log₁₀(2)	0.363	0.143
7	[0.400,0.500)	2*a3	b3-log₁₀(2)	0.436	0
8	[0.500,0.600)	a1	b1	0.120	0.076
9	[0.600,0.800)	a2	b2	0.362	0.137
10	[0.800,1.000)	a3	b3	0.427	0.041

Thus the methodology of selecting partition ranges such that they are 2 times the ranges of other partitions results in minimum storage requirements for the coefficients, with a minimum hardware overhead for generation of remaining coefficients.

The coefficients for the linear approximation of $1+ log_{10}\left(m\right)$ for the regions 8, 9 and 10 are given as

Region-10: $a3 = 0.48, b3 = 0.5180$
Region-9: $a2 = 0.62, b2 = 0.4075$
Region-8: $a1 = 0.79, b1 = 0.3047$

Figure 3(a) shows the 10-region approximation with the region bounds indicated by the arrows. It is seen that the selection of regions results in more number of partitions as m tends to '0' and less number of partitions as m tends to '1'.

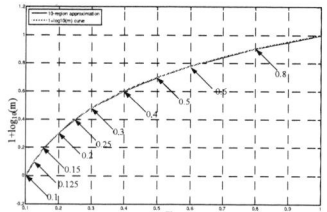

Figure 3.   10-region Approximation of $1+log_{10}(m)$

The proposed 10-region approximation has been simulated with 1,000,000 7-digit decimal numbers. The maximum absolute error obtained ranges over $-0.0015 \le E_{absolute} \le 0.00445$. The region-wise maximum absolute positive and negative errors are indicated in Table I. The error simulation result for the proposed decimal logarithm is shown in figure 4. Though the absolute error obtained is greater than the one presented in [6], it has advantage in

978-1-4577-1608-9/11 $26.00 © 2011 IEEE   104

terms of the coefficients stored. In the algorithm presented in [6], 32 coefficients of 4-digit length have been stored, whereas in the proposed algorithm only 6 coefficients (3 coefficients of 2-digit length and 3 coefficients of 4-digit length) need to be stored.

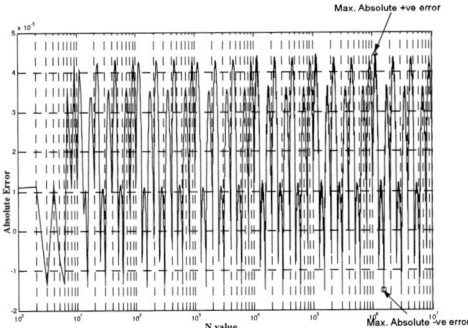

Figure 4.   Error Simulation for 10-region approximation

The proposed 10-region approximation can be scaled to 20-region approximation. This can be done by dividing each region of 10-region approximation in to two regions. For 20-region approximation 12 coefficients are to be stored and used for generating other coefficients. Initial results show that a 20 region approximation has maximum absolute error in the range $-0.000948 \leq E_{absolute} \leq 0.000811$, which is comparable to that of [6], but with reduced storage requirements.

IV.   HARDWARE IMPLEMENTATION

Hardware implementation of the proposed 10-region approximation shown in Figure 5 has two stages. The first stage consists of the leading zero detector (LZD) and a 16-bit shift register. The characteristic of logarithm and the 4-digits of the MSB are achieved in the first stage. The second stage consists of the error correction logic. Implementation of the error correction logic is shown in figure 5.

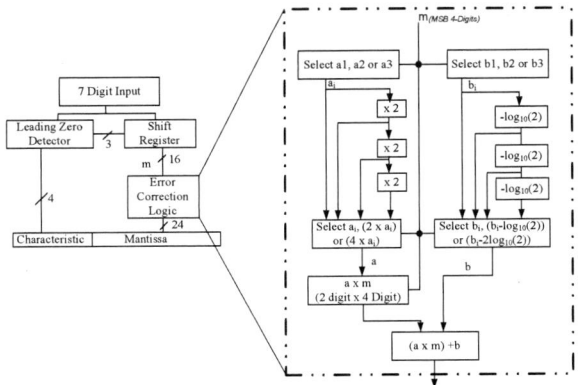

Figure 5.   Implmentation of 10-region Approximation

First a 7-digit fixed point number which is represented by 28-bits BCD is sent to the LZD. In the first stage the characteristic of the decimal logarithm output is obtained by detecting leading zero. Depending on the characteristic the 16-bit shift register obtains MSB 4-digits, which act as inputs

to the error correction logic. The error correction logic uses the 4-MSB digits to select the corresponding slope and intercept coefficients at different stages. The slope coefficient i.e. *a (3-digit)*, is multiplied by 4-digit MSB of *m* and the result is added to intercept coefficient i.e. *b (4-digit)* to get the mantissa part of the output. The characteristic along with the mantissa forms the final output.

V.   CONCLUSION

A novel approach for partition region selection, which can be applied for decimal logarithm conversion, is proposed. This approach is applied to a 10-region linear approximation, giving rise to an absolute error range of $-0.0015 \leq E_{absolute} \leq 0.00445$. It can also be extended to a 20-region linear approximation for decimal logarithm, which results in an error range of $-0.000948 \leq E_{absolute} \leq 0.000811$. The proposed algorithm results in reduced storage requirements than the existing linear approximation algorithm.

REFERENCES

[1]   Hyejung Kim, Byeong-Gyu Nam, Ju-Ho Sohn, Hoi-Jun Yoo, "A 231-MHz, 2.18-mW 32-bit Logarithmic Arithmetic Unit for Fixed-Point 3-D Graphics System," *IEEE Journal of Solid State Circuits*, vol. 41, no. 11, 2373-2381, November 2006.

[2]   The IEEE Standard for Floating-Point Arithmetic (IEEE 754 2008), *IEEE Computer Society*, Aug 2008.

[3]   J. A. Pineiro, M. D. Ercegovac, and J. D. Bruguera, "High-radix logarithm with selection by rounding: Algorithm and implementation," *J.VLSI Signal Process. Syst.* vol. 40, pp. 109–123, May 2005.

[4]   S. Paul, N. Jayakumar,S.P. Khatri,"A Fast Hardware Approach for Approximate, Efficient Logarithm and Antilogarithm Computations", *IEEE Transactions on Very Large Scale Integration (VLSI) Systems*, vol.17, no. 2, 269-277, February 2009.

[5]   M. J. Schulte and J. E. Stine, "Approximating elementary functions with symmetric bipartite tables," *IEEE Trans. Computers*, vol. 48, no. 8, pp. 842–847, Aug. 1999.

[6]   D. Chen and Y. Choi and D. Teng and K. Wahid and S. Ko,"A novel decimal-to-decimal logarithmic converter," *Proc.IEEE Symp. on Circuit and System (ISCAS'08)*, pp. 688–691, May 2008.

[7]   Dongdong Chen, Yu Zhang, Younhee Choi, Moon Ho Lee, Seok-Bum Ko, "A 32-bit Decimal Floating-Point Logarithmic Converter," *Proc. of the IEEE 19th Symposium on Computer Arithmetic*, pp. 195-203, June 2009.

[8]   Ramin Tajallipour, Md. Ashraful Islam, Khan Wahid, "Fast Algorithm of A 64-bit Decimal Logarithmic Converter", *Journal of Computers*, Vol 5, No 12, 1847-1855, Dec 2010.

[9]   J. N. Mitchell, "Computer multiplication and division using binary logarithms," *IRE Trans. Electron. Computers*, vol. 11, pp. 512–517, Aug.1962.

[10]   K. H. Abed and R. E. Siferd, "CMOS VLSI implementation of a lowpower logarithmic converter," *IEEE Trans. Computers*, vol. 52, no. 11, pp. 1421–1433, Nov. 2003.

[11]   S. L. SanGregory, C. Brothers, D. Gallagher, and R. E. Siferd, "A fast low-power logarithm approximation with CMOS VLSI implementation," in *Proc. IEEE Midw. Symp. Circuits Syst.*, vol. 1, pp.388–391, Aug. 1999.

[12]   A.Vázquez, E.Antelo, P.Montuschi, "A New Family of High-Performance Parallel Decimal Multipliers", *IEEE 18th Symposium on Computer Arithmetic (ARITH 18)*, pp. 195-204, June 2007.

[13]   J. M. Muller, "Elementary Functions, Algorithms and Implementation." Boston, USA: Birkha¨user Verlag, 2005.

Reconfigurable Adders for Binary/BCD addition/Subtraction

Syed Ershad Ahmed, Sreehari Veeramanchaneni, Moorthy Muthukrishnan N, M.B Srinivas

Department of Electrical Engineering
Birla Institute of Technology and Science –Pilani, Hyderabad Campus
Hyderabad –500078, India

Abstract—In this paper a new reconfigurable architecture for efficient Binary coded decimal (BCD) addition/subtraction is presented. The architecture is mainly designed keeping in mind the signed magnitude format. The proposed architecture avoids the usage of additional 2's complement and 10's complement circuitry, for correcting the results to sign magnitude format. The architecture is run-time reconfigurable to facilitate both BCD and Binary operations. Simulation results show that the proposed architecture is 13.6% better in terms of delay than the existing designs.

Keywords-BCD; Unified; adder/subtractor; reconfigurable

I. INTRODUCTION

Hardware support for Decimal arithmetic is receiving an increased attention, due to the growing importance of decimal arithmetic in commercial, financial and internet based applications. Specifications for decimal floating point arithmetic have been added to the final version of the IEEE 754-2008 standard for floating point arithmetic [1]. The decimal computation is required not only when numbers are presented for human inspection, but is also often a necessity when fractions are involved. Decimal fractions cannot be represented by binary fractions. The value 0.1, for example, requires an infinitely recurring binary number. If a binary approximation is used instead of an exact decimal fraction results can be incorrect even if subsequent arithmetic is correct [2].

To facilitate even binary applications on the same hardware a reconfigurable approach needs to be adopted. This paper presents a reconfigurable architecture, which can perform both binary and BCD addition/subtraction. The operands are assumed to be in signed magnitude format and the result obtained from the reconfigurable adder structure is also in the signed magnitude representation.

The rest of the paper is organized as follows: Section 2 gives a brief explanation of the existing architectures. The proposed design for the unified binary and BCD adder/subtractor is given in section 3. In section 4, a brief overview on different carry generation networks is presented. Section 5 presents, simulation results for the proposed and existing architectures are given and comparisons are carried out. Conclusion is presented in section 5. Section 6 presents the future work.

II. EXISTING ARCHITECTURES

There have been many contributions on decimal arithmetic especially adders/subtractors. Some of the initial contributions came from Schoomkler et al. [3] and Adiletta

et al. [4]. The first BCD sign-magnitude adder/subtractor was designed by Grupe [5]. An area efficient sign-magnitude adder was later developed by Hwang [6]. In this approach two additional conversions are introduced before and after the binary addition.

An adder similar to the carry select adder was presented in [7]. The design concurrently calculated two results, one assuming the presence of the input carry and the other in its absence. It then selects the appropriate results as the carry is computed. Fischer et al. [8] later came up with the improved version of this design where only a single adder was used to reduce the area overhead.

The approach to construct BCD architectures in many IBM processors is based on the work presented by Haller et al. in [9]. This architecture operated in a single cycle, though requiring corrections in some cases. In the case of subtraction there is need for the computation of the complement after the subtraction to obtain the correct result, hence increasing the latency.

Another improvement in the above architecture was the optimization of the carry chain resulting in slight delay improvement with increased area.

A universal adder design was proposed by D.R.Humberto et al. [10] (figure 1). This design uses effective addition/subtraction operations on unsigned/ sign-magnitude and various complement representations. This design overcomes the limitations of the previously reported approaches that produce some of the results in complement representation when operating on sign-magnitude numbers.

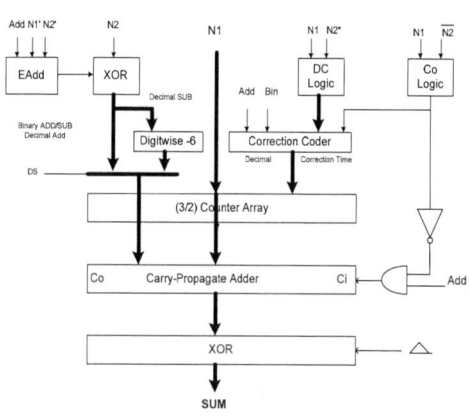

Figure 1. Humberto's proposal [10]

978-1-4577-1608-9/11 $26.00 © 2011 IEEE

III. DESCRIPTION OF THE PROPOSED ADDER

The proposed architecture performs both BCD and binary addition/subtraction on both signed and unsigned numbers. The main algorithm used for performing binary and BCD addition/ subtraction is modification of architecture proposed by Fischer et al. [8] The modifications are done so as to make the proposed algorithm work for signed magnitude representations also. In Fischer's [8] and Haller's [9] proposal, when subtrahend is greater than the minuend, the result is in its complement form and an extra 2's complement circuit or 10's complement circuit is needed for correcting the result. In Humberto's [10] proposal, this is avoided by introducing the comparator circuit and a complement correction circuit at the end of subtraction operation. In our proposal the adder itself is used to find out which of the operands is greater and the correspondingly a complement operation is done using the end around carry method. In the following section the detailed implementation of the proposed unified binary and BCD adder/subtractor is discussed.

Let us assume X and Y being two n+1-bit signed magnitude numbers such that $X = [X_n X_{n-1} X_{n-2}...... X_0]$ and $Y = [Y_n Y_{n-1} Y_{n-2}...... Y_0]$, where X_n and Y_n are sign bits of binary and BCD numbers. The *Eop* (Effective Operation) given by the equation (1) will determine the effective operation. The logic one value of *Eop* indicates effective addition operation, while logic zero indicates effective subtraction operation.

$$Eop = (X_n \odot Y_n) \oplus op \quad (1)$$

Where input $op = 0$ indicates addition operation and $op=1$ indicates subtraction operation. To indicate the binary or BCD operation an additional control signal *Bin* is used. *Bin* =1 indicates binary operation and *Bin* = 0 indicates BCD operation.

After the effective operation *Eop* is found using the equation 1, the sign of the final result is computed using the sign of the first operand X i.e. X_n and the carry-out. If the final effective operation is addition then the sign of the final result is equal to the sign of the first operand i.e. X. But if the effective operation is subtraction the final sign depends on the sign of X and also the carry-out of the adder circuit, which indicates if $X > Y$ or $X \leq Y$. The sign of the final result is given by

Final Sign $S_n = X_n$ if *Eop* = '1' i.e. addition

$S_n = X_n \oplus (Cout)'$ if *Eop* = '0' i.e. subtraction. The implementation is shown in the figure 2

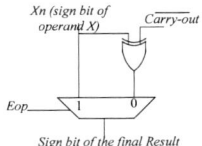

Figure 2. Final Sign computation logic

The architecture of the proposed unified binary/BCD adder/subtractor is shown in the figure 3. In the following

section the different parts of hardware active during the binary or BCD addition subtraction is explained.

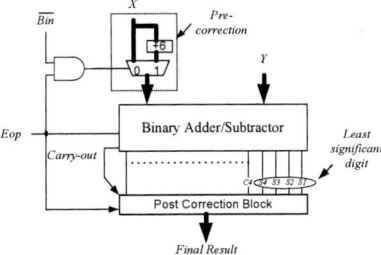

Figure 3.   Architecture of unified BCD and binary adder / subtractor

A. Binary operation

The function of the proposed unified adder is fairly straightforward for binary addition/subtraction operation. A 32-bit prefix adder, based on Sklansky [11] structure is used to carry out addition. When the effective operation is subtraction i.e. *Eop* = 0, operand Y is inverted at the input side. The normal addition operation is carried out to result in $X + (Y)'$. The resulting carry out of this addition indicates whether $X > Y$ or $X \leq Y$. Based on this the following operations i.e., addition of '1' or inverting operation is done to get the correct result. The addition of '1' is done by end-around carry method [12]

If $X > Y$(*Eop* = 0 and carry-out =1)then result $= X + (Y)' +1$

If $X \leq Y$(*Eop* = 0 and carry-out = 0)then result $= (X + (Y)')'$

The binary adder structure including the final sum computation block and the above implementation is shown in the figure 4.

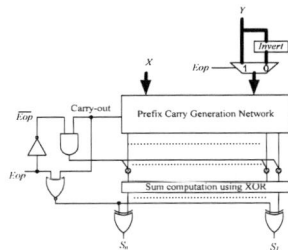

Figure 4.   Implementation of Binary Adder/subtractor of operands in signed magnitude

B. Decimal Operation

The BCD addition operation is performed by pre-correction block where digit-wise addition of 6 is done. After Pre-correction, the binary adder is used for addition operation. The final correction block includes conditional subtraction of 6 depending up on the carry out at each digit stage. The carry out of '1' at digit stage indicates no correction, while the carry out of '0' at the digit stage indicates a correction by subtraction of 6. The subtraction of 6 i.e. 0110 binary is accomplished by addition of 1010 (2's complement of 0110). The following example illustrates the above decimal addition operation.

978-1-4577-1608-9/11 $26.00 © 2011 IEEE

Let $X = 5\ 5\ 6$

$Y = 2\ 3\ 9$

In BCD format: $X = 0101\ 0101\ 0110$

$Y = 0010\ 0011\ 1001$

Addition of digit-wise 6 i.e. $(0110)_2$ to X results in new X,

$X = 0101\ 0101\ 0110$

$+6$ $0110\ 0110\ 0110$

Hence,   new $X = 1011\ 1011\ 1100$

Now the 'new X' is added to Y and correction is applied.

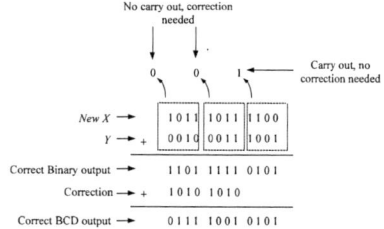

Figure 5.   Illustration of BCD addition algorithm

The BCD subtraction in the first part of the circuit is treated as binary subtraction and the difference is obtained by 2's complement technique. The same method followed in the binary subtraction is followed at this stage. After the operation is completed the post correction block is used to get the correct BCD result.

In the first case when $X>Y$, if the digit of the minuend is greater than that of subtrahend (digit wise carry =1), the binary output of that digit is the correct BCD output and there is no need for any correction. But if a digit of subtrahend is greater than the minuend (digit wise carry = 0), then $(1010)_2$ has to be added to the binary output for that digit to get the correct result. The following example illustrates the above decimal subtraction operation.

Let $X = 5\ 5\ 6$

$Y = 2\ 3\ 9$

In BCD format: $X = 0101\ 0101\ 0110$

$Y = 0010\ 0011\ 1001$

Taking 1's complement of Y(treating it as normal binary subtraction) results in new Y,

New $Y = 1101\ 1100\ 0110$

Next the subtrahend is added to the minuend with carry-in of '1' which is actually the carry out (implemented using the end around carry method as in binary subtraction method) and the correction is done if needed.

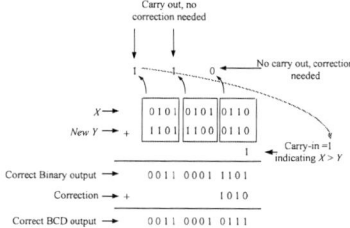

Figure 6.   Illustration of BCD Subtraction algorithm (when $X>Y$)

In second case when $X \leq Y$, if the digit of the minuend is greater than that of subtrahend (digit wise carry = 1),then $(1010)_2$ has to be added to the binary output for that digit to get the correct result. But if a digit of subtrahend is greater than the minuend (digit wise carry = 0), then the binary output of that digit is the correct BCD output and there is no need for any correction. The following example illustrates the above decimal addition operation.

Let $X = 2\ 3\ 9$

$Y = 5\ 5\ 6$

In BCD format: $X = 0010\ 0011\ 1001$

$Y = 0101\ 0101\ 0110$

Taking 1's complement of Y(treating it as normal binary subtraction) results in new Y,

New $Y = 1010\ 1010\ 1001$

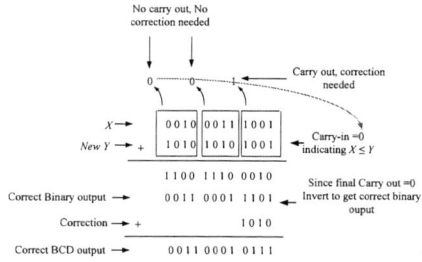

Figure 7.   Illustration of BCD Subtraction algorithm (when $X \leq Y$)

As seen in the illustrative examples in all cases the correction is done by adding $(1010)_2$. The optimized implementation of the pre-correction block (which implements the +6 circuit) and the post correction block (which implements the subtraction of 6 or addition of $(1010)_2$) are shown in figure 8 (a) and (b) respectively. The control signal for the pre correction circuit is given as

$$Cnt1 = Eop.\ (Bin)'$$

which indicates that the addition of $(0110)_2$ is activated only for BCD addition operation.

Similarly the control signal for post correction block is given as

$$Cnt2 = ((Carry\text{-}out + Eop) \oplus C4).\ (Bin)'$$

which indicates BCD subtraction operation. It also takes into consideration both the cases of $X>Y$ and $X \leq Y$ and also carry at each digit stage.

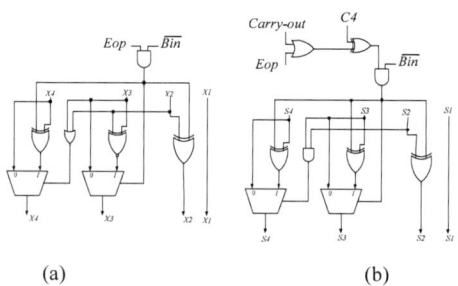

(a)   (b)

Figure 8.   (a) Pre Correction block (b) Post Correction block for BCD

978-1-4577-1608-9/11 $26.00 © 2011 IEEE   108

IV. Different Carry Generation Networks

The prefix carry generation network forms a very important block of the proposed unified binary/BCD adder subtractor. Many implementations of the prefix carry generation block are possible. In the current work we have used skalansky[11] carry generation network. The other prefix structures that can be used for the implementation of the carry generation block are analyzed in [13]. Higher radix prefix networks that can also be used. One such higher radix prefix carry generation network was presented in [14].

All the above mentioned carry generation networks use weinberger [15] carry recurrence relation for designing the networks. The use of an alternative carry recurrence relation can also be explored, which were presented by ling [16]. A comparitive analysis of all the carry recurrence relations was presented in [17].

The design of proposed adder with all the different types of carry generation blocks mentioned above can be explored.

V. Simulation Results

All the adder architectures have been structurally described using Verilog HDL and simulated using Cadence Incisive Unified Simulator (IUS) v6.1 covering all functional combinations. These adders were mapped on the TSMC $180nm$ Technology Typical library (operating conditions 1.8 V, 25°C), using Cadence RTL Compiler v7.1. Inputs were set to have a toggle rate of 50% and a frequency of 1GHz for calculating dynamic power.

Comparison was done with Humberto architecture [10] , and not with the architecture described in [6-9], as it is the only unified adder/subtractor architecture which supports Two's complement signed, unsigned, signed magnitude operands.

TABLE I. SIMULATION RESULTS

	Humberto[10]	Proposed
Delay (nS)	4.004	3.460
Power (mW)	13.37	14.50
Power-Delay (pJ)	53.53	50.17
Area (um²)	12068	10498

It is clear from the Table I that the proposed design has an improvement of 13.6% in terms of delay and around 6.3% improvement in power-delay product when compared with existing architecture [10].

VI. Conclusion

This paper presented a modified architecture for fast BCD addition/subtraction that performs binary addition/subtraction. The proposed architecture avoids the usage of additional 2's complement and 10's complement circuitry, for correcting the results to sign magnitude format. The proposed design is simulated and synthesized using TSMC $180nm$ Technology. The proposed architecture shows an improvement of 13.6% in delay when compared with the existing architecture.

VII. Future Work

As part of future work we are exploring the possibility of extending the work presented in this paper to a multi-precision binary/BCD adder/subtractor. Such a unit can be used in SIMD (Single Instruction Multiple Data) arithmetic, which finds applications in media processing.

Design of high performance and energy efficient reconfigurable multioperand adders, which can operate on binary or decimal operands, is also currently being explored. These multioperand adders find application in multipliers, which form the basic arithmetic blocks of anydigital signal processors.

References

[1] IEEE Standard for Floating-Point Arithmetic, Aug 29, 2008

[2] Michael F. Cowlishaw,"Decimal Floating-Point: Algorism for Computers", IEEE Symposium on Computer Arithmetic 2003: 104-111, June 2003

[3] M.S.Schmookler and A. Weinberger. "Decimal Adder for Directly Implementing BCD Addition Utilizing Logic Circuitry", *International Business Machines Corporation, US patent 3629565*, pages 1 – 19, Dec 1971.

[4] M. J. Adiletta and V. C. Lamere. "BCD Adder Circuit". *Digital Equipment Corporation, US patent 4805131*, pages 1 – 18, Jul 1989.

[5] U. Grupe."Decimal Adder", *Vereinigte Flugtechnische Werke-Fokker gmbH, US patent 3935438*, pages 1 – 11, Jan 1976.

[6] S. Hwang. "High-Speed Binary and Decimal Arithmetic Logic Unit", *American Telephone and Telegraph Company, AT&T Bell Laboratories, US patent 4866656*, pages 1-11, Sep 1989.

[7] Flora, Laurence P., "Fast BCD/Binary Adder", US Patent 5007010.

[8] H. Fischer andW. Rohsaint. "Circuit Arrangement for Adding or Subtracting Operands Coded in BCD-Code or Binary-Code", *Siemens Aktiengesellschaft, US patent 5146423*, pages 1 – 9, Sep 1992.

[9] W. Haller, U. Krauch, and H. Wetter. Combined Binary/Decimal Adder Unit. *International Business Machines Corporation, US patent 5928319*, pages 1-9, Jul 1999.

[10] D.R.Humberto Calderón, G. N. Gaydadjiev, S. Vassiliadis, "Reconfigurable Universal Adder", Proceedings of the IEEE International Conference on Application-Specific Systems, Architectures, and Processors (ASAP 07), pages 186-191, July 2007.

[11] J. Sklansky, "Conditional-sum addition logic," *IRE Trans. Electronic Computers*,vol. EC-9, pages 226-231, June 1960.

[12] E. M. Schwarz, "High-Performance Energy-Effcient Microprocessor Design,"ch. 8. Binary Floating-Point Unit Design: The Fused Multiply-Add Dataflow,pp. 189-208. Dordrecht, The Netherlands: Springer, 2006.

[13] Harris, D.; , "A taxonomy of parallel prefix networks," *Signals, Systems and Computers, 2003. Conference Record of the Thirty-Seventh Asilomar Conference on* , vol.2, no., pp. 2213- 2217 Vol.2, 9-12 Nov. 2003.

[14] R. Zlatanovici, S. Kao, B. Nikolić, "Energy – delay optimization of 64-bit carry-lookahead adders with a 240ps 90nm CMOS design example," IEEE Journal of Solid-State Circuits, vol. 44, no.2, pp.569-583, February 2009.

[15] Weinberger, J.L. Smith, "A Logic for High-Speed Addition," Nat. Bur. Stand. Circ., 591:3-12, 1958.

[16] H. Ling, "High-Speed Binary Adder," IBM Journal of Research and Development, vol. 25, no.3, pp. 156-166, May 1981.

[17] R. Zeydel, T. Kluter, V. G. Oklobdzija, *Efficient Mapping of Addition. Recurrence Algorithms in CMOS*, 17th IEEE Symposium on Computer Arithmetic, June 2005, pp. 107-113.

Conditional Sum Block for High Sparse Adders

Sai Phaneendra P, Sreehari Veeramachaneni, N Moorthy Muthukrishnan, M.B. Srinivas

Department of Electrical Engineering

Birla Institute of Technology and Science-Pilani, Hyderabad Campus

Hyderabad, India

Abstract—**Sparse tree adders are used for energy efficient and high speed addition operation in microprocessors. However, when the operand length increases, sparse tree adder will have the problem of fan-out and/or wiring in the carry generation network as in prefix adders. Increase in the sparseness will be a solution for this problem. But, Adders with higher sparseness have high fan-out and complex conditional sum computation blocks, which acts as a bottle neck. This paper presents a modified prefix adder with late carry-in which has less fan-out and can be used as sum computation blocks in sparse adders. Sparse tree adders using the proposed sum computation block and with other existing sum computation blocks have been implemented and compared. The proposed design results in reduced power-delay product and area when compared to existing designs.**

I. INTRODUCTION

Addition is the basic arithmetic operation in general purpose processors, digital signal processor and application specific processors. Adders form a critical functional element in processors, as a part of ALU or in address generation unit (AGU). In 1958 Weinberger [1] proposed the widely known carry recurrence algorithm for addition. Based on this algorithm, several parallel prefix adders are proposed to improve the performance of addition [2-8]. These prefix adders differ in their logic cells, fan-out and wiring tracks. A study of different prefix adders is presented in [9]. Ling modified Weinberger algorithm resulting in reduced carry computation complexity and increased sum computation complexity [10]. In [11] Doran performed an analysis to determine the set of recurrences, which have recurrence properties and are similar to Weinberger's and Ling's recurrence.

Different adder structures like carry-select adder, carry-skip adder, condition sum adder and prefix adders along with their performance comparison was explained in [12]. For high speed addition prefix adders are preferable. For higher operand lengths, the increase in logic cells, fan-out and lateral wiring complexity limits the usage of prefix adders. Hence, sparse tree adders are seen as alternative for wide adders with reduced wiring and fan-out.

Many sparse tree based designs are proposed in recent years [13-15][19]. S. Naffziger presented a sparse adder in [13], which uses Ling's algorithm. Sanu Mathew proposed a 32 bit sparse-tree adder core for address generation unit, to reduce the thermal hot spots [14]. Energy-Delay comparison between sparse tree adder and prefix adder was presented in [15]. Analysis for design trade-offs was done and a set of guidelines for the design of 64-bit adders was presented in

[16]. Different techniques for energy efficient adder were discussed and a 64-bit energy efficient adder was presented in [17].

As the operand length increases furthermore, these sparse adders will also suffer with high fan-out and lateral wiring in carry generation network as in prefix network. This paper will discuss a method to reduce this high fan-out and wiring complexity problem.

The paper is organized as follows. Section II discusses about the previous work done and section III presents the proposed design and implementation. Section IV shows the simulation results and comparisons. Section V concludes the paper.

II. RELATED WORK

Any prefix adder can be considered as a three stage circuit. The first stage computes the carry propagate and generate terms. These terms are used to compute the final carries in the second stage. In the final stage sums are generated using these carries.

Carry computation can be transformed to a prefix problem [3] using the associative operator $\circ$, which associates pairs of generate and propagate bits as in below:

$$(g,p)\circ(g',p') = (g + p.g', p.p') \quad (1)$$

Where g and g' represent generate terms and p and p' represent propagate terms.

Using the operator $\circ$ consecutive propagate and generate pairs can be grouped to generate i^{th} carry as follows:

$$C_i = (g_i, p_i) \circ (g_{i-1}, p_{i-1}) \circ \ldots (g_1, p_1) \circ (g_0, p_0) \quad (2)$$

Here C_i represents final carry for generation of $(i+1)^{th}$ sum. g_i and p_i represents generate and propagate terms for i^{th} bit respectively.

Figure 1 shows implementation of different blocks used in this paper. Figure 1(a) is a PG block which generates propagate and generate terms of ith bit. Figure 1(b) shows XOR which generate sum. Figure 1(c) represents buffer. Figure 1(d) is a carry merge block which functions as associate operator. Figure 1(e) is a sum computational block with late carry-in.

978-1-4577-1608-9/11 $26.00 © 2011 IEEE

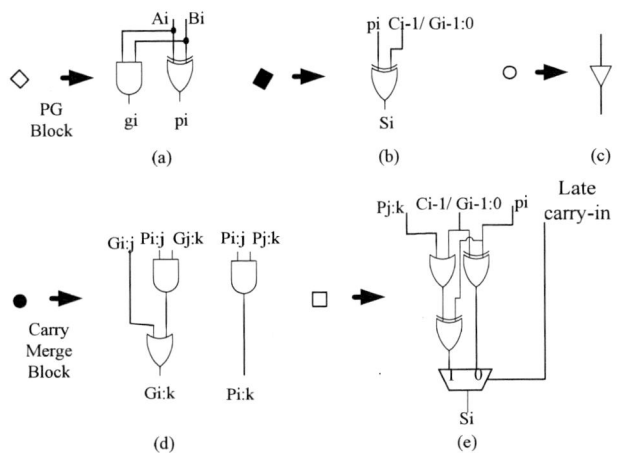

Figure 1. Implementation of different blocks

Among prefix adders, Kogge-Stone adder is a well-known high performance prefix adder because of its minimum logic depth, constant fan-out and regular structure. But when operand length increases, the lateral wiring complexity will increase. The long wire lengths and wiring complexity will result in delay and power consumption. Thus to overcome this problem, sparse tree adders are preferable [14].

Sparse tree adder consists of two segments, one is carry generation block (CGB) and other is conditional sum computation block (SCB). In CGB all the carries are not computed as in prefix adders. Only few carries are computed depending on degree of sparseness. Degree of sparseness is the number of sums selected conditionally. For example degree of sparseness 2 means that the carry will select two sum bits conditionally. Hence, only alternate carries are generated in CGB. In SCB, the sums are generated in two sets assuming input carry as 0 and 1. Optimized sparse 4 SCB was presented in [15]. Sparse 2 SCB was presented in [17]. The carries obtained in CGB will select the sums in SCB.

Figure 2 and 3 shows 64 bit Sklansky and Kogge-Stone based CGB respectively for a sparse 4 adder. As seen from figures, for higher operand lengths, Sklansky based CGB has high fan-out in critical path and Kogge-Stone based CGB has wiring complexity and long wiring tracks.

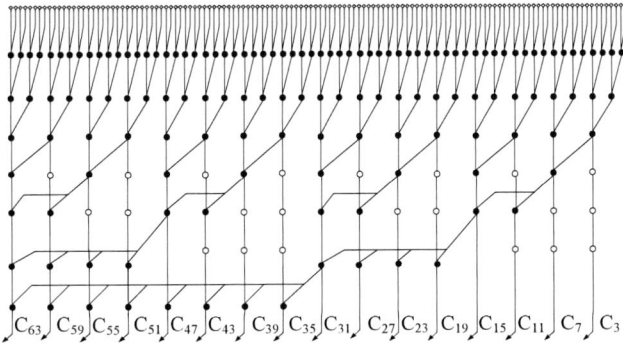

Figure 2. 64 bit Sklansky based CGB for a sparse 4 adder

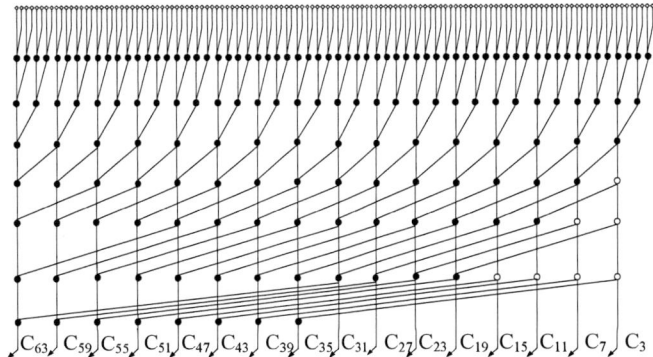

Figure 3. 64 bit Kogge-Stone based CGB for a sparse 4 adder

This can be overcome by increasing the degree of sparseness. As degree of sparseness increases the SCB complexity will increase. Moreover, the loading on carry signal will increase as number of sum bits in conditional block increase.

Sklansky proposed a fast carry-in or late carry-in concept where the input carry was merged with prefix carries at the end of the prefix carry network. Figure 4 shows Sklansky prefix adder with late carry in. Tyagi proposed a reduced area scheme carry select adder which can be used as SCB [18]. Detailed explanation and applications of late carry-in was presented in [12].

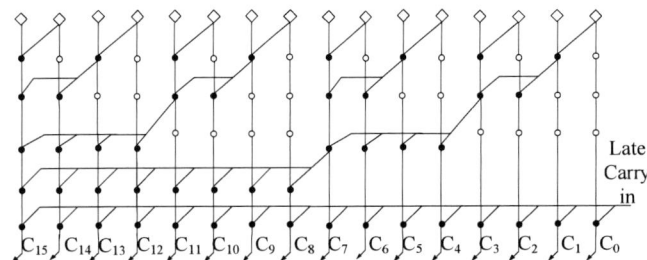

Figure 4. Sklansky parallel prefix adder with late carry-in

III. PROPOSED SUM COMPUTATIONAL BLOCK

The concept of late carry–in and problem of sparse 4 adders for higher operand lengths motivate us to design a higher sparseness SCB. As the sparseness increases, compound adder proposed by Tyagi [18] or any prefix adder structures with late carry-in can be used. But the fan-out of carry signals will increase. For an 8-bit SCB, the carry signal has to drive eight MUXes to select the sum. This limits the usage of direct higher bit carry select adders or prefix adders with late carry-in as SCB. To avoid this, loading on the carry signal should be reduced.

In order to reduce the loading on carry signal multiple carry select stages are used in proposed design. Though it will increase delay by one stage, it gives flexibility to use higher sparseness SCBs and also reduces fan-out and wiring tracks in CGB.

The proposed SCB has two segments. The first segment is a prefix block which generates intermediate carries. These intermediate carries are combined with the carry signal of the CGB block to generate final carries in the second segment.

978-1-4577-1608-9/11 $26.00 © 2011 IEEE   111

Finally, these final carries are used to select the sum conditionally in the n-bit SCB. Here *n* represent number of conditional sum bits selected.

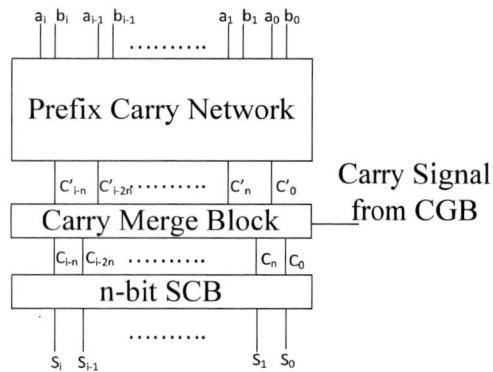

Figure 5. Block diagram of prposed SCB

In this paper we used a sparse 2 SCB as n-bit SCB. Thus, it reduces the generation of carries by half and only alternate carries are generated in prefix network. This also reduces the loading on carry signal coming from CGB.

Figure 6 shows the proposed 8-bit SCB with Kogge-Stone and Sklansky based prefix network. The thick dotted lines are generate signal and the thin dotted lines are propagate signal that are generated in PG block. From the figure, the loading of the carry signal is 4 for an 8-bit SCB with an increase in extra stage in critical path.

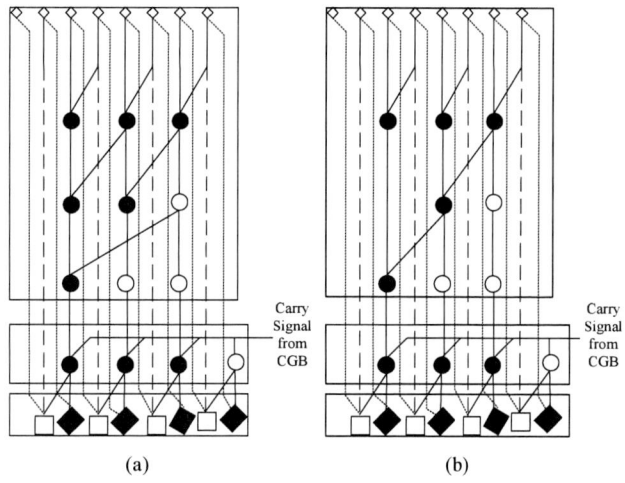

Figure 6. Proposed SCB with (a) Kogge-Stone and (b) Sklansky based prefix adder

The loading effect can further reduce by using sparse 4 SCB in n-bit SCB block. Thus the generalized fan out on the carry signals coming from CGB is given as

$$Fan-out\ of\ Carry\ Signal = \frac{Degree\ of\ Sparseness}{n}$$

Where 'n' is the degree of lower sparse SCB used.

Figure 7 shows the CGB for 64-bit Sklansky based sparse-8 adder. As seen from figure 7, the fan-out of the critical carry

generation path of sparse-8 CGB is decreased when compared to that of sparse-4 CGB shown in figure 2.

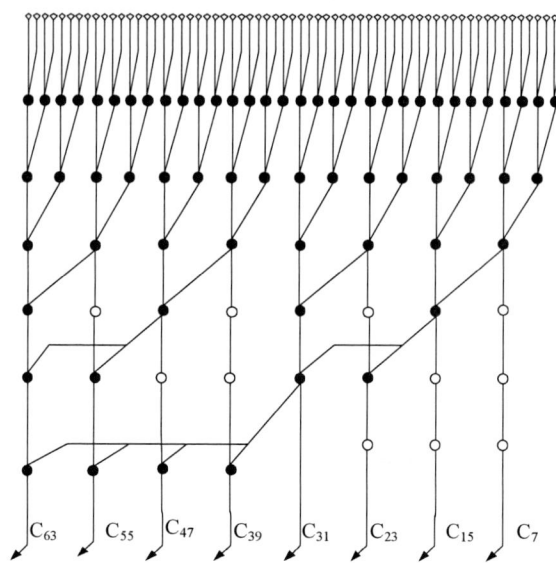

Figure 7. CGB for sparse 8 64-bit Sklansky based sparse adder

Since some of the group Generate and Propagate terms have already been found in the carry generation stage, the same terms can be used in the sum computation block. If the intermediate propagate and generate terms generated at the end of the second stage that is $(G_{[1:0]},P_{[1:0]})$, $(G_{[3:2]},P_{[3:2]})$..etc are used for sum computation, due to which there will be reduction in the number of carry merge blocks at the sum computation stage, which in turn results in power and area.

Figure 8 shows the modified SCB of 8-bit Sklansky and Kogge-Stone based prefix adders by reusing the generate and propagate terms in CGB.

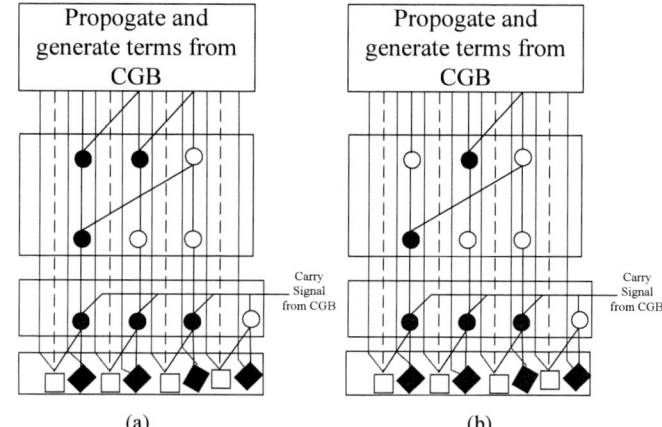

Figure 8. Proposed SCB of 8-bit (a) Kogge-Stone and (b) Sklansky based prefix adders after reusing propagate and generate terms in CGB

In the proposed implementation the max fan-out is four for a 64 bit sparse adder. Even sparse 16 adders can also be implemented using this model with a fan-out of 8 in the critical path.

Figure 9 shows the implementation of 64 bit sparse-16 adder. The CGB is less complex and have minimum wiring and fan-out. But loading on carry signal will increases when compared to 64 bit sparse-8 adder.

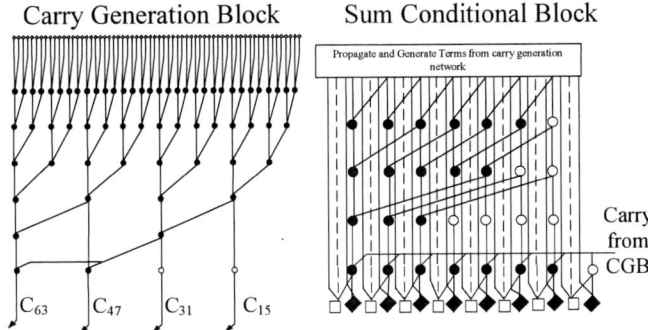

Figure 9. 64-bit Sparse 16 adder with proposed SCB

As the sparseness increases more will be the loading effect on carry signal in proposed model also but less compared to conventional SCB.

IV. RESULTS AND COMPARISON

Each prefix adder has been described in Verilog HDL and simulated using Cadence Incisive Unified Simulator (IUS) v6.1. All the adders were mapped on to the Synopsys 90*nm* generic Technology library, using Cadence RTL Compiler v7.1. The derived netlist was then passed to Cadence First Encounter XL v7.1 for floor-planning and routing. All the simulations are done with single drive strength.

Table I presents area, delay, power and power-delay product for proposed SCB for 64-bit sparse-8, 16, 32, and 64-bit sparse-4 with conditional sum block [15]. Sparse-8 adder with proposed SCB is having less delay compared to sparse-4 adder. Simulation results shows that adding a stage to reduce the loading effect on carry signal takes less delay than having a high fan-out on carry signal.

Table I Area, power, delay and power-delay product for 64-bit adder with various sparseness

64-Bit adder	Sparse-4 with conditional sum[15]	Sparse 8 adder with Proposed SCB	Sparse 16 adder with Proposed SCB
Area (um²)	4316 [100%]	3998 [92.63%]	4383 [101.55%]
Power (mW)	0.286 [100%]	0.247 [86.36%]	0.262 [91.60]
Delay (ns)	1.101 [100%]	0.969 [88.01%]	1.03 [93.55%]
Power-Delay Product (pJ)	0.315 [100%]	0.2394 [76%]	0.271 [86.03%]

64-bit sparse-16 adder has a fan-out of 2 in CGB block and fan-out of 8 in SCB block. 64-bit sparse-8 adder has uniform fan-out of 4 in both CGB and SCB. Hence, 64-bit sparse-16 adder has more delay compared to 64-bit sparse-8 adder.

V. CONCLUSION

This paper presents a sum computational block used for high sparseness. Sparse adders with varying sparseness using the proposed approach have been implemented. Results show an improvement of 13-24% efficiency in terms of energy when compared to existing 64-bit sparse-4 conditional sum adder. Also, the proposed implementation results in better performance in terms of area, power and power-delay product, when compared with existing sparse adders.

REFERENCE

[1] Weinberger, J.L. Smith, "A Logic for High-Speed Addition," Nat. Bur. Stand. Circ., 591:3-12, 1958.

[2] P.M. Kogge and H.S. Stone, "A Parallel Algorithm for the Efficient Solution of a General Class of Recurrence Equations," IEEE Trans. Computers, vol. 22, no. 8, pp. 786-792, Aug. 1973.

[3] R.P. Brent and H.T. Kung, "A Regular Layout for Parallel Adders," IEEE Trans. Computers, vol. 31, no. 3, pp. 260-264, Mar. 1982.

[4] T. Han and D. Carlson, "Fast Area-Efficient VLSI Adders," Proc. Symp. Computer Arithmetic, pp. 49-56, May 1987.

[5] R.E. Ladner and M.J. Fisher, "Parallel Prefix Computation," J. ACM, vol. 27, no. 4, pp. 831-838, Oct. 1980.

[6] S. Knowles, "A Family of Adders," Proc. 14th Symp. Computer Arithmetic, pp. 30-34, Apr. 1999.

[7] J. Sklansky, "Conditional-sum addition logic," RE Trans. Electronic Computers, vol. EC-9, pp. 226-231, June 1960.

[8] N. Burgess, *"New Models of Prefix Adder Topologies"*, J. VLSI Sig. Proc., 40, pp. 125-141 (2005).

[9] D. Harris, "A taxonomy of parallel prefix networks," 37th Asilomar Conf. on Comp. Circ and Systems, Nov. 2003, pp. 2213 – 2217.

[10] H. Ling, "High-Speed Binary Adder," IBM Journal of Research and Development, vol. 25, no.3, pp. 156-166, May 1981.

[11] Doran, R.W.; , "Variants of an improved carry look-ahead adder," *Computers, IEEE Transactions on* , vol.37, no.9, pp.1110-1113, Sep 1988.

[12] R. Zimmermann, *Binary Adder Architectures for Cell-Based VLSI and their Synthesis*, PhD thesis, Swiss Federal Institute of Technology (ETH) Zurich, Hartung-Gorre Verlag, 1998.

[13] Naffziger, S.; , "A sub-nanosecond 0.5 μm 64 b adder design," *Solid-State Circuits Conference, 1996. Digest of Technical Papers. 42nd ISSCC., 1996 IEEE International* , vol., no., pp.362-363, Feb 1996.

[14] Mathew, S.; Anders, M.; Krishnamurthy, R.K.; Borkar, S.; , "A 4-GHz 130-nm address generation unit with 32-bit sparse-tree adder core," *Solid-State Circuits, IEEE Journal of* , vol.38, no.5, pp. 689- 695, May 2003.

[15] Oklobdzija, V.G.; Zeydel, B.R.; Dao, H.; Mathew, S.; Ram Krishnamurthy; , "Energy-delay estimation technique for high-performance microprocessor VLSI adders," *Computer Arithmetic, 2003. Proceedings. 16th IEEE Symposium on* , vol., no., pp. 272- 279, 15-18 June 2003.

[16] Zlatanovici, R.; Sean Kao; Nikolic, B.; , "Energy–Delay Optimization of 64-Bit Carry-Lookahead Adders With a 240 ps 90 nm CMOS Design Example," *Solid-State Circuits, IEEE Journal of* , vol.44, no.2, pp.569-583, Feb. 2009.

[17] Zeydel, B.R.; Baran, D.; Oklobdzija, V.G.; , "Energy-Efficient Design Methodologies: High-Performance VLSI Adders," *Solid-State Circuits, IEEE Journal of* , vol.45, no.6, pp.1220-1233, June 2010.

[18] Tyagi, A.; , "A reduced-area scheme for carry-select adders," *Computers, IEEE Transactions on* , vol.42, no.10, pp.1163-1170, Oct 1993.

[19] Dimitrakopoulos, G.; Nikolos, D.; , "High-speed parallel-prefix VLSI Ling adders," *Computers, IEEE Transactions on* , vol.54, no.2, pp. 225-231, Feb. 2005.

978-1-4577-1608-9/11 $26.00 © 2011 IEEE

A 3.9 pJ/Pulse Differential IR-UWB Pulse Generator in 90 nm CMOS

Kin Keung Lee, Øivind Næss and Tor Sverre "Bassen" Lande
Department of Informatics, University of Oslo, N-0316 Oslo, Norway
E-mail: kklee@ifi.uio.no

Abstract—A low-power impulse radio (IR) ultra wideband (UWB) pulse generator (PG) for low-data-rate (LDR) applications is presented. Different constraints on LDR IR-UWB systems are analyzed. A new differential IR-UWB PG structure is proposed, it utilizes polarization property of antennas to increase the output power. The PG is power-efficient since it does not have precise high-frequency carrier generators. Also, it contains only digital circuits, which is very beneficial for modern CMOS technology. The PG is realized in a TSMC 90 nm CMOS process. Measurements show the energy consumption from a 1.2 V supply to be 3.9 pJ/pulse for 5–80 MHz pulse repetition frequency (PRF). The core area is 0.0011 mm^2 (26 µm × 42 µm).

Index Terms—IR, UWB, RFID, CMOS, pulse generator, wireless sensor network

I. INTRODUCTION

Since Federal Communication Commission (FCC) released a large spectral mask (i.e. 3.1–10.6 GHz) for unlicensed use [1], UWB technology has been an attractive field of research. The huge spectral mask not only gives the benefit of higher date rate, but also reduced power consumption of transmitting circuits by using IR technology. Also, the impulse signaling nature allows turning off most circuits inside the transmitter between transmissions, which makes the IR-UWB transmitters power-efficient. Because of these features, IR-UWB technology is widely adopted on power-harvesting LDR wireless applications, for example, wireless sensor networks (WSN) and RFID tags. One of the most challenging tasks is to design low-power IR-UWB PG which is usually the most power-consuming component in the UWB transmitters.

Besides the power efficiency, the output power spectral density (PSD) of the PG has to meet the FCC part 15 requirements, which is usually a main constraints for high-date-rate UWB applications. The impacts of FCC requirements on power harvesting LDR applications are analyzed in this paper, the main constraints we found, however, are the output power of PG and the available harvesting energy. Base on these considerations, a new differential structure is proposed in this paper, it utilizes polarization property of antennas to increase the output swing, which can be as large as two V_{DD}.

Delay-line based IR-UWB PGs are commonly used since they do not need precise high-frequency carrier generators and, thus, can be very low-power (5.2 pJ/pulse was measured in [2]). However, they usually require additional components to bias the output DC voltage. This limitation is eliminated in the proposed PG, which greatly reduces the circuit complexity and power consumption.

The PG is realized in a TSMC 90 nm CMOS process. It uses digital gate-delay for timing, which enables good power efficiency and acceptable spectral filling. Also, it only contains digital circuits, which is beneficial for modern CMOS technology. Measurements show the energy consumption from a 1.2 V supply to be 3.9 pJ/pulse for 5–80 MHz PRF. The core area is 0.0011 mm^2 (26 µm × 42 µm). The PG performs very well, especially in term of power consumption and chip area, and is competitive to other recently published UWB PGs [2]–[5].

II. CONSTRAINTS ON LDR IR-UWB PGs

The FCC spectral masks for indoor and hand-held UWB systems are shown in Fig. 1a, the out of band emission mask is very conservative, especially for the hand-held devices ($\leq$ −61.3 dBm/MHz). In addition, FCC requires that the −10 dB bandwidth (BW_{-10dB}) of these UWB systems lies inside the UWB band (i.e. 3.1–10.6 GHz). As a result, most UWB PGs have a center frequency (f_{center}) around 7 GHz (the middle of the UWB band) to use the spectral mask efficiently. The trade-offs are the larger propagation loss[1] and the energy consumed by the high speed output drivers, this makes it difficult to apply them on power-harvesting applications. For such applications, instead of the FCC masks, the main constraints are actually the PG output power and the energy harvested. To prove this, we will use a frequency-shifted Gaussian pulse (multiplying a

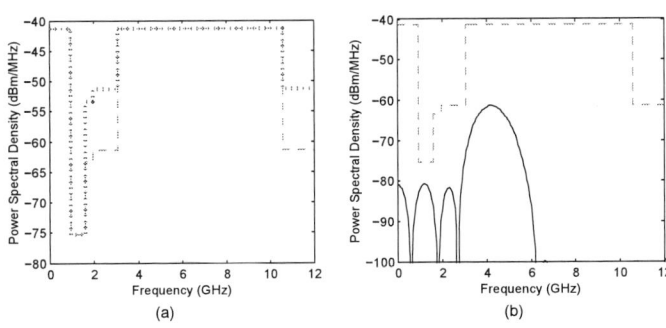

Fig. 1. a) FCC masks for (Dash) hand-held and (Dotted) indoor devices. b) (Dashed) FCC mask for hand-held device and (Solid) example PSD for LDR applications.

---

[1]For example, the free space propagation loss is inversely proportional to the square of carrier frequency.

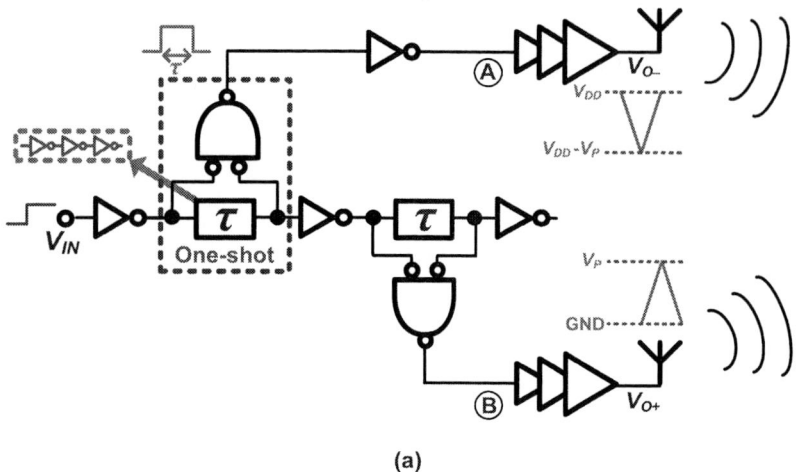 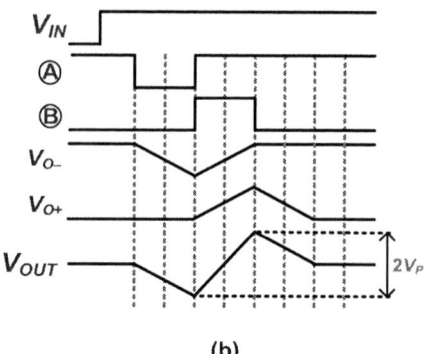

(a)   (b)

Fig. 2.   a) Proposed PG. b) Timing diagram.

first-order Gaussian and a high-frequency sine waveforms) in the calculations here, which gives:

$$V_o(t) = V_p \cdot e^{-\frac{t^2}{2\alpha}} \cdot \cos(\omega_c t) \qquad (1)$$

The corresponding frequency component is:

$$V_o(\omega) = V_p \sqrt{2\alpha^2 \pi} \cdot \left[e^{-\frac{\omega^2}{8\alpha^2}} * \sin(\omega_c t) \right] \qquad (2)$$

Consider a 50 Ω unity-gain antenna is connected to the PG, the peak power can be expressed as:

$$P_{\max, 1\text{-Hz}} = 2\pi\alpha^2 \cdot \frac{V_p^2}{50} \qquad (3)$$

Assume the peak is flat for 1 MHz, we get:

$$P_{\max, 1\text{-MHz}} = 10^6 \cdot 2\pi\alpha^2 \cdot \frac{V_p^2}{50} \qquad (4)$$

One of the FCC requirements is that the radiated emission is lower than –41.3 dBm EIRP with a 1 MHz resolution bandwidth inside the UWB band [1], which results:

$$-41.3 = 10 \cdot \log \frac{P_{\text{FCC,max}}}{1\text{m}} \qquad (5)$$

$$\rightarrow P_{\text{FCC,max}} = 74.1 \text{ nW} \qquad (6)$$

Assume α = 0.86 ns ($BW_{-10\text{dB}}$ = 2 GHz) and V_P = 0.5 V, using (4), we can find:

$$P_{\max, 1\text{-MHz}} = 93 \text{ fW} \qquad (7)$$

From (6) and (7), we know that a maximum of 8k pulses can be transmitted. Similarly, if we want to transmit 1k pulses per second, which is usually limited by the energy harvested and memory [6], with α equals to 0.86 ns (i.e. $BW_{-10\text{dB}}$ = 2 GHz), V_P can be as high as 28 V [2].

---

[2]In this case, V_P is actually limited by another FCC requirement — the peak emission level contained within a 50 MHz bandwidth centered on f_{center} must be lower than 0 dBm EIRP (P.78 of [1]).

Clearly, for power-harvesting LDR applications, the performance is mainly limited by the output power (the output swing) and the energy harvested, not the FCC masks. Also, because the duty cycle is very low (with the order of 0.1%)[3], the emission energy level is usually lower than the FCC masks, the situation is illustrated in Fig. 1b. It would be a trade-off between energy consumption, propagation loss and antenna size to set f_{center} to lower frequencies, for example around 4 GHz.

III. PROPOSED DIFFERENTIAL IR-UWB PG

The schematic of the proposed differential PG is depicted in Fig. 2a. The structure is similar to the PGA-PG in [4] and [7], but the idea is different. The designed PG does not have any Gaussian pulse generators. The timing diagram is shown in Fig. 2b. The one-shot circuit is triggered by a falling edge and creates a fixed-width pulse, τ, for each input edge. If two identical antennas with the same polarization plane are connected to the outputs, the EM signals will be added together. The pulse width (τ_{OUT}) is expected to be around 260 ps (i.e. f_{center} is around 4 GHz) and $BW_{-10\text{dB}}$ is set by the antenna bandpass property. Notice that the new different structure can also be extended to generate multi-cycles IR-UWB pulses easily by cascading and scaling them with different driving strengths, this enables us to decrease the $BW_{-10\text{dB}}$ with a trade-off of energy consumption.

One of the main advantages of this structure is the large output power, the PG output swing can be as large as two V_{DD}. This provides a potential increase in communication distance. Also, the PG contains only combinational CMOS gates which do not consume static power.

Several delay-line based IR-UWB PGs have been reported in [2], [4] and [7]. Since there is no precise high-frequency

---

[3]The duty cycle here is defined as the ratio between the energy-harvesting time and the total transmission time. The energy-harvesting time (with the order of 10 ms) is often much longer than the total transmission time (for example, 2.56 μs for sending 256 pulses with 10 ns chip size).

978-1-4577-1608-9/11 $26.00 © 2011 IEEE   116

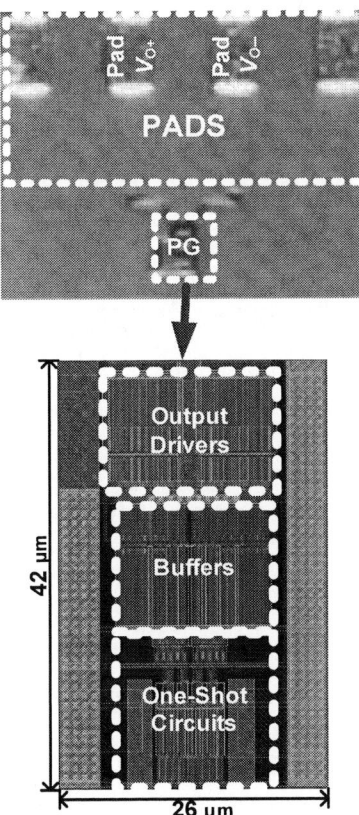

Fig. 3. (Top) Die photo. (Bottom) Circuit layout.

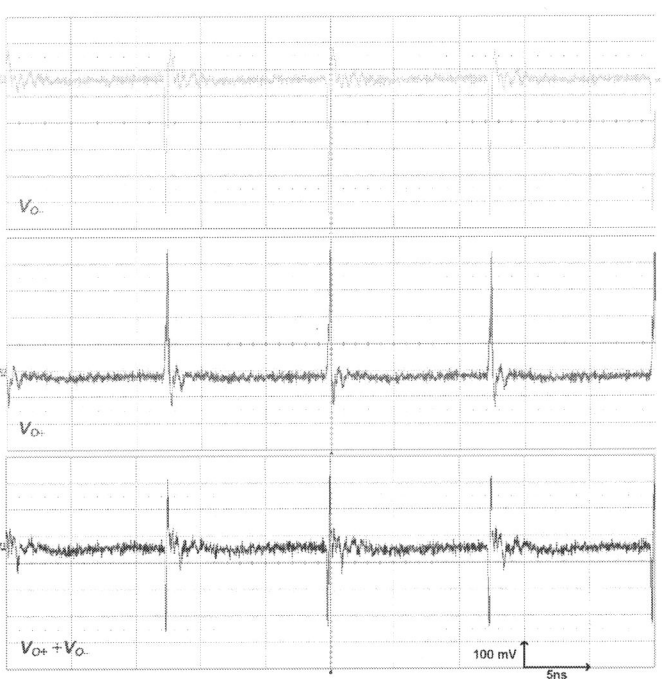

Fig. 4. Output waveform for an 80 MHz PRF.

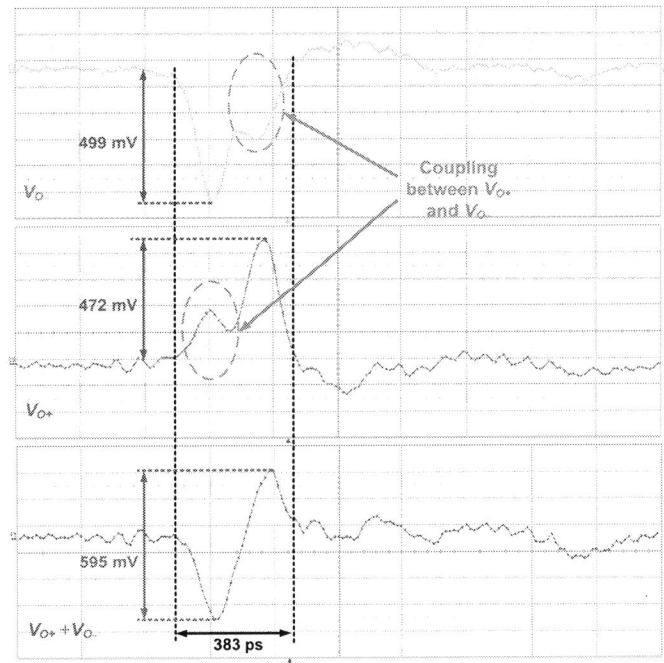

Fig. 5. Zoom-in of Fig. 4.

carrier generator required, they are shown to be very low-power (5.2 pJ/pulse was measured in [2]). One of the problems of such PGs is to bias the output DC voltage of the output drivers, In [4], the leakage current of the output transistors is used to perform the biasing, which may be sensitive to the process variation. Also, it is difficult to bias the output DC voltage to $V_{DD}/2$ [4], hence the maximum achievable output swing is smaller. In [2], a resistive divider was used, however, the divider drives static current and this makes the PG not suitable for LDR applications. In the proposed PG, the DC output voltages of the output drivers are biased to V_{DD} and GND by design. We do not need any additional components to bias the DC output voltage, which greatly reduces the circuit complexity and power consumption.

IV. EXPERIMENTAL RESULTS

The PG is realized in a TSMC 90 nm CMOS process. A die photo is shown in Fig. 3. A small silicon area of 0.0011 mm² for the PG core is achieved because the PG contains only digital circuits. The die is packaged in a 48-leads QFN package. The PG outputs are connected to LeCroy WaveMaster 830Zi with 100 nF coupling capacitors in between. Measurements show the energy consumption from a 1.2 V supply to be

---

[4]In order to bias the output DC voltage to $V_{DD}/2$, the leakage current of the PMOS and NMOS output transistors should be the same, which is difficult to control.

3.9 pJ/pulse for 5–80 MHz PRF and goes up when the PRF is smaller than 5 MHz because of the circuit leakage current. However, for power-harvesting LDR applications, the PG will be turned on a very short time only because of the low duty cycle and the leakage current usually does not have a significant contribution.

Fig. 4 shows the output waveforms for an 80 MHz PRF

978-1-4577-1608-9/11 $26.00 © 2011 IEEE

TABLE I
SUMMARY OF MEASUREMENT RESULTS AND COMPARSION
TO OTHER PUBLISHED IR-UWB PGS

Design	Tech. (CMOS)	$V_{O,P-P}$ (mV)	τ_{OUT} (ns)	Energy Cons. (pJ/Pulse)	Area (mm^2)
This work	90 nm	595	0.38	<u>3.9</u>	<u>0.0011</u>
[2]	90 nm	414	0.96	5.2	0.0015
[3]	180 nm	70	N/A	12	0.045
[4]	130 nm	84	0.75	5.6	0.02
[5]	180 nm	673	0.6	14	0.11

and a zoom-in version is shown in Fig. 5. The swing of V_{O+} and V_{O-} are 499 mV and 472 mV respectively, which results a maximum possible swing of almost 1 V. However, the V_{O+} and V_{O-} pads are misplaced next to each other (see Fig. 3) and the coupling through the pads and bondwire, which is circled in Fig. 5, reduce the swing to 595 mV. The problem can be remedied by arranging a GND pad in between the V_{O+} and V_{O-} pads to avoid the coupling.

The τ_{OUT} is around 383 ps, which is longer than the expected value. It is believed that the error is due to the mismatch between the signal paths off-chip (PCB, bondwire etc.). With the on-chip or stacked antennas, we can control the matching better. A summary of the measurement results and comparison to some recently published IR-UWB PGs [2]–[5] are given in Table I.

V. CONCLUSION

Different constraints on power-harvesting LDR IR-UWB systems were analyzed. Because of the low duty cycle, the performance were constrained by the output power and the energy harvested, unlike the conventional designs which is mainly constrained by the FCC masks. Based on these considerations, a new differential IR-UWB PG structure was proposed, it utilized polarization property of antennas to increase the output power. Also, it was power-efficient since it did not have precise high-frequency carrier generators. The PG was implemented in a TSMC 90 nm process, occupying 0.0011 mm^2. The energy consumption was 3.9 pJ/pulse for 5–80 MHz PRF. The performance was competitive to other recently published UWB PGs, especially in term of power consumption and chip area.

ACKNOWLEDGMENT

The authors would like to thank Novelda AS for their useful suggestions on IR-UWB PG design.

REFERENCES

[1] Federal Communication Commission, *Revision of Part 15 of the Commission's Rules Regarding Ultra-Wideband Transmission Systems*, adopted Feb. 2002, released Apr. 2002.

[2] K. K. Lee, M. Z. Dooghabadi, H. A. Hjortland, Ø. Næss, and T. S. Lande, "A 5.2 pJ/pulse impulse radio pulse generator in 90 nm CMOS," in *Proc. IEEE International Symposium on Circuits and Systems*, May 2011, pp. 1299–1302.

[3] V. V. Kulkarni, M. Muqsith, K. Niitsu, H. Ishikuro, and T. Kuroda, "A 750 Mbit/s, 12 pJ/b, 6-to-10 GHz CMOS IR-UWB transmitter with embedded on-chip antenna," *IEEE Journal of Solid-State Circuits*, vol. 44, no. 2, pp. 394–403, Feb 2009.

[4] X. Wang *et al.*, "FCC-EIRP-aware UWB pulse generator design approach," in *Proc. IEEE International Conference on Ultra-Wideband*, Sep 2009, pp. 592–596.

[5] S. Sim, D.-W. Kim, and S. Hong, "A CMOS UWB pulse generator for 6–10 GHz applications," *IEEE Microwave and Wireless Components Letters*, vol. 19, no. 22, pp. 83–85, Feb 2009.

[6] G. Cimatti, R. Rovatti, and G. Setti, "Chaos-based spreading in DS-UWB sensor networks increases available bit rate," *IEEE Transactions on Circuits and Systems I: Regular Papers*, vol. 54, no. 6, pp. 1327–1339, Jun 2007.

[7] H. Hie *et al.*, "A varying pulse width 5th-derivative Gaussian pulse generator for UWB transceivers in CMOS," in *Proc. IEEE Radio and Wireless Symposium*, Jan 2008, pp. 171–174.

A Novel Digital Predistortion Technique for Class-E PA with Delay Mismatch Estimation

U-Wai Lok, Pui-In Mak, Wei-Han Yu, R.P. Martins [1]

State-Key Laboratory of Analog and Mixed-Signal VLSI, University of Macau, Macao, China
1 – On leave from Instituto Superior Técnico (IST) / TU of Lisbon, Portugal

Abstract—This paper proposes a polynomial-based adaptive digital predistortion (DPD) method with delay mismatch estimation for the class-E power amplifier (PA), which can compensate the AM-AM and the strong AM-PM distortion produced by the switching operation of class-E PA. On the other hand, due to the strict requirement of timing alignment in polar transmission, a low-complexity correlation-based scheme for delay mismatch estimation and compensation filter is developed. Simulation results confirmed that the proposed DPD can improve the EVM performance from 5.67% to 0.72%, leading also to 18 dB rejection of spectrum re-growth under the condition of perfect timing alignment. The results also reveal that our proposed method of digital predistortion with delay mismatch estimation, as well as delay mismatch compensation has 1% EVM improvement as compared with applying DPD method directly without timing alignment.

I. INTRODUCTION

Class-E power amplifier (PA) with polar transmission offers high power efficiency, but suffers from AM-AM and strong AM-PM distortions. The distortion causes spectral re-growth and devastates the constellation diagram. As shown in Fig. 1, especially for the AM-PM distortion, it varies abruptly at low input amplitude. The main cause of this effect is the gate-drain feedthrough of the MOS switch, as depicted in Fig. 2.

In order to obtain high power efficiency while ensuring good linearity to fit high order QAM signal numerous linearization techniques have been proposed for class-E PA. One of the most low-cost techniques is digital pre-distortion (DPD). DPD with look-up table (LUT)-based method was widely used as described in [1] and [2], which simply use the error feedback signal to construct the inverse function and generate LUT for each entity. However, the accuracy of the inverse function depends on the number of entities of the LUT, and a large amount of entities will result in large memory size. On the other hand, polar transmission requires precise timing alignment between amplitude and phase paths [3]. A small timing mismatch (e.g., 2 ns) can degrade the error-vector magnitude (EVM) by 1% and produces obvious spectrum re-growth. This effect will affect the training of DPD and leads to spectrum re-growth even after DPD. Timing mismatch estimation has been proposed in [4], where the complicated least mean squares method makes it quite problematic in terms of implementation.

Here, we propose a DPD scheme using a polar-based polynomial method with recursive least squares algorithm. The advantage of the proposed method is to compute the required coefficients to generate inverse functions of AM-AM

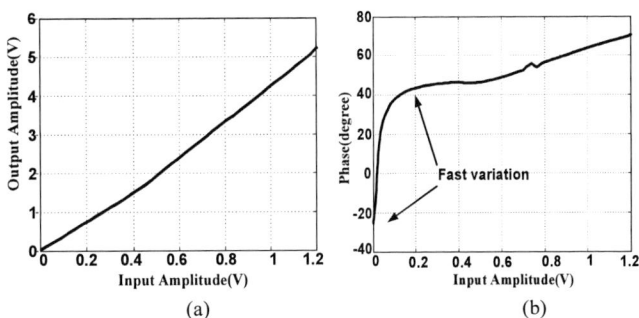

Fig. 1 (a) AM-AM distortion, and (b) AM-PM distortion of Class-E PA.

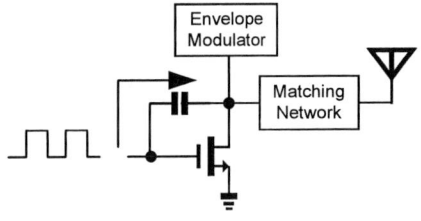

Fig. 2 Simplified Class-E PA.

and AM-PM distortion of class-E PA. We also propose a piecewise AM-PM predistortion scheme against the strong AM-PM effect. This method saves memory when compared with the LUT-based method.

Section II presents the proposed DPD algorithm. In Section III we propose a simple method for delay mismatch estimation for polar transmission scheme. Simulation results will be shown in Section IV and the conclusions will be drawn in Section V.

II. ALGORITHM OF DIGITAL PREDISTORTION

A. Mathematical Models of Power Amplifier and Polar-Based Polynomial Digital Pre-distortion

Figure 3 shows the block diagram of the pre-distortion structure applied in a transmitter with a class-E PA. M-QAM OFDM baseband signals are transmitted, the ith transmitted sample is expressed as $s_{in}^{(i)} = a_{in}^{(i)} exp(j\theta_{in}^{(i)})$, where $a_{in}^{(i)}$ and $\theta_{in}^{(i)}$ represent the amplitude and phase of the ith transmitted sample respectively. With AM-AM and AM-PM distortion, the ith pre-distorted amplitude $a_{pd}^{(i)}$ and the ith pre-distorted phase $\theta_{in}^{(i)}$ can be expressed as:

$$a_{pd}^{(i)} = \mathbf{P}(a_{in}^{(i)})$$
$$= P_1 a_{in}^{(i)} + P_2 \left(a_{in}^{(i)}\right)^2 + ... + P_5 \left(a_{in}^{(i)}\right)^5 \quad (1)$$

and

$$\theta_{pd}^{(i)} = \mathbf{Q}(a_{in}^{(i)})$$
$$= Q_0 + Q_1 a_{in}^{(i)} + Q_2 \left(a_{in}^{(i)}\right)^2 + ...Q_5 \left(a_{in}^{(i)}\right)^5 \quad (2)$$

where pre-distortion functions $\mathbf{P}(.)$ and $\mathbf{Q}(.)$ are the AM-AM and AM-PM inverse function of the class E power amplifier. The ith sample of the received signal from the feedback path will be

$$s_o^{(i)} = a_o^{(i)} \exp(j\theta_o^{(i)})$$
$$= \mathbf{G}(a_{pd}^{(i)}) \exp(j(\theta_{pd}^{(i)} + \theta_{in}^{(i)} + \varphi(a_{pd}^{(i)}))) \quad (3)$$

where $\mathbf{G}(.)$ and $\varphi(.)$ denote the AM-AM and AM-PM distortion functions of the Class E power amplifier, which depend on the pre-distorted signal's amplitude $a_{pd}^{(i)}$.

B. Digital Pre-distortion with Recursive Least Squares

The fundamental idea of the AM-AM pre-distortion algorithm is to obtain the coefficients as defined in (1) to minimize the defined function

$$e_A = \sum_{i=1}^{M} \left| a_{pd}^{(i)} - \mathbf{P}(a_o^{(i)}) \right|^2$$
$$= \sum_{i=1}^{M} \left| a_{pd}^{(i)} - \mathbf{P}(\mathbf{G}(a_{pd}^{(i)})) \right|^2 \quad (4)$$

if function $\mathbf{P}(.)$ equals the inverse of the function $\mathbf{G}(.)$, then $\mathbf{PG}(a_{pd}^{(i)})=a_{pd}^{(i)}$, which means that the cost function is equal to 0 when the exact inverse function of AM-AM distortion is obtained. To minimize the cost function, we can apply the least squares method directly. However, a large amount of computation to invert the matrix is required; so the recursive least squares (RLS) algorithm is employed to find the AM-AM predistortion vector $[P_1 \quad P_2 \quad .... \quad P_5]^T$, which can find the least squares solution recursively with small amount of memory requirement and fast convergence [5]. The complexity of this algorithm for multiplications and additions are $O(k^3)$ and $O(k^2)$ ($O(.)$ represents the Big O function), where k is the order of polynomial function. It means that the complexity is dominated by the matrix computation as shown in (6). The algorithm of RLS is organized as below:

The initial conditions are:

$$\mathbf{K}^{(0)} = \beta * \mathbf{I}, \quad (5)$$

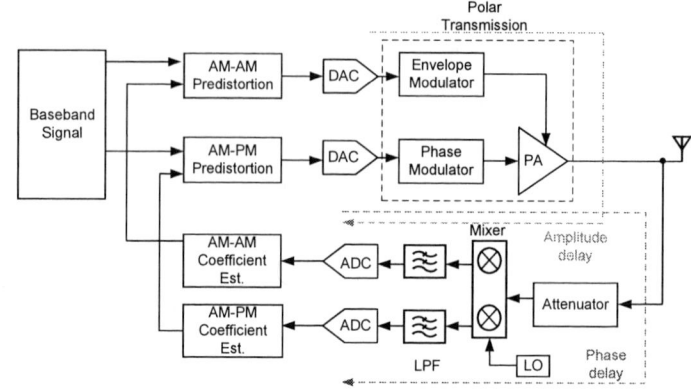

Fig .3 Block diagram of a transmitter with class-E PA, showing amplitude and phase delays.

$$\mathbf{P}^{(0)} = [1 \quad 0 \quad .... \quad 0]^T$$

Where $\mathbf{I}$ is an identity matrix and β is an arbitrary large number.

The tracking stage is operated as:

$$\mathbf{K}^{(n)} = \mathbf{K}^{(n-1)} - \frac{\mathbf{K}^{(n-1)} \mathbf{A}_{n,o} \mathbf{A}_{n,o}^T \mathbf{K}^{(n-1)}}{1 + \mathbf{A}_{n,o}^T \mathbf{K}^{(n-1)} \mathbf{A}_{n,o}}$$

$$\mathbf{P}^{(n)} = \mathbf{P}^{(n-1)} + \mathbf{K}^{(n)} \mathbf{A}_{n,o} (a_{pd}^{(n)} - \mathbf{A}_{n,o}^T \mathbf{P}^{(n-1)}) \quad (6)$$

$$\mathbf{A}_{n,o}^T = \left[a_o^{(n)} \quad \left(a_o^{(n)}\right)^2 \quad .... \quad .... \quad \left(a_o^{(n)}\right)^5 \right]$$

For AM-PM pre-distortion the RLS algorithm is applied to obtain the coefficients vector $[Q_0 \quad Q_1 \quad .... \quad Q_5]^T$ which minimizes the cost function e_θ of AM-PM pre-distortion as follows:

$$e_\theta = \sum_{i=1}^{M} \left| \theta_o^{(i)} - \theta_{in}^{(i)} - \mathbf{Q}(a_{in}^{(i)}) \right|^2 \quad (7)$$

Since $\theta_o^{(i)} = \varphi(a_{pd}^{(i)}) + \theta_{in}^{(i)}$, if the function $\mathbf{Q}(.)$ equals to $\varphi(a_{pd}^{(i)})$, then the cost function e_θ is 0. The AM-PM with RLS process can be expressed as:

The initial conditions are:

$$\mathbf{D}^{(0)} = \alpha * \mathbf{I}$$
$$\mathbf{Q}^{(0)} = [0 \quad 0 \quad .... \quad 0]^T \quad (8)$$

Tracking stage:

$$\mathbf{D}^{(n)} = \mathbf{D}^{(n-1)} - \frac{\mathbf{D}^{(n-1)} \mathbf{A}_{n,in} \mathbf{A}_{n,in}^T \mathbf{D}^{(n-1)}}{1 + \mathbf{A}_{n,in}^T \mathbf{D}^{(n-1)} \mathbf{A}_{n,in}} \quad (9)$$

$$\theta_{diff}^{(n)} = \theta_o^{(n)} - \theta_{pd}^{(n)}$$

$$\mathbf{Q}^{(n)} = \mathbf{Q}^{(n-1)} - \mathbf{D}^{(n)} \mathbf{A}_{n,in} (\theta_{diff}^{(n)} - \mathbf{A}_{n,in}^T \mathbf{Q}^{(n-1)}) \quad (10)$$

where $\mathbf{A}_{n,in} = [1, a_{in}^{(n)}, (a_{in}^{(n)})^2, ...(a_{in}^{(n)})^5]^T$

C. Piece-wise AM-PM Digital Pre-distortion

As shown in Fig. 1(b), AM-PM distortion of the Class E amplifier varies abruptly with low input amplitude. The Least Squares method cannot accurately find the inverse function of AM-PM distortion with limited values of polynomial orders and it may cause the divergence of coefficients tracking. To mitigate this issue, we equally divide the output voltage into three regions and find the inverse AM-PM curve separately with RLS algorithm. Consequently, a better inverse AM-PM curve can be obtained.

III. DELAY MISMATCH ESTIMATION FOR POLAR TRANSMISSION IN CLASS-E AMPLIFIER

The existence of a time difference between the amplitude and phase signals for polar transmission can be exemplified by the diagram of Fig. 3. This mismatch induces spectrum re-growth and degrades the EVM performance, as well as degrades the performance of the DPD. In order to mitigate this effect, amplitude and phase delay mismatches should be estimated and compensated. In addition, during the training process of digital pre-distortion, the timing difference between received signals from the feedback path and the feedforward signal should be aligned, otherwise, the digital pre-distortion will fail (diverge) and spectrum re-grown will become even worse than without applying digital pre-distortion. This path delay can be determined by finding the maximum delay of amplitude and phase delay. Then, the task would be to estimate three delay parameters associated with amplitude, phase and path, and to compensate them. Since the proposed estimation method for amplitude and phase delay is identical, it will only be mentioned the concept of the integer and fractional delay estimation, and afterwards the compensation of these delays with delay filters.

A. Correlation-Based Integer Delay Estimation

To train coarse / integer timing delay within 1 sample period error training signals with high auto-correlation and small cross-correlation are transmitted and will be received from the RF feedback path. The cross-correlation between the transmitted signals and feedback signals is expressed as

$$C_\tau = \sum_{n=1}^{N} \left(A_{in}(n) \right) \left(A_{out}(n - \tau) \right) \qquad (11)$$

where τ represents the estimated integer delay (in sample period), $A_{out}(n-\tau)$ is the $(n-\tau)^{th}$ sample period of feedback path, $A_{in}(n)$ is the nth transmitted signal period and C_τ is the cross-correlation with delay τ. Due to the nature of auto-correlation there will be a maximum value, and the integer delay can be obtained by looking for the value of τ to maximize the function in (11).

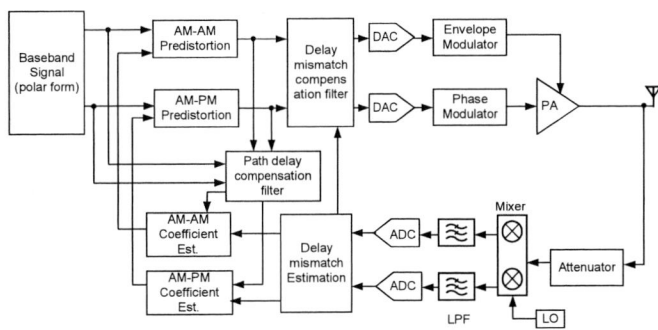

Fig. 4. Block diagram of digital pre-distortion with delay mismatch estimation and compensation.

B. Fractional Delay Estimation

Fractional delay is located in the range of [-0.5 0.5] of the sampling period. This fractional delay signal can be reconstructed by using a Farrow structure interpolation function as:

$$\begin{cases} a_2(\tau_f) = (\tau_f^2 - 1)(\tau_f + 2)\tau_f / 24 \\ a_1(\tau_f) = (\tau_f^2 - 4)(\tau_f + 1)\tau_f / 6 \\ a_0(\tau_f) = (\tau_f^2 - 1)(\tau_f - 4)\tau_f / 4 \\ a_{-1}(\tau_f) = (\tau_f^2 - 4)(\tau_f - 1)\tau_f / 6 \\ a_{-2}(\tau_f) = (\tau_f^2 - 1)(\tau_f - 2)\tau_f / 24 \end{cases} \qquad (12)$$

and the reconstructed signal with fractional delay τ_f is expressed as

$$y(n - \tau_f) = \sum_{i=-2}^{2} a_i(\tau_f) A_{out}(n + i) \qquad (13)$$

where τ_f is the fractional delay within interval [-0.5 0.5], $A_{out}(n-i)$ is the $(n-i)^{th}$ sample period of received signal from the feedback path. After fractional delay signals are constructed with interval of 0.1 sample period, the correlation method needs to be applied to find the τ_f to maximize the cross-correlation function as

$$C_{\tau_f} = \sum_{n=1}^{N} \left(A_{in}(n) \right) \left(y(n - \tau_f) \right) \qquad (14)$$

After finding the amplitude and phase delay, we need to compensate amplitude, phase and path delays before starting the training of the digital pre-distortion. The path delay is equal to the maximum delay between amplitude and phase delays. As shown in Fig. 4, the first step would be to estimate amplitude, phase and path delays from the feedback path, then the delay mismatch compensation filter will compensate the time difference between amplitude and phase delay to align amplitude and phase signals for the polar transmission. On the other hand, the path delay compensation filter will compensate the time delay between feed-forward path and feedback path, which is used for the training process of DPD. The hardware complexity is mainly based on the number of samples for correlation computations in (13) and (14). The

978-1-4577-1608-9/11 $26.00 © 2011 IEEE

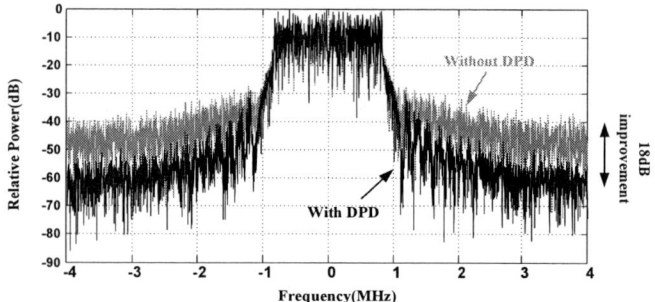

Fig. 5 PA model output power spectrum for OFDM 64-QAM signal.

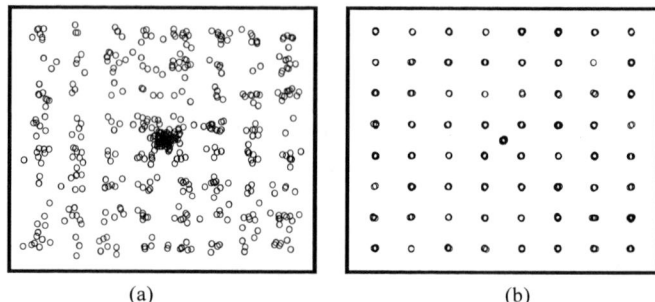

Fig 6. 64-QAM constellation diagram (a) before digital pre-distortion (b)after digital pre-distoriton.

Big O functions for multiplications and additions are O(N^2) and O(N) respectively, where N is the number of samples for calculating correlation functions.

IV. SIMULATION RESULTS

The MATLAB/Simulink is employed as the simulator to verify the proposed method and compare the performances with the requirement of the IEEE 802.11a standard. Fig. 5 shows the PA output spectrum of OFDM 64-QAM signals with and without digital pre-distortion under the condition of perfect timing alignment. Spectrum re-growth has an overall 18 dB improvement after applying digital pre-distortion. Comparing with the IEEE 802.11a standard, the spectrum mask requires -20/-28/-40 dBc at 11/20/28 MHz. It can be observed that at 30 MHz, the spectrum re-growth is close to -40 dBc without digital predistortion. As a result, the Class E PA will be suitable for the 802.11a system after DPD. Fig. 6 shows the simulated constellation diagram of decoded OFDM 64-QAM signals. The constellation diagram without any DPD is shown in Fig. 6(a), the EVM output in this case is 5.67%. The constellation diagram with DPD is shown in Fig. 6(b). The EVM performance in this case is improved to 0.72%. This EVM can fulfill the requirement of 802.11a standard, which specifics 3% EVM.

Fig. 7 shows the constellation diagram of decoded OFDM 64-QAM with and without delay mismatch compensation for 1.5 ns mismatch while applying DPD. As shown in Fig. 7(a), the presence of distortions produced from class-E PA is compensated by digital predistortion, however, delay mismatch degrades the EVM performance which is 2.52%. In Fig. 7(b), after estimated and compensated delay mismatch, as well as applying DPD, the EVM performance is improved from 2.52% to 1.49%.

V. CONCLUSIONS

A novel DPD with delay mismatch estimation for Class-E PA has been proposed. It allows class-E PA to reach high power efficiency while ensuring linearity suitable for high-tier wireless applications such as IEEE 802.11a Wireless LAN. The proposed piecewise AM-PM predistortion method can effectively compensate the abrupt variation of AM-PM

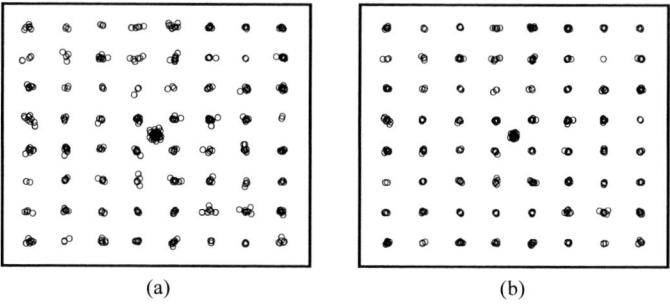

Fig 7. 64-QAM constellation diagram with and without delay mismatch compensation for 1.5 ns while applying digital predistortion for class E PA.

distortion generated by the Class E PA. The polynomial-based RLS method saves memory when compared to the look-up table method. In addition, a simple method for delay mismatch estimation has been proposed. It mitigates the effect of timing mismatch in the polar transmission scheme, and provides timing alignment for the training process of the DPD. EVM and constellation diagram results verified the feasibility of the entire DPD scheme.

ACKNOWLEDGMENTS

This work was supported by the Research Committee of University of Macau, and Macao Science and Technology Development Fund (FDCT).

REFERENCES

[1] K .J .Muhonen, M. Kavehrad, and R. Krishnamoorthy, "Look-up table techniques for adaptive digital predistortion: A development and comparison" IEEE Trans. Veh. Technol., vol. 49, no. 9, pp. 1995-2002, Sept. 2000.

[2] Y .Y . Woo, J. Kim, J. Yi, S. Hong, I. Kim, J. Moon, and B. Kim, "Adaptive digital feedback predistortion technique for linearizing power amplifier" IEEE Trans. Microwave Theory and Techniques, vol. 55, no. 5, May 2007.

[3] K. Waheed, R. B. Staszewski, S. Rezeq,"Curse of digital polar transmission: Precise delay alignment in amplitude and phase modulation paths" in Proc. IEEE ISCAS, pp. 3142-3145, May 2008.

[4] J-F. Bercher and C. Berland,"Envelope/phase delays correction in an EER radio architecture" in Proc. IEEE Int. Electron., Circuits, Syst. Conf., pp. 443-446, 2006.

[5] S. Haykin, Adaptive Filter Theory, 4th Ed. Upper Saddle River, NJ: Prentice-Hall, 2001.

978-1-4577-1608-9/11 $26.00 © 2011 IEEE

A 0.13-μm CMOS 0.8-10.6GHz Low Noise Amplifier with Active Balun for Multi-Standard Applications

Kaichen Zhang[1], Wei Li[1], Fan Ye, Ning Li[1] and Junyan Ren[1, 2]

1. State Key Laboratory of ASIC and System; 2. Micro-/Nano-Electronics Science and Technology Innovation Platform

Fudan University, Shanghai, P. R. China

E-mail: jyren @fudan.edu.cn

Abstract—A CMOS 0.8-10.6GHz Low Noise Amplifier (LNA) with an active balun is proposed for multi-standard applications. A resistive negative feedback stage is adopted for wideband input impedance matching and an active balun is proposed to realize single-to-differential (S2D) conversion. This LNA was designed in a 0.13-μm RF CMOS process with an active area of 0.33mm^2. The post simulation results show that the balun-LNA achieves a minimum NF of 2.8 dB, S11 less than -10 dB, a maximum gain of 16.3 dB and a maximum IIP3 of -3.2 dBm with a power consumption of 13.7 mW from 1.2-V supply.

I. INTRODUCTION

For different applications, many wireless communication standards have emerged, such as GSM, WCDMA, LTE, Bluetooth, 802.11a/b/g/n and UWB. In recent years, some works have been reported towards software-defined-radio (SDR) and multi-standard receivers [1] [3]. As a critical module in receiver, low noise amplifier (LNA) is challenged to work for multi-standards. One structure to realize such a wideband LNA is based on multi-narrow-band LNAs, as shown in Fig. 1(a). The other is directly a single broadband LNA, as shown in Fig. 1(b). Obviously, the second scenario is easy to realize and has an advantage in power and area efficiency [4]. Meanwhile, as the first radio frequency (RF) block after antenna, LNA has to face an issue on single-to-differential (S2D) conversion. Traditionally, off-chip baluns are used for this function. But the insertion loss seriously affects the LNAs' noise figure. Furthermore, off-chip baluns are not suitable for monolithic integration. On-chip transformers overcome the integration problem but still suffer from insertion loss and poor area-efficiency. LNAs with on-chip active baluns seem to be an attractive solution. In traditional active balun [5], a pair of common source (CS) and common gate (CG) stage are adopted to achieve S2D conversion. However, the parasitic capacitor at gate of the CS stage limits its bandwidth.

In this paper, a broadband LNA with on-chip active balun is proposed. In order to eliminated the interaction between wideband performance and balun function, the resistive negative feedback topology is adopted as the input stage for wideband input impedance matching and a modified wideband balun is proposed for S2D functionality.

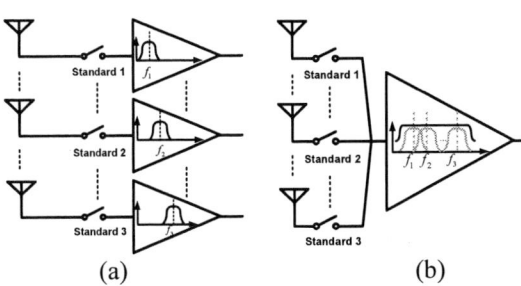

Figure 1. (a) Multiple switching narrowband LNAs. (b) Broadband multi-standard LNA.

Current reuse technology is used to save power, increase the gain and reduce noise. The imbalance of gain and phase of the balun are within 0.4dB and 4.3 degrees. The post simulation results show that a minimum NF of 2.8 dB, a S11 less than -10 dB, a maximum gain of 16.3 dB and a maximum IIP3 of -3.2 dBm are achieved.

II. ARCHETECTURE AND CIRCUIT DESIGN

The biggest challenge of designing such a broadband multi-standard balun-LNA is to achieve good input impedance matching, high gain and low noise figure, while realizing the functionality of single-to-differential conversion.

A. Resistive Negative Feedback Topology

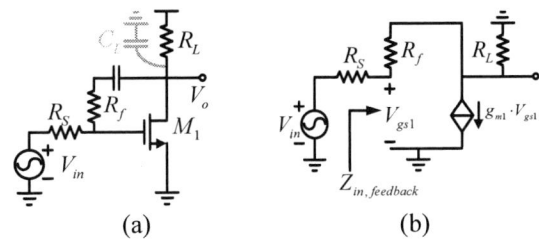

Figure 2 (a) Topology of classic resistive feedback LNA. (b) Small signal equivalent model.

The resistive negative feedback LNA has been widely used [6] [7], mainly because of its good wideband input

---

This work is co-sponsored by National Science & Technology Major Project of China under grant 2009ZX03006-009, 2011ZX03004-001-02 and State Key Laboratory of ASIC & System under Grant MS20080207.

978-1-4577-1608-9/11 $26.00 © 2011 IEEE

matching, high gain and the low noise performance. The topology of a typical resistive negative feedback LNA is shown in Fig. 2 (a), and its small signal equivalent model is shown in Fig. 2 (b). The input impedance $Z_{in,feedback}$ and gain $A_{V,feedback}$ is derived as follows.

$$Z_{in,feedback} = \frac{1}{1+g_{m1}R_L}(R_F + R_L). \quad (1)$$

$$A_{V,feedback} = -\left(g_{m1} - \frac{1}{R_F}\right)(R_F // R_L). \quad (2)$$

R_F and R_L are feedback resistor and load resistor respectively, g_{m1} is the transconductance of M_1. The expressions above are obtained under ideal conditions. In reality, R_L should be replaced by $R_L // (1/sC_L)$, when the circuit frequency response is considered. C_L is the parasitic capacitor at the output node. According to (1) and (2), it is evident that the load parasitic capacitor C_L will reduce the circuit bandwidth and affect input impedance matching at the same time. Therefore, C_L should be paid more attention to especially in the design of high frequency wideband LNAs. The noise figure of the classic resistive negative feedback LNA is derived as [4]

$$NF \approx 1 + \frac{\gamma g_m}{R_S g_m} + \frac{1}{R_S R_L g_m^2} + \frac{4R_S}{R_F}\left(\frac{-1}{1+\left(\frac{R_F + R_S}{(1+g_m R_S)R_L}\right)}\right)^2. \quad (3)$$

The noise figure can be decreased by increase the feedback resistor R_F, however, the input impedance $Z_{in,feedback}$ is changed simultaneously. The trade-off between input power matching and noise matching should be carefully considered.

B. Active Balun Topology

The on-chip active balun-LNA [5] has been increasingly used because its area consumption is less than the traditional transformer while realizing the same single-to-differential (S2D) functionality. A typical balun-LNA topology is shown in Fig. 3[5].

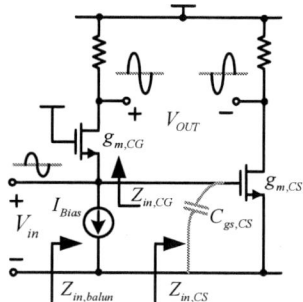

Figure 3. Traditional active balun topology.

$$Z_{in,balun} = Z_{in,CG} // Z_{in,CS} = \frac{1}{g_{m,CG}} // \frac{1}{C_{gs,CS} \cdot S} \quad (4)$$

$C_{gs,CS}$ is the parasitic capacitor between gate and source of M_2, and $g_{m,CG}$ is the transconductance of M_1. From (4), it can be inferred that the input impedance matching will be deteriorated significantly with the increasing frequency. In other words, this kind of active balun structure brings constraint to input impedance matching at high frequency, which means that making the balun stage as input stage of a broadband LNA is not the best choice.

III. PROPOSED BROADBAND BALUN-LNA

A novel broadband balun-LNA is proposed in Fig. 4. A resistive feedback LNA is used as the first stage, while an active balun works as the second stage of LNA. By rearranging the input matching stage and balun stage, constrains between broadband input impedance matching and S2D functionality are reduced.

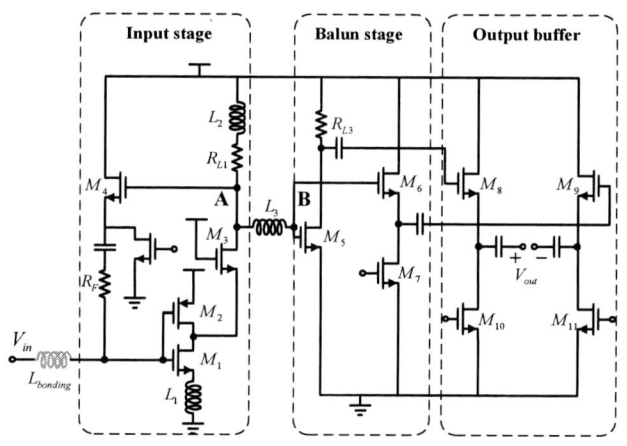

Figure 4. The proposed broadband balun-LNA.

A. Input Matching Stage

The input matching stage is designed based on resistive negative feedback topology due to its good wideband performance. Current reuse technology, consisting of transistor M_1 and M_2, is employed to increase the transconductance of the LNA and save the DC current simultaneously. L_1 is used as a source degeneration inductor to resonate with the gate-source parasitic capacitor C_{gs}. The bonding inductor at the input node is exploited as gate peaking inductor [6] to expand the bandwidth of LNA at high frequency. And it is implied from (1) that the parasitic capacitor C_L at the output node of the input stage will deteriorate the bandwidth performance significantly. Therefore, inductor L_2 is introduced to compensate the attenuation of gain at high frequency. The effect of L_2 is shown in Fig 5. The L_3 is introduced between input stage and balun stage to resonate with the parasitic capacitors at node A and B and expand the bandwidth. The effect of L_2 is shown in Fig 6.

As the first stage of the LNA, the input matching stage contributes the dominant noise to the LNA. According to (3),a reasonable NF can be achieved under the input impedance matched condition by choosing the parameter carefully.

B. Active Balun Stage

From (4), the traditional active topology is not suitable for broadband input impedance matching design due to the gate parasitic capacitor of the CS stage. Therefore, the active balun is arranged at latter stage to perform the single-to-differential functionality, while input matching being realized by resistive

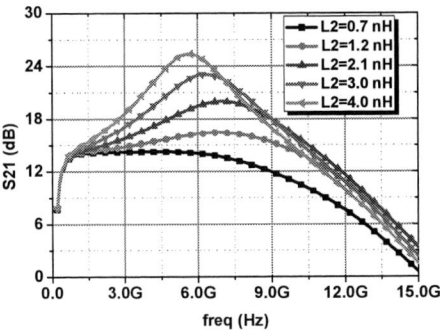

Figure 5. The compensation effect of L_2.

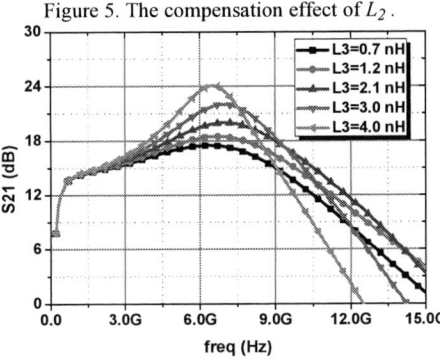

Figure 6. The compensation effect of L_3.

feedback stage. Thus, the limit of bandwidth in traditional active balun is eliminated. However, according to (4), the input impedance of balun stage ($Z_{in,balun-stage}$) is usually small, which is in parallel with the output impedance of the first stage. By replacing the parameter R_L in (1) and (2) with $Z_{Load,first}$ (total load impedance of the first stage, $Z_{Load,first}=R_L$ // $Z_{in,balun-stage}$), it is evident that the small $Z_{in,balun-stage}$ will lower the $Z_{Load,first}$. Consequently, the gain and input matching performance are affected.

A novel active balun topology is proposed in Fig. 7. A source follower is used to replace the common-gate stage in active balun, which could also provide a positive phase at the

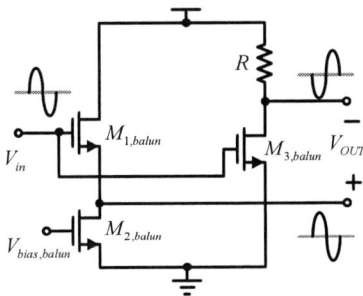

Figure 7. The improved active balun for the proposed LNA

output node. By doing this replacement, the input impedance

of balun stage increases due to the gate input of both the source follower and the common-source amplifier. Therefore, the problem discussed above becomes less noticeable.

The simulation results of the modified active balun are shown in Fig.8. The gain imbalance is within 0.4 dB, and the phase imbalance is restricted to 4.3 degrees from 0.8-10.6GHz.

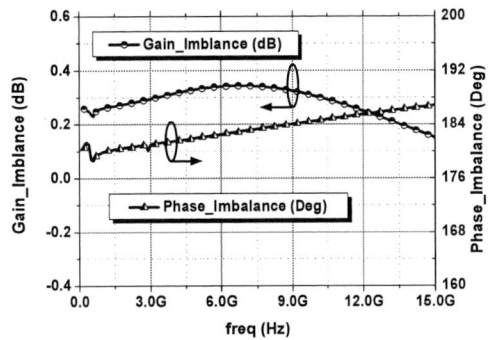

Figure 8. Simulation results of gain and phase imbalance.

IV. POST SIMULATION RESULTS

The proposed LNA is designed in 0.13-μm, 1P8M RF CMOS process. The layout of the LNA is shown in Fig. 9. The core size of the LNA is 0.5mm × 0.66mm. The source followers consisting of M_9-M_{12} are employed as output test buffer to realize 50Ω output impedance matching, as shown in Fig. 4. The DC current of the output buffer is 11.4mA. The power consumption of the core circuit is 13.7mW (de-embedded the power consumption of the output buffer) from a 1.2-V supply. The post simulation results show that the LNA achieves the minimum NF of 2.8 dB, the maximum gain (S21) of 16.3 dB. The simulated -3dB bandwidth is from 0.8-to-10.6 GHz. The S-parameter and NF simulation results are shown in Fig. 10(a)-(b). The output buffer leads to an at least 6dB gain attenuation. Fig. 10(b) shows the real gain of the LNA de-embedded the loss due to the test buffer. The IIP3 simulated from 0.8-to-10.6 GHz is shown in Fig 10(c).

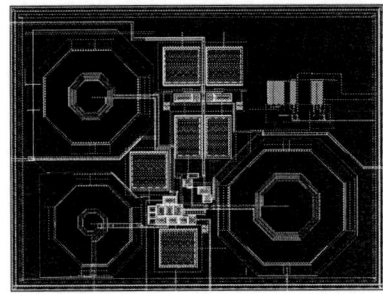

Figure 9. Layout of the proposed broadband multi-standard balun-LNA

The figure-of-merit (FOM) value we used is introduced in [8]. It is expressed in (5).

$$FOM = \frac{Gain \cdot IIP3(mW) \cdot Bandwidth(GHz)}{(F-1) \cdot P_{DC}(mW)}. \qquad (5)$$

The Gain and F are the absolute values, and P_{DC} is the DC power consumption. The performance comparison among the proposed broadband multi-standard balun LNA and recently published LNAs is summarized in TABLE I. The area and power consumption of the single-end LNAs listed in TABLE I will be doubled if calculated as differential circuits.

reduce power consumption. The modified active balun is proposed for broadband application. The proposed LNA achieves excellent FOM value among the recently published works.

TABLE I.   PERFORMANCE SUMMARY & COMPARISON

	This Work	Ref. [2]	Ref. [4]	Ref. [7]	Ref. [9]
Freq. (GHz)	0.8-10.6	0.1-14	0.5-7	0.2-9	4.5-11
Gain (dB)	16.3	12.4	22	10	12.5
S11 (dB)	<-10	<-7.3	<-7	<-10	<-8.2
NF (dB)*	2.8	2.7	2.3	4.2	3.9
IIP3 (dBm)	-3.2	-3.8	-10.5	-8	N/A
Topology	Active balun	Single-End	Single-End	Single-End	Active balun
VDD (V)	1.2	1.8	1.2	1.2	1.2
Power (mW)	22.8(13.7)[a]	14.4[c]	12[c]	20[b]	20.4
FOM	2.54	1.95	0.87	0.14	N/A
CMOS Process	**130-nm**	90-nm	90-nm	90-nm	90-nm

[a] Power consumption with (without) test buffer   * Minimun Noise Figure
[b] Power consumption of Differential Amplifier   [c] Power consumption of Single-end Amplifier

References

[1] R. Bagheri, A. Mirzaei, S. Chehrazi, M.E. Heidari, L. Minjae, M. Mikhemar, T. Wai, A. A. Abidi, "An 800-MHz–6-GHz Software-Defined Wireless Receiver in 90-nm CMOS," *IEEE J. Solid-State Circuits*, vol. 41, no. 12, pp. 2860-2876, Dec. 2006.

[2] C. Po-Yu, S. S. H. Hsu, "A Compact 0.1–14-GHz Ultra-Wideband Low-Noise Amplifier in 0.13-μm CMOS," *IEEE Trans. Microw. Theory Tech.*, vol.58, no.10, pp.2575-2581, Oct. 2010.

[3] Z. Kaichen, L. Wei, Z. Feng, L. Ning, and R. Junyan, "A CMOS 0.5–10.6 GHz inductor feedback low noise amplifier for multi-standard application," *IEEE Int. Conf. Solid-State and Integrated Circuit Technol.* Nov.2010 pp.764-766.

[4] B. G. Perumana, J. -H. C. Zhan, S. S. Taylor, B. R. Carlton, L. Laskar, "Resistive-Feedback CMOS Low-Noise Amplifiers for Multiband Applications," *IEEE Trans. Microw. Theory Tech.*, vol.56, no.5, pp.1218-1225, May 2008.

[5] S. C. Blaakmeer, E. A. M. Klumperink, D. M. W. Leenaerts, and B. Nauta, "Wideband Balun-LNA With Simultaneous Output Balancing, Noise-Canceling and Distortion-Canceling," *IEEE J. Solid-State Circuits*, vol. 43, no. 6, pp.1341-1350, June. 2008.

[6] C. Hsien-Ku, C. Da-Chiang, J. Ying-Zong, L. Shey-Shi Lu, "A Compact Wideband CMOS Low-Noise Amplifier Using Shunt Resistive-Feedback and Series Inductive-Peaking Techniques," *IEEE Microw. Wireless Compon. Lett.*, vol.17, no.8, pp.616-618, Aug. 2007.

[7] C. Tienyu, C. Jinghong, L. A. Rigge, J. Lin, "ESD-Protected Wideband CMOS LNAs Using Modified Resistive Feedback Techni-ques With Chip-on-Board Packaging," *IEEE Trans. Microw. Theory Tech.*, vol.56, no.8, pp.1817-1826, Aug. 2008.

[8] A. Amer, E. Hegazi, H. Ragai, "A Low-Power Wideband CMOS LNA for WiMAX," *IEEE Trans. Circuits Syst. II, Exp. Briefs*, vol.54, no.1, pp.4-8, Jan. 2007.

[9] A. Bevilacqua, C. Sandner, M. Tiebout, A. Gerosa, and A. Neviani, "A 6–9-GHz programmable gain LNA with integrated balun in 90-nm CMOS", in *IEEE Int. Conf. Ultra-Wideband*. 2008. p. 25-28.

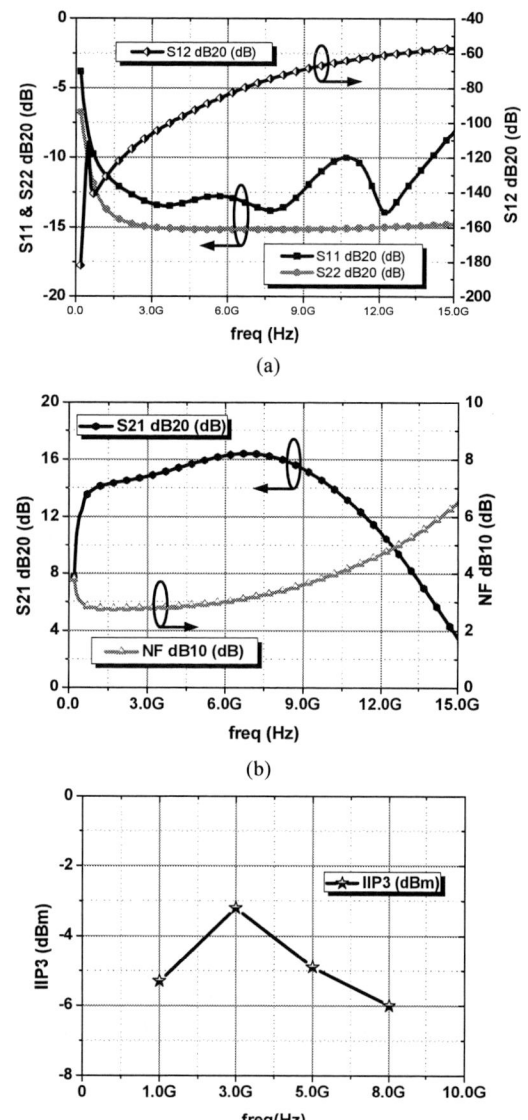

Figure 10.Post Simulation Results of the LNA (a) S11&S22 &S12
(b) S21&NF (c)IIP3

V. CONCLUSION

A 0.8-10.6GHz broadband balun-LNA has been presented. Standards in this frequency range can be compatible, such as WCDMA, WLAN and UWB. The resistive negative feedback topology is utilized for wideband input matching. Current ruse technology is introduced to increase the transconductance and

Memory efficient LDPC decoder design

Yuan Yao, Wei Liang and Fan Ye
State Key Laboratory of ASIC and System
Fudan University
Shanghai , China
fanye@fudan.edu.cn

Junyan Ren, *Member, IEEE*
State Key Laboratory of ASIC and System, and Micro-/Nano-
Electronics Science and Technology Innovation Platform
Fudan University
Shanghai , China

Abstract—**Hardware complexity of LDPC decoders, which is caused by storage and processing of massive information, is the major reason that encumbers LDPC codes from widely application. Reducing the quantization word length of decoding information can effectively decrease the hardware complexity. But for the absolute value of information keeps increasing during decoding, short word length with finite quantization ranges will lead to serious saturation errors and damage decoding performance. Two quantization schemes is proposed in this paper to reduce the number of memory bits required by decoder design by using short word length while guarantee bit-error-rate (BER) performance. Results shows that these two quantization schemes can simplify the hardware complexity with very little loss of decoding performance.**

I. INTRODUCTION

Low density parity check (LDPC) codes [1] have been investigated extensively in recent years. LDPC codes have several advantages: more inherent parallelism, high throughput potentials and no error floors at high SNRs. LDPC codes are specified by sparse parity check matrices and can be represented by Tanner graph with two sets of nodes, called the check nodes and the variable nodes.

LDPC codes have been demonstrated to perform very close to the Shannon limit [2-4]. They have been adopted as optional forward error correction (FEC) codes in the emerging communication standards such as IEEE 802.11n, 802.16e (WiMax), and DVB-S2.

Layered decoding architecture (LDA) [5] is proposed for LDPC decoder design to reduce the number of iterations. Denote α the scaling factor, Λ the message of variable node unit (VNU), r the message passed from check node unit (CNU) to VNU, and q the message passed from VNU to CNU. LDA with min-sum algorithm [6-8] can be described as:

$$q_{j,n_j}[k] = \Lambda_{n_j}[k] - r_{j,n_j}[k] \ (n_j \in N(j)) \quad (1)$$

$$new\_r_{j,n_j}[k] = (\prod_{n_{j'} \in N(j) \backslash n_j} sign(q_{j,n_{j'}}[k])) \quad (2)$$
$$\times(\min_{n_{j'} \in N(j) \backslash n_j} \{|q_{j,n_{j'}}[k]|\} \times \alpha)$$

$$new\_\Lambda_{n_j}[k] = q_{j,n_j}[k] + new\_r_{j,n_j}[k] \quad (3)$$

The details can be referred to [5]. In LDA the parity check matrix of LDPC code is divided into several horizontal layers. In each iteration, the layers are processed one by one. Equations (1), (2) and (3) are performed in CNU and VNU with each layer. In LDPC decoder design, quantization word length of decoding information is a tradeoff between BER and hardware complexity. Different number of bits can be used to denote values of Λ, q and r. If more bits are used, the decoder can achieve a better BER performance with larger chip area. If fewer bits are used, the decoder can be constructed with smaller chip area and the BER performance will be worse. In this paper, two new dynamic quantization schemes based on layered decoding architecture are proposed. These schemes can reduce decoder's chip area by using fewer bits to denote values of Λ, q and r while guarantee BER performance.

The rest of this paper is organized as follows. Section 2 introduces the proposed dynamic quantization schemes. BER performance of these schemes are simulated and shown in section 3. Section 4 gives hardware comparison between decoder design with and without dynamic quantization. The 1/2-rate LDPC code in 802.16e is used during comparison. Finally, conclusions are presented in section 5.

II. PROPOSED DYNAMIC QUANTIZATION SCHEMES

The matrix for 1/2-rate quasi cyclic [9], [10] (QC) LDPC code in 802.16e is used to describe proposed quantization method. The matrix is shown in figure 1. The size of the matrix is 1152×2304 and the size of the submatrix is 96×96. This matrix contains twelve layers and each layer has 24 submatrixes. A non-zero submatrix denoted by a marked value S is obtained by circularly shifting an identity matrix by S towards right.

This work was supported by National Science & Technology MajorProject with No. 2011ZX03004-001-02 and No. 2009ZX03006-009.

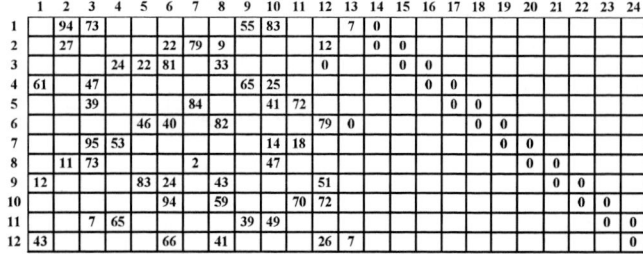

	1	2	3	4	5	6	7	8	9	10	11	12	13	14	15	16	17	18	19	20	21	22	23	24
1		94	73						55	83			7	0										
2		27				22	79	9				12		0	0									
3				24	22	81		33				0			0	0								
4	61		47						65	25						0	0							
5		39					84			41	72						0	0						
6					46	40		82				79	0					0	0					
7			95	53						14	18								0	0				
8		11	73				2				47									0	0			
9	12				83	24		43				51									0	0		
10						94		59			70	72										0	0	
11			7	65					39	49													0	0
12	43					66		41				26	7											0

Figure 1. Matrix for the rate-1/2 QC LDPC code in 802.16e

During decoding iterations, values of Λ and r will become larger and larger [11]. The details are shown in Figure2. The 1/2-rate QC LDPC code in 802.16e is decoded with layered decoding architecture. In Figure 2, $Q(M,N)$ denotes that in fixed point decoder design M bits are used to represent values of Λ and N bits are used to represent values of r. $E(|\Lambda|)$ denotes the average absolute value of Λ. $E(|r|)$ denotes the average absolute value of r. It can be seen that in decoder design with fixed point, absolute values of Λ and r are limited by the finite number of bits which are used to represent Λ and r. It leads to saturation errors and damage BER performance. This figure also shows that in better SNR conditions absolute values of Λ and r grow faster.

We propose two dynamic quantization schemes to solve this problem. These two schemes are based on layered decoding architecture.

The first scheme, which can be called scheme 1, reduce the absolute values of Λ and r during certain iterations. The details of this method are listed in Table 1. In this table *IteMax* denotes the maximum number of iteration. The k denotes the current iteration. The *ite* denotes the set of iterations after which absolute values of Λ and r shall be reduced. Denote β the scaling factor which is used to reduce absolute values of Λ and r. When value of *IteMax* equals to twenty, *ite* can be set to five, ten and fifteen. When *ite* equals to these values, values of Λ and r will be reduced by a factor of β after this iteration.

The second scheme, which can be called scheme 2, reduce absolute values of Λ and r when these values satisfy a certain condition. The details of this method are listed in Table 2. In this table *IteMax* denotes the maximum number of iteration. The k denotes the current iteration. The β denotes the scaling factor which is used to reduce absolute values of Λ and r. Suppose six bits are used to represent values of Λ and four bits are used to represent values of r. Denote l_v the maximum absolute value which can be represented by six bits. The P denotes the number of Λ whose absolute value is above $0.9l_v$. Value of P will be calculated after each iteration. The N denotes the total number of Λ. If value of P is above $N \times \eta\%$ after one layered decoding iteration, Λ and r will be reduced by a factor of β.

Table 1. Details of dynamic quantization scheme 1

for $k = 1$ to *IteMax*
one layered decoding iteration
if $k \in ite$
for each Λ
$\Lambda^{(k)} = \Lambda^{(k)}/\beta$
end for
for each r
$r^{(k)} = r^{(k)}/\beta$
end for
end if
end for

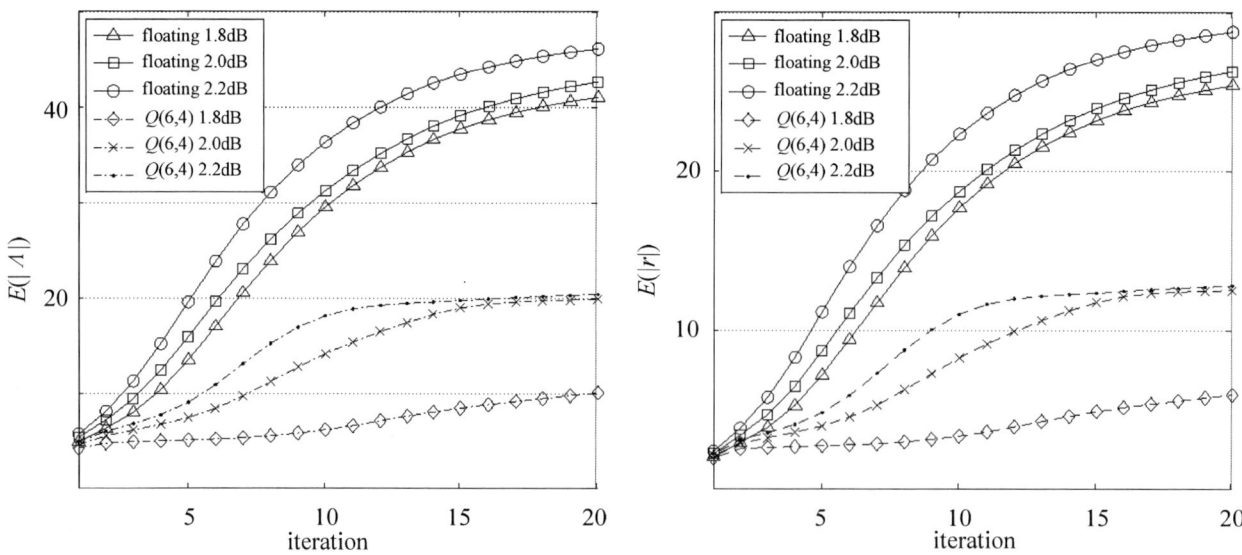

Figure 2. Values of $E(|\Lambda|)$ and $E(|r|)$ during decoding procedure with different SNR per bit (AWGN, BPSK)

Table 2. Details of dynamic quantization scheme 2

for $k = 1$ to $IteMax$
one layered decoding iteration
for each $\varLambda$
If $|\varLambda^{(k)}| > 0.9l_v$
$P^{(k)} = P^{(k)} + 1$
end if
end for
if $P^{(k)} > N \times \eta\%$
for each $\varLambda$
$\varLambda^{(k)} = \varLambda^{(k)}/\beta$
end for
for each r
$r^{(k)} = r^{(k)}/\beta$
end for
end if
end for

In these two dynamic quantization schemes, the value of β is set to be 8/5 and the value of $\eta\%$ is set to be 10%. These values are selected based on a lot of simulation results. They are selected to simplify the hardware implementation while guarantee a good BER performance.

III. SIMULATION RESULTS

Figure3 shows the comparison of BER between decoders with and without dynamic quantization. The 1/2 rate QC LDPC code in 802.16e and layered decoding architecture are used during simulation. BER of $Q(8,6)$ without dynamic quantization is very close to floating point. BER of $Q(6,4)$ with dynamic quantization scheme 1 is close to the floating point with less than 0.2 dB loss. BER of $Q(6,4)$ with dynamic quantization scheme 2 is close to the floating point with less than 0.1 dB loss. Both of the two dynamic quantization schemes have a better BER performance than $Q(6,4)$ without dynamic quantization.

IV. HARDWARE COMPARISON

Comparison of memory cost between Q(6,4) with scheme 2 and Q(8,6) without dynamic quantization is listed in table 3. The matrix for 1/2-rate LDPC code in 802.16e and layerd decoding architecture are used to construct the decoder during comparison. In $Q(6,4)$ with scheme 2, values of r in each row of the matrix require three bits to store the smallest value, three bits to store the second smallest value, three bits to describe the location of the smallest value [12] and six bits to store signs of the values. In $Q(8,6)$ without dynamic quantization,

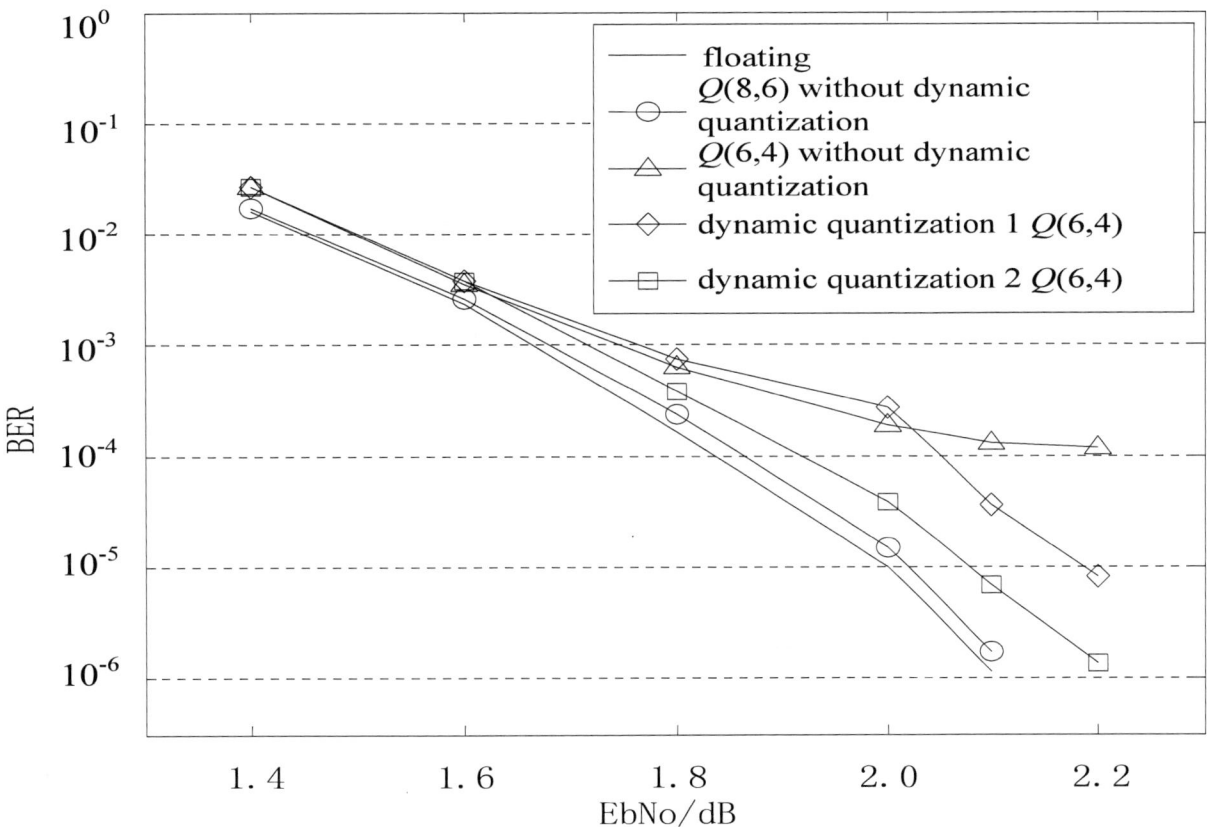

Figure 3. Comparison of BER between LDPC decoder design with and without dynamic quantization

values of r in each row of the matrix require five bits to store the smallest value, five bits to store the second smallest value, three bits to describe the location of the smallest value and six bits to store signs of the values. The table shows that compared to Q(8,6) without dynamic quantization, Q(6,4) with scheme 2 reduces the memory bits by 25%. Although dynamic quantization requires additional logic circuits, hardware complexity of all the subblocks of the decoder are closely related with the number of bits used to represent Λ and r. So by using Q(6,4) with scheme 2 instead of Q(8,6) without dynamic quantization, the chip area can be reduced by around 25% while leading to a very small loss of BER as shown in section 3 and figure 3.

Table 3 Comparisons of memory units

	Memory bits used to store r	Memory bits used to store Λ	Total bits
Q(8,6) without dynamic quantization	$(5+5+3+6) \times 1152$	8×2304	40320
Q(6,4) with scheme 2	$(3+3+3+6) \times 1152$	6×2304	31104

V. CONCLUSIONS

This paper proposes two dynamic quantization schemes for LDPC decoder design with layered decoding. This work can reduce the hardware complexity of LDPC decoders while guarantee BER performance.

REFERENCES

[1] R. Gallager, "Low-density parity-check codes," IRE Trans. Inf. Theory, vol. 8, no. 1, pp. 21–28, 1962.

[2] D. J. C. MacKay and R. M. Neal, "Near Shannon limit performance of low density parity check codes," Electron. Lett., vol. 33, no. 6, pp. 457–458, Mar. 1997.

[3] D. J. C. MacKay, "Good error-correcting codes based on very sparse matrices," IEEE Trans. Inf. Theory, vol. 45, pp. 399–431, Mar. 1999.

[4] T. J. Richardson and R. L. Urbanke, "The capacity of low-density parity-check codes under message-passing decoding," IEEE Trans. Inf. Theory, vol. 47, pp. 599–618, Feb. 2001.

[5] D. Hocevar, "A reduced complexity decoder architecture via layered decoding of LDPC codes," in Proc. IEEE Workshop Signal Process.Syst. (SiPS), Oct. 2004, pp. 107–112.

[6] M. Fossorier, M. Mihaljevic, and H. Imai, "Reduced complexity iterative decoding of low density parity check codes based on belief propagation," IEEE Trans. Commun., vol. 47, no. 5, pp. 673–680, May 1999.

[7] E. B. Guilloud and J. Danger, " λ -Min Decoding Algorithm of Regular and Irregular LDPC codes," in Proc. 3rd Int. Symp. Turbo Codes and Related Topics, Sep 2003, pp. 451–454.

[8] J. Chen, A. Dholakia, E. Eleftheriou, M. Fossorier, and X.-Y. Hu, "Reduced Complexity Decoding of LDPC Codes," IEEE Trans. Commun., vol. 53, no. 8, pp. 1288–1299, Aug. 2005.

[9] M. Fossorier, "Quasicyclic low-density parity-check codes from circulant permutation matrices," IEEE Trans. Inf. Theory, vol. 50, no. 8, pp.1788–1793, Aug. 2004.

[10] S. Myung, K. Yang, and J. Kim, "Quasi-cyclic LDPC codes for fast encoding," IEEE Trans. Inf. Theory, vol. 51, no. 8, pp. 2894–2901, Aug. 2005.

[11] S. Y. Chung, T. J. Richardson. "Analysis of sum product decoding of low density parity check codes using a Gaussian approximation", IEEE Trans on Information Theory, pp. 657-670, 2001.

[12] D. Bao, B. Xiang, X. Y. Zeng. Programmable Architecture for Flexi-Mode QC-LDPC Decoder Supporting Wireless LAN/MAN Applications and Beyond", IEEE Trans on Circuits and Systems-I: regular papers,2010, 57(1):125-138.

978-1-4577-1608-9/11 $26.00 © 2011 IEEE

A Double Active-Decoupling Technique for Reducing Package Effects in a Cognitive-Radio Balun-LNA

Miao Liu, Pui-In Mak, Yao-Hua Zhao and R. P. Martins [1]

State-Key Laboratory of Analog and Mixed-Signal VLSI, University of Macau, Macao, China (E-mail: pimak@umac.mo)
1 – On leave from Instituto Superior Técnico (IST) / TU of Lisbon, Portugal

Abstract—**Package effects degrade the performances of high frequency circuits such as wideband low-noise amplifiers (LNAs). This paper describes an area-saving technique to improve the performance of a balun-LNA covering the range of 50 MHz to 10 GHz in the presence of package effects. A double active-decoupling technique based on a feedback inverter amplifier is proposed. Optimized in GP 65-nm CMOS, the new technique shows 2.3-dB maximum gain loss compensation, 6.6GHz bandwidth recovery for input matching, and 22-dB maximum noise filtering in both supply rails. The added active circuitry draws 16 mW from a single 1.2-V supply, and the required physical capacitor is reduced to 4 pF.**

I. INTRODUCTION

Cognitive radios (CRs) have become the RF design trend for its efficiency and flexibility of utilizing any unoccupied channels in a wide frequency range [1]. However, it is an enormous challenge to balance all levels of abstraction. The low noise amplifier (LNA) is the first stage in the receiver path, and its behavior significantly affects the performance of the whole system. Two key design issues of LNA are: (1) broadband characteristic, i.e., the relatively flat noise figure (NF) and gain, and adequate input matching across the signal band, 50MHz to 10GHz for our case, and (2) nonlinearity.

A LNA topology introduced in [1] has employed a three-stage common source (CS) amplifier in cascade with a resistive feedback. It exploits inductive input impedance of the negative feedback to cancel the input capacitance, which has achieved an optimized trade-off between the input matching, NF and bandwidth (BW). As described in [1], for CRs, the wideband LNA is likely to be the bottleneck in terms of the linearity. IP2 would be the worst among all, since the other IM products of the system can be superposed to the desired channel. Differential outputs may cancel the even order harmonics, thus improving the IP2; however for the case when the antenna is single-ended, a balun would be needed. In [1], the output is sensed between the last two stages. Based on this pseudo-differential sensing, the signal is partially cancelled, and the gain and IP2 are both increased. However, the nonlinearity cancelation relies on the gain and phase matching between the two outputs and they always vary with frequency.

Based on our previous work [2], an RC degeneration circuit adding to the last CS stage is proposed as shown in Fig. 1. The RC degeneration can lower down the gain of the third stage, as well as provides better matching between the two output nodes, besides it can improve the linearity [2]. The degeneration capacitor C_{deg} also effectively boosts up the BW of the balun LNA.

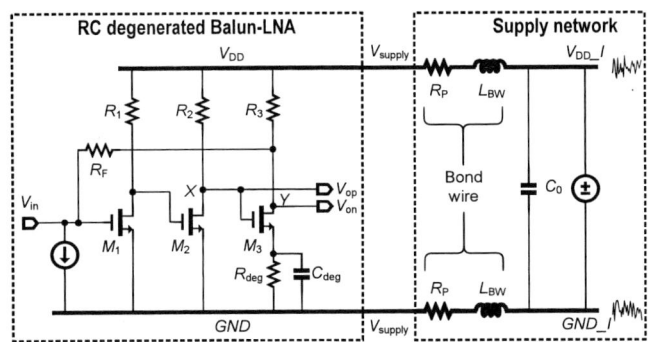

Fig. 1. RC-degenerated balun-LNA and practical supply network model.

This RC degenerated balun LNA has achieved favorable performance, however, only in "ideal" circumstances. With package parasitics involved, signal loss will occur in high-frequency, eventually introducing a NF decrease and an input matching BW reduction. Besides, as the transistor dimension scales down, the supply voltage and the noise might give rise to a drawback in terms of circuit design [3]. A double active-decoupling technique (active decaps) is proposed in this paper to reduce the sensitivity of the balun-LNA to package effects. Specifically, this technique improves the gain loss and input matching, and offers noise filtering to both supply rails.

II. PROPOSED DOUBLE ACTIVE-DECOUPLING TECHNIQUE

The power supply including package parasitics can be modeled by the "supply network" of Fig. 1. V_{DD}_I and GND_I are the power supply terminals without package parasitics which may contain noise. V_{DD} and GND represent the power supply with package parasitics. V_{supply} denotes either V_{DD} or GND, which is the supply rail voltage. L_{BW} is the inductance resulting mainly from the bond wires, R_P is the parasitics resistance of the supply network, and C_0 denotes the on-chip intrinsic circuit capacitance. L=0.4 nH, R=0.2 Ω and C_0=200 pF are chosen as in [3].

For the LNA operating at very high frequency, the output signals V_{op} and V_{on} will vary with the input signal frequency, so it does the signal current flowing through nodes X and Y, consequently. With the package effects taken into account, the V_{supply} voltage will vary with frequency. For simplicity, the package effects will be analyzed by just considering the bond wire inductance L_{BW}.

As shown in Fig.2 (a), an ideal voltage source connected by the bond wire inductance represents the voltage source with package parasitics V_{supply}, and the ac current source

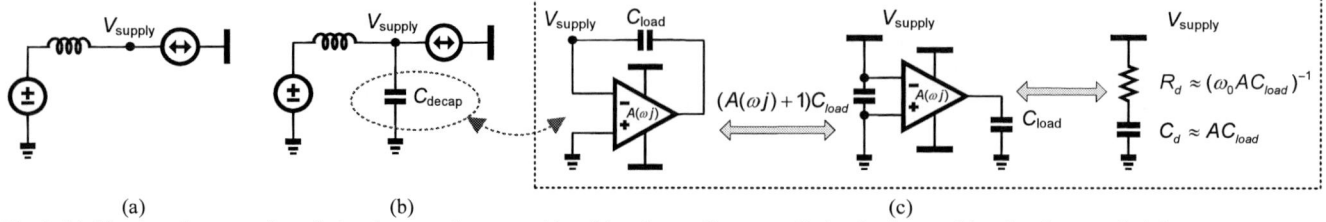

(a)   (b)   (c)

Fig. 2. (a) Circuit performance degradation due to package parasitics, (b) package effects cancellation by decaps, (c) active decaps principle.

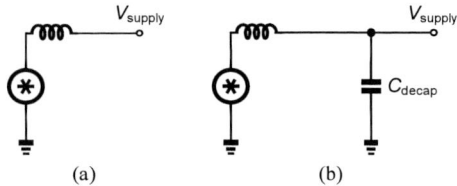

(a)   (b)

Fig. 3. (a) Noisy power supply in the presence of package inductance, (b) noise suppression by applying decaps.

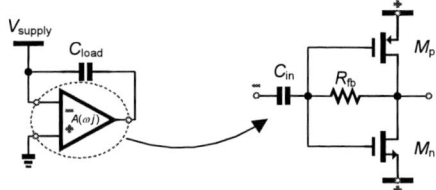

Fig. 4. Schematic of proposed active decap circuit.

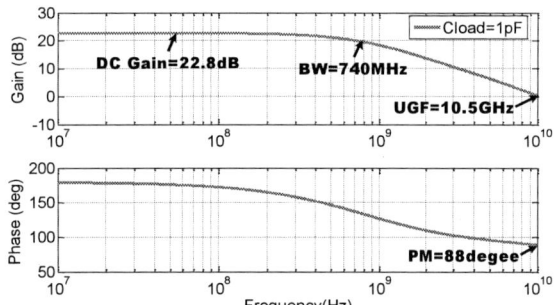

Fig. 5. Simulated bode plot of the active decap opamp.

denotes the signal current. A small capacitance can be seen from V_{supply} in any practical circuit, together with bond wire inductance L_{BW}, they will form a LC circuit with a time constant $\tau=1/(LC)$. As the frequency of the signal current increases, the effective supply voltage will decrease, and this will lead to gain degradation of the LNA in high frequency.

In order to stabilize the frequency-varying power supply, a capacitor is required to be applied to V_{supply} for decreasing the time constant, and thus to cancel the package inductance. Conventionally, passive decoupling capacitor (passive decaps) is used as shown in Fig. 2 (b). It is predicted that the variation of V_{supply} will be suppressed down very quickly if a relatively large capacitor is used.

Active decoupling capacitor (active decaps) circuit has been introduced in [3] and [4] to suppress the substrate noise in mixed-signal integrated circuits, which employs the Miller effect to boost the effective decap value. Fig. 2 (c) shows

the principle of using Miller effect for decap boosting. The negative input and the output terminals of an operational amplifier (opamp) are connected with a capacitor C_{load}, and the positive input of the opamp is connected to virtual ground. The two voltage supplies of the opamp are generated locally, and are supposed to be precisely 1.2V and 0V, respectively. Then the equivalent capacitance seeing from V_{supply} will be boosted by a factor of $(1 + A(j\omega))$, where $A(j\omega)$ represents the gain of the opamp, and varies with frequency ω.

Assuming that the opamp used in the active decap circuit is single stage with a dc gain of A and a dominant pole at ω_0, then the transfer function of this opamp can be formulated by $A(\omega j)=A/(1+\omega j/\omega_0)$. It would be more intuitive to analyze the active decaps by modeling it through its passive equivalent. Due to the band-limited property of the active opamp, the gain will drop beyond its BW, thus the active decaps can be replaced by a series connected resistor R_d and capacitor C_d as shown in Fig. 2 (c). The effective capacitance is $A \cdot C_{load}$. And the effective resistance R_d can be found by equating the impedance of $A(\omega j) \cdot C_{load}$ with the impedance of the series connected R and C as in (1):

$$Z(A(\omega j) \cdot C_{load}) = 1/[(A/(1+\omega j/\omega_0) \cdot C_{load} \cdot \omega j)]$$
$$= Z(R_d) + Z(A \cdot C_{load})$$
$$= R_d + 1/(A \cdot C_{load} \cdot \omega j) \qquad (1)$$

Thus the effective resistance is $R_d = 1/(\omega_0 \cdot AC_{load})$. The active decaps technique will avoid large peaking caused by bond wire inductance and decaps due to this resistance, a big advantage compared to passive decaps. And, active decaps saves more chip area than passive decaps; with a much smaller capacitor when comparing it with its passive counterpart. However, the opamp used in the active decaps consumes power. So, if a small capacitor and a reasonable power consumption opamp is used, a large unpractical passive decaps could be avoided. Besides, as mentioned in [3], a large passive capacitor may introduce significant leakage problems. And with the downscaling of technology, active decaps will be more attractive.

As mentioned in [2], the balun-LNA with RC degeneration is very sensitive to supply line noise, especially for ground rail noise. And it will be proved that applying decaps also helps to suppress the supply line noise in the presence of package effects. Fig. 3(a) shows the supply line with package inductance. For a noisy power supply, there is no noise suppression at node V_{supply}. In Fig. 3(b), after

978-1-4577-1608-9/11 $26.00 © 2011 IEEE

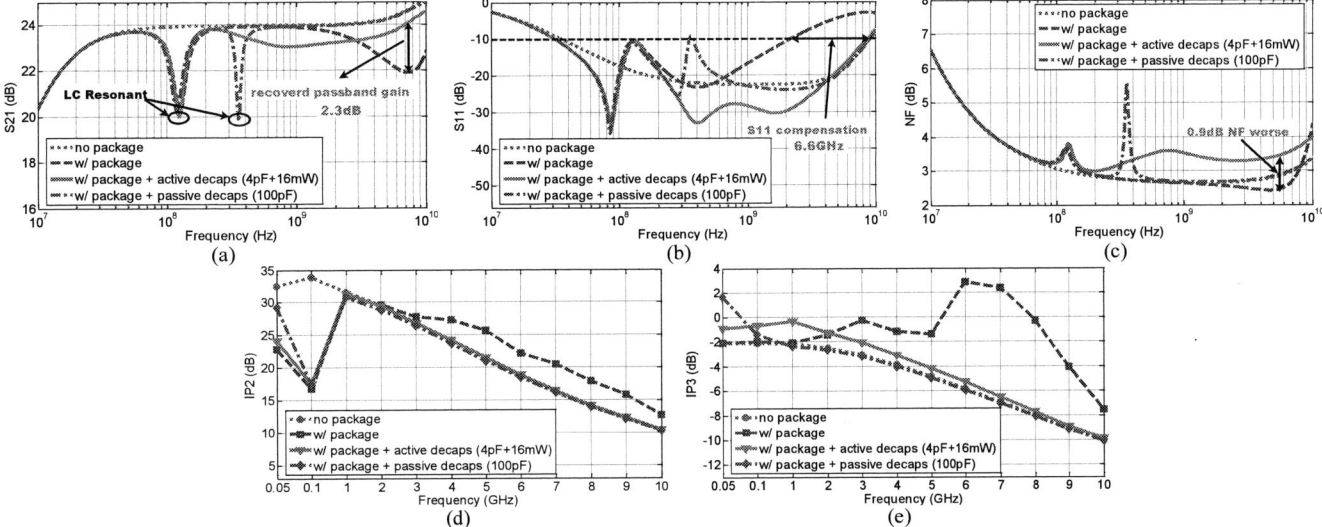

Fig. 6. Simulation results for (a) S21, (b) S11, (c) NF, (d) IP2, (e) IP3.

applying decaps to the supply line, noise will be suppressed down very effectively. In order to comprehensively reduce the supply line noise, a double active-decoupling technique is applied to the balun-LNA. One is for the V_{DD}, the other is for the GND; both are to cancel the parasitic inductance and improve the noise filtering to achieve better performances.

For the opamp used in the active decaps, there will be the following considerations: (1) very wide BW to reduce package parasitics effectively across the wideband signal range, e.g. 10 GHz GBW, (2) large gain to boost up the decaps more effectively and to save more area, (3) reasonable power consumption, and (4) simple structure. Finally, the inverter with resistive feedback is proposed as the opamp selection shown in Fig. 4. The negative input terminal will be connected to the V_{supply} to sense the supply line noise, and the positive input terminal will be connected to the two supplies of the opamp, namely the source terminals of Mp and Mn, as the virtual ground terminal. An input capacitor C_{in} is required to couple the supply signal to the opamp. The value of C_{in} should be much larger than the sum of C_{gg_NMOS} and C_{gg_PMOS}, in order to avoid signal loss due to voltage dividing. Here C_{in} is chosen to be 1 pF.

The gain and the GBW of the opamp can be formulated as in (2) and (3). This opamp has large gain and GBW due to the relatively large equivalent transconductance G_m, and it does not need any bias circuit. It is simple, effective and stable. The above properties are just adequate for the design.

$$A_v = \frac{1 - R(g_{mn} + g_{mp})}{1 + R / (r_{on} \| r_{op})} \approx -(g_{mn} + g_{mp}) \times (r_{on} \| r_{op}) \quad (2)$$

$$GBW = \frac{G_m}{C_{load}} = \frac{g_{mn} + g_{mp}}{C_{load}} \quad (3)$$

Then, an opamp design example was implemented. It has a DC gain of 22.8 dB, bandwidth of 740 MHz, unity gain frequency of 10.5 GHz and phase margin of 88° with 1 = pF load capacitor under 8 mW of power. The resulting

performances are shown in Fig. 5. As defined by (1) and (2), better performances can be obtained with larger power budget, so this opamp reveals itself quite adequate for the related applications.

III. SIMULATION RESULTS

The results based on circuit-level simulations are presented in Fig. 6(a)-(e) and Fig. 7(a) and (b): i) "no package" represents the simulation results for the LNA under the power supply without any package effects; ii) "w/ package" stands for the power supply with package effects. Referring back to Fig. 1, the package effects are modeled by the simple RLC circuits with L = 0.4 nH, R = 0.2 Ω, and C_0 = 200 pF; iii) "w/ package + active decaps (4 pF + 16 mW)" is the case where the power supply with package effects and with active decaps are applied. The double active decaps draws a total power of 2x8 mW and consumes a total capacitance of 2 x (1+1) pF for the input and load; iv) "w/ package + passive decaps (100 pF)" is the case where the power supply with package effects and with passive decaps are applied, which consumes a total capacitance of 2 x 50 pF. To avoid large peaking at the resonant frequency caused by the passive decaps a 2 x 0.5 Ω resistor is utilized.

A. The gain of the balun-LNA (S21)

As mentioned before, when the frequency increases the impedance of the supply rails will follow it, which will cause gain degradation. Active decaps cancels the inductance effects of the supply lines, which, on the other hand, compensates the gain degradation of the LNA, as shown in Fig. 6(a). The gain performance has no degradation without supply package parasitics, and degrades in high frequency with the package parasitics taken into account. After applying decaps to the supply lines, the signal loss is compensated, and there is a peaking at approximately 140 MHz for all the cases with supply package considered. This, corresponds to a resonant frequency caused by supply

978-1-4577-1608-9/11 $26.00 © 2011 IEEE

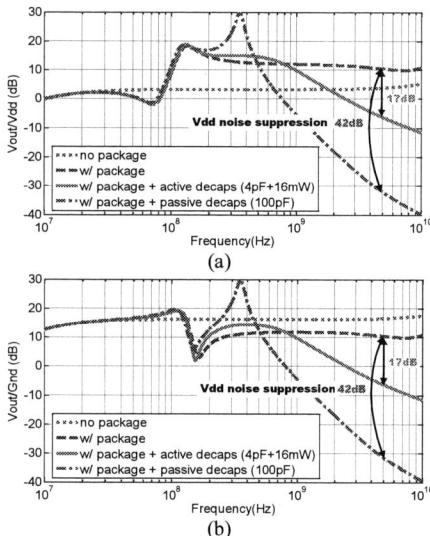

(a)

(b)

Fig. 7. Noise filtering performances for (a) voltage supply, (b) ground.

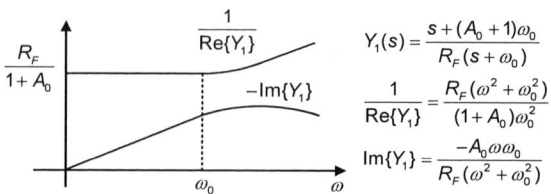

Fig. 8. Behavior of Y_1 components with frequency.

package parasitics, and can be reduced by adding a resistor to the supply lines. However the resistor may divide the supply voltage, thus leading to power supply degradation. There is another large peaking at around 350 MHz for passive decaps only. It is the LC resonant frequency caused by the supply line inductance and passive decaps, which can also be suppressed by adding resistors. The resistor may degrade the performance of decoupling and add noise. The peaking is avoided in the proposed active decaps technique due to its intrinsic series resistance.

B. The input matching (S11)

The impedance of the antenna is assumed to be 50 Ω. It is challenging to match this over a wide range of frequency. If the open loop transfer function of the LNA is modeled as a one-pole system for simplicity, the input admittance will behave like in Fig. 8. It is shown that the gain enhancement of the LNA will lower down $1/\text{Re}\{Y_1\}$ and boost up $-\text{Im}\{Y_1\}$. So, better input matching will be obtained with gain enhancement. Fig. 6(b) shows the input matching performances for the above four types of supplies. As predicted, after package effects are involved, the effective BW is reduced to 1.5 GHz. This is unacceptable in our LNA design. By applying decaps, the BW will recover and due to the large peaking caused by the supply inductance and decaps, passive decaps shows a worse performance when compared with its active counterpart at mid frequency.

C. Noise figure (NF) and linearity

With the gain improved, the NF will be suppressed down theoretically, since the input referred noise is equal to the total noise divided by the gain. However, the opamp used in the active decaps will introduce additional noise, so for better performance, the noise of the opamp should be minimized. The simulated NF performances are shown in Fig. 6(c). Active decaps add some noise to the whole circuit,

but since the gain has been improved by decaps, there is only 0.9-dB maximum NF degradation.

As shown in Fig. 6(d) and (e), after applying decaps, the linearity is quite unaffected.

D. Noise filtering

Noise from outside supplies will be suppressed effectively in high frequency by decaps due to inductance canceling for both supply lines. The noise filtering performances are shown in Fig. 7. Since the same decaps is applied to both of the supply lines, the noise suppression performance is the same for V_{DD} and GND.

IV. CONCLUSIONS

A double active-decoupling technique based on an inverter resistive feedback opamp is proposed to reduce package effects for cognitive radio balun-LNA in GP 65-nm CMOS process. The design example employs two active decaps that have achieved a 2.3-dB maximum gain loss compensation, 6.6-GHz BW recovery for input matching, 22-dB noise suppressions for both supply lines with 4-pF capacitance and 16-mW power consumption. The simulation results are compared to passive decaps, considering a 100-pF capacitance. It is shown that the active decoupling technique is much more area-efficient and can avoid large peaking which is harmful to circuit performances.

ACKNOWLEDGEMENT

This work was supported by the Research Committee of University of Macau, and Macao Science and Technology Development Fund (FDCT).

REFERENCES

[1] B. Razavi, "Cognitive Radio Design Challenges and Techniques," *IEEE J. Solid-State Circuits*, vol. 45, no. 8, Aug. 2010.

[2] W.-F. Cheng, K.-F. Un, P.-I. Mak and R. P. Martins, "A Highly-Linear Ultra-Wideband Balun-LNA for Cognitive Radios," *IEEE International Conference on Computer as a Tool (EUROCON)*, pp. 1-4, April 2011.

[3] J. Gu and R. Harjani, "Design and Implementation of Active Decoupling Capacitor Circuits for Power Supply Regulation in Digital ICs," *IEEE Tran. On Very Large Scale Integration (VLSI) Systems*, vol. 17, no. 2, pp. 292-301, Feb. 2009.

[4] T. Tsukada *et al.*, "An on-chip active decoupling circuit to suppress crosstalk in deep-submicron CMOS mixed-signal SoCs," *IEEE J. of Solid-State Circuits*, vol. 40, no. 1, pp. 67–79, Apr. 2005.

978-1-4577-1608-9/11 $26.00 © 2011 IEEE

An Effective Buffer Space Management in Serial RapidIO Endpoint

Wu Fengfeng, Jia Song, Wang Yuan

Key Laboratory of Microelectronics Devices and Circuits (MOE), Institute of Microelectronics
Peking University
Beijing, China

Abstract—RapidIO is a high performance open standard for the next-generation embedded interconnection technology. In this paper, an improved pivotal buffer core which plays a crucial role in the RapidIO packet transmission is described in detail. It uses the four-isolated-queue as the outbound framework for better quality of service and a certain amount of blocks with sharing and binding attempts mechanism as the main inbound structure for higher space utilization. It supports the advanced transmitter-controlled flow control so as to effectively manage the buffer space. Simulation results show that the valid link data utilizations of the proposed design with transmitter-controlled flow control could be kept over 95%, comparing to the fact that those of the buffers with basic receiver-controlled flow control gradually decrease to lower than 50% in the same condition.

I. INTRODUCTION

As the performances of microprocessors have been optimized steadily, the improvement of interconnection becomes the immediate problem in the embedded systems. Traditional bus architectures such as PCI have reached their performance limits and been proven to be deficient in modern applications [1]. The need for high-speed, reliable connectivity for next-generation system is the driving force behind the development of the RapidIO interconnection architecture [2].

RapidIO defines a point-to-point and packet-switched fabric for processor-to-processor and board-to-board interconnection. It supports all needed microprocessor and I/O transactions and is transparent to existing applications and operation system software [3]. It adopts a three-layer architectural hierarchy including the logical layer, transport layer and physical layer. The topmost logical layer defines the overall protocol and packet formats, and provides the information necessary for endpoints to initiate and complete transactions. The middle transport layer provides the route information for packets to move from endpoint to endpoint. The lowest physical layer describes the link level interface details including packet transport mechanisms, flow control, electrical characteristics and link-level error management [4]. Both serial and parallel physical protocols are described. The Serial RapidIO (SRIO) can be used in connected micro-

processors, memory mapped I/O devices that operate in networking equipment, memory subsystems and general purpose computing [5]. Different baud rates of 1.25Gbaud, 2.5Gbaud and 3.125Gbaud are available for these applications. To prevent the loss of packets due to a lack of buffer space in the link receiver, the serial link level flow controls including basic receiver-controlled flow control and advanced transmitter-controlled flow control are defined [6]. Fig.1 shows common hardware structures of SRIO endpoints and the simple direct-connected model. In this paper, we focus on the design of the pivotal buffer core including outbound and inbound buffers. It acts as the joint between the logical and physical cores and accounts for packet buffering to provide normal packet transmission and retransmission. Furthermore, the buffer space management mechanism directly determines the flow control ability and the link data utilization.

II. RELATED WORK

Motorola and Mercury collaborated on the initial specification of RapidIO. Now the main reference is the set of formal specifications, published by the RapidIO Trade Organization [6]. An overall frame and top layer design of SRIO endpoint implemented by Xilinx can be found in [7]. Another excellent SRIO peripheral block diagram proposed by Texas Instruments is described in [8]. Reference [9] presents a relatively detailed design and introduces the four-isolated-queue buffer. Our group previously proposed a buffer design with different queue depths for smooth transmission flow [10]. In this paper, an improved buffer core which can optionally work in receiver-controlled or transmitter-controlled flow control mode is proposed.

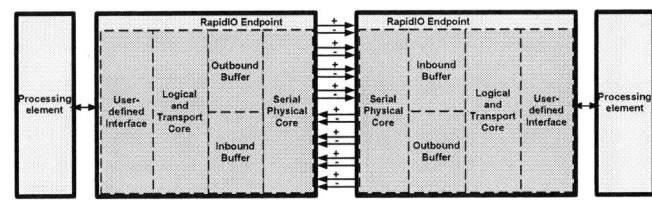

Figure 1. The structure diagram of SRIO endpoints

III. PROPOSED DESIGN

As the hinge inside a RapidIO endpoint, the proposed buffer core consists of outbound buffer in the transmitting channel and inbound buffer in the receiving channel. The former is responsible for sorting and backing up the outgoing packets and managing the packet flow over the link. The later takes charge of storing incoming packets and forwarding them to logical core. Co-operating with the control symbols from physical core, these buffers are able to manage the packet transmission and control the packet flow between the two link partners.

A. Outbound Buffer Design

The outbound buffer consists of the Packet FIFO Queues, Outbound Buffer Manager and Inbound Buffer Manager, as shown in Fig 2. The Packet FIFO Queues are responsible for storing packets from logical core and sending them to physical core. To achieve better quality of service and deadlock avoidance, the four-isolated-queue structure is adopted as the main framework but the working principle is different from the design in [9]. Incoming packets would be led into the corresponding queues according to their priorities and always sent out from the existent highest to lowest priority. Only when the packet-accepted control symbols have been received, indicating that the related packets had been successfully received by the link partner, could the corresponding occupied space be set free. The Sending Multiplexer selects the chosen output queue and limits the amount of outgoing packets according to the current capacity status and the control signals. The Outbound Buffer Manager takes charge of managing the normal packet transmission and retransmission. All outgoing packets would be identified through their AckIDs assigned by the AckID Status Manager. Meanwhile the current read pointers of the four queues would be saved correspondingly in the Pointers Backup module so that packet retransmissions could be easily established by looking up these read pointers. The functionality of Inbound Buffer Manager is to control the amount of outgoing packets in the transmitter-controlled flow control mode. According to the protocols, the opposite link partner should provide control symbols carrying reports about the amount of its free inbound buffer space. Each of these reports would update the Received Report Register. The Outstanding Packet Counter maintains a total count of outstanding packets which have not yet been acknowledged by the link partner. The counter value increases when a new packet has been sent out and decreases when a packet-accepted control symbol has been received. In the Free Buffer Counter, the actually available space of the link partner can be estimated by subtracting the current outstanding packet amount from the latest reported available amount. To progressively limit the packet priorities that can be transmitted due to the decrease of the actually available space, watermarks are adopted as the suggested implementation [6]. Therefore, the control signals for the Sending Multiplexer are finally determined by the estimated actually available space and the watermarks.

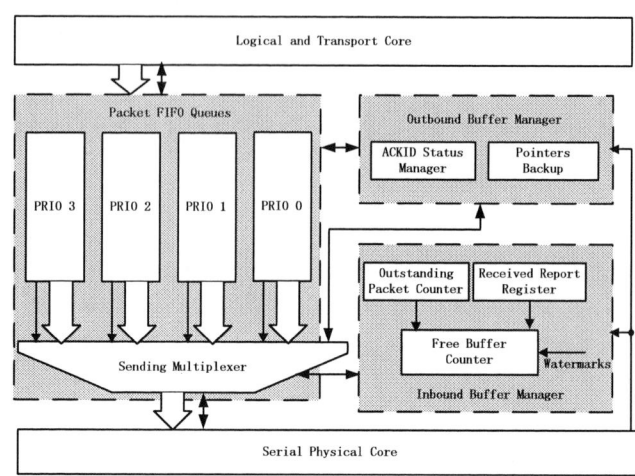

Figure 2. Structure diagram of outbound buffer

B. Inbound Buffer Design

The Inbound Buffer is divided into three main components as illustrated in Fig 3: the Packet FIFO Blocks, Block Manager and Free Buffer Calculator. Instead of the four queues, the main receiving FIFO adopts a certain amount of blocks, each of which can store a maximum-length packet. The amount of these blocks depends on the design capacity. For example, it could be set to be 32 so that the inbound buffer could hold at most 32 maximum-length packets. Inside the Block Manager, the Inlet Block ID FIFO stores all free block IDs. These free blocks have no priority attributes and they might flexibly serve for any of the four priorities. Compared with the four-isolated-queue structure, this design can achieve a better space utilization by adapting to any combination of the packet priorities. The Inlet Block Selector is responsible for receiving a new packet and writing it into the free block which is currently chosen by the read pointer in the Inlet Block ID FIFO. Once a free block is occupied by a packet with priority i ($i = 0,1,2,3$), its block ID would be removed from the Inlet Block ID FIFO and then moved into the corresponding Outlet Block ID FIFO i. Thus this block is enslaved and forced to have a priority attribute until it is set free again. The Outlet Block ID FIFOs have priority attributes and adopt the four-isolated-queue structure so as to distinguish the priorities of the occupied blocks. When the logical core is ready to receive packets, the Outlet Block Selector would fetch a Block ID in Outlet Block ID FIFOs from existent highest to lowest priority. Then the Receiving Multiplexer would sequentially output the packets stored in this block. After this block is completely read, its ID would be removed from the Outlet Block ID FIFO and then moved into the Inlet Block ID FIFO. In this way, the block becomes free again. If both ports work in transmitter-controlled flow control mode, the Free Buffer Calculator would output the total amount of free blocks (the maximum value is 30). Otherwise it should output all 1s for the receiver-controlled flow control mode according to the protocols.

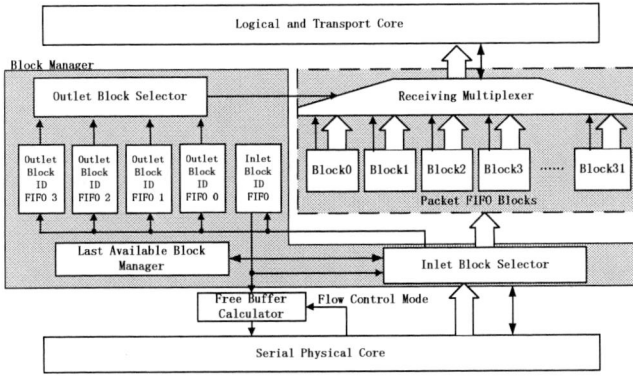

Figure 3. Structure diagram of inbound buffer

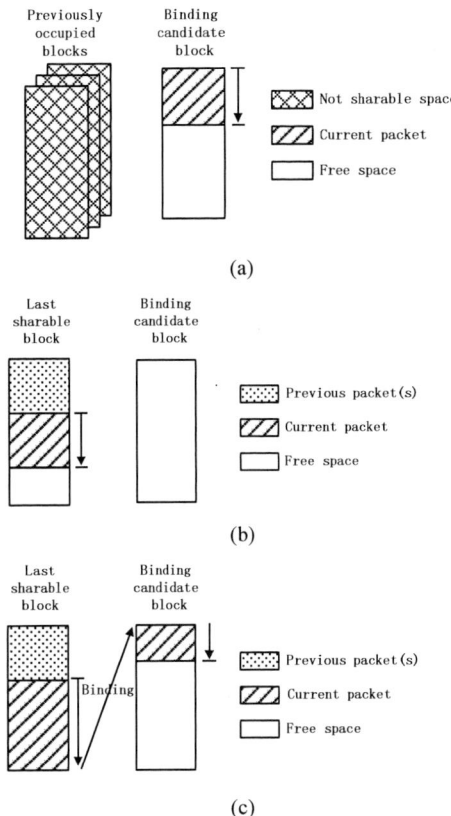

Figure 4. Different results of the block sharing and binding attempts

C. Block Sharing and Binding Attempts Mechanism

Though each block can store a maximum-length packet, the space utilization would be low if each block only stores one packet, especially when the packet is carrying a short transaction like Doorbell. To solve this problem, we proposed the block sharing and binding attempts mechanism. To block N, if an incoming packet with priority i is not long enough to occupy the whole block space, then a following packet with the same priority could have the chance to use the remaining block space. We name block N the last sharable block of priority i if it satisfies the following conditions. Firstly, block N should be unread and its block ID is still stored in Outlet Block ID FIFO i. Secondly, block N should be not full so as to store a new packet or part of a new packet. Furthermore, there is at most only one last sharable block corresponding to each priority. All these judgments would be executed by the Last Available Block Manager.

According to the reports about the amounts of free blocks from link partner, a port could control the amounts of outgoing packets without concern that one or more of the packets shall be forced to retry. In the normal transmission, if a new packet is coming in, there should be at least one free inbound block. In other word, the Inlet Block ID FIFO should not be empty. In this case, we name the free block currently pointed by the read pointer in Inlet Block ID FIFO "the binding candidate". Once a packet with priority i arrives, the Inlet Block Selector would firstly try to find out whether there is a last sharable block of priority i. If no, this packet would be written into the binding candidate, as shown in Fig. 4(a). Then both sharing and binding attempts are abandoned. If yes, this packet would be written into the remaining space of that block. If the whole incoming packet could not fill up the remaining space, as shown in Fig. 4(b), the binding candidate is useless and waits for the next packet. The sharing attempt is completed and the binding attempt is abandoned. If the remaining space of the last sharable block is not enough for this packet, as shown in Fig. 4(c), the remaining part of the packet would be successively written into the binding candidate. Both sharing and binding attempts are completed. Each time after a packet with priority i has been stored, if the currently written block still has remaining space, it would become the new last sharable block of priority i.

IV. PERFORMANCE AND ANALYSIS

The verification model was established by directly connecting two endpoints through their 4x electrical interfaces, as shown in Fig. 1. One endpoint played the role of the transaction transmitter and the other endpoint acted as the receiver. After all initializations had been completed, the tests were executed through the normal link transmission.

A. The utilization of the inbound buffer space

To analyze the improvement of the space utilization, the proposed design with the block sharing and binding attempts was compared with the same structure without this mechanism. In each test, a total number of 32 transactions with a certain function type and the highest priority were generated. Meanwhile the output enable signal of the receiver was always set invalid. Then all the 32 packets were obstructed and forced to be stored in the inbound buffer of the receiver. The utilizations of the inbound blocks without and with the sharing and binding attempts are shown in Fig. 5(a) and Fig. 5(b), respectively. We can see that no matter how long the packet was, each packet occupied one block in the former and no more packets could be received. However, in the later, some blocks were shared and totally occupied meanwhile the remainders were still free and able to hold more packets. The amount of the free blocks became greater as the packet length went shorter.

978-1-4577-1608-9/11 $26.00 © 2011 IEEE

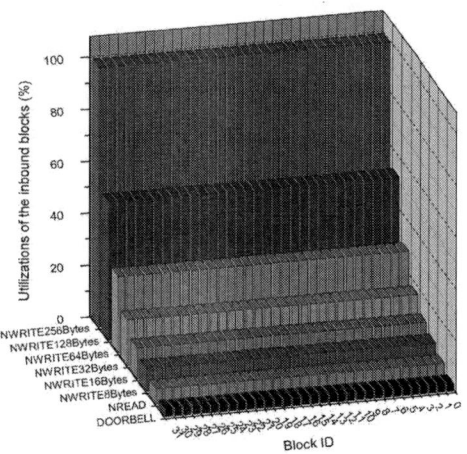

(a) Design without block sharing and binding attempts mechanism

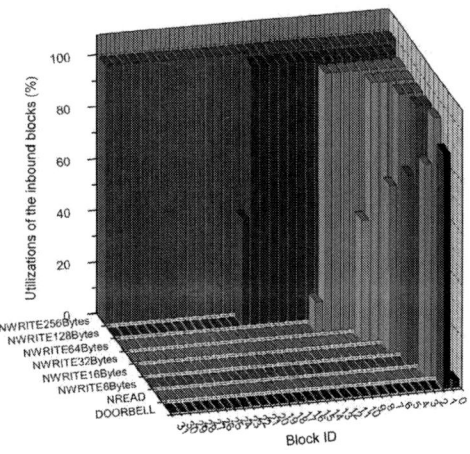

(b) Design with block sharing and binding attempts mechanism

Figure 5. Different groups of utilizations of inbound blocks storing 32 packets with the same function type

B. The utilization of valid link data during transmission

The output enable signal of the receiver could be set valid after a certain delay to restore the normal unobstructed transmission. During the jam time, the flow control ability could be reflected by the utilizations of the valid link data. The proposed buffer core with block sharing and binding attempts was compared with the traditional four-isolated-queue outbound and inbound buffer design with the same maximum capacity. A total number of 100 NWRTIE transactions with 256-byte data payload were generated as the test stimulus. The utilizations were calculated by measuring the total amounts of data frames sent into PCS (excluding the idle sequences) in the transmitter and the total amounts of valid packet frames written into inbound buffer in the receiver. We can see from Fig. 6 that as the jam delay increased, the utilizations of the designs with receiver-controlled flow control gradually dropped to lower than 50% due to the packet retransmissions.

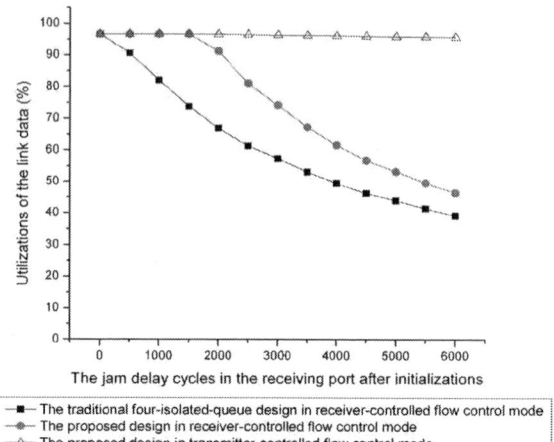

Figure 6. Utilizations of valid link data of different designs with different flow control modes

However, the proposed buffer core with transmitter-controlled flow control could still keep an over 95% high utilization by effectively avoiding the packet retransmissions.

V. CONCLUSIONS

An improved buffer core of SRIO is proposed in this paper. It contributes to provide an effective buffer space management based on transmitter-controlled flow control. It also helps to increase link utilization and makes packet transmission more reliable. It is considered to be applicable as part of high performance SRIO endpoint in high speed embedded system communications.

REFERENCES

[1] T. Scheckel, "Serial RapidIO: Benefiting System Interconnects," SOC Conference, Herndon, USA, pp. 317-318, September 2005.

[2] S. Fuller, T. Cox, "Anatomy of a forward-looking open standard," Computer, IEEE Computer Society, Vol. 35, Issue. 1, pp. 140 – 141, 2002

[3] J. Adams, C. Katsinis, W. Rosen, et al., "Simulation experiments of a high-performance RapidIO-based processing architecture," Network Computing and Applications, Cambridge, USA, pp. 336-339, October 2001.

[4] S. Fuller, "RapidIO: The embedded system interconnect," UK: John Wiley & Sons, Ltd, 2005, pp.14-44.

[5] J. Zhang, H. B. Su, Q. Z. Wu, "Research and Implement of Serial RapidIO Based on Mul-DSP," Computational Intelligence and Software Engineering, Wuhan, China, pp. 1-4, December 2009.

[6] "RapidIO Interconnect Specification Rev 2.1," RapidIO Trade Association, August 2009.

[7] Xilinx, "LogiCORE IP Serial RapidIO v4.3," product specification, March 2008.

[8] Texas Instruments, "TMS320C645x DSP Serial RapidIO," product specification, March 2006.

[9] Y. W. Lin, M. Lu, "Research on a novel interconnect technology for B3G system," Wireless Communications, Networking and Mobile Computing, Beijing, China, pp. 1-4, September 2009

[10] X. B. Zhao, S. Jia, Y. Wang, et al., "Buffer design based on flow control in RapidIO," PrimeAsia 2010, Shanghai, China, pp. 150-153, September 2010.

978-1-4577-1608-9/11 $26.00 © 2011 IEEE

AUTHOR INDEX

Ahmed, Syed Ershad 106
Chan, Chi-Hang 9
Chen, Chunchun 33, 53
Chen, Lei 78
Chen, Xuewei 33
Chimpleekul, Puttachai 90
Chio, U-Fat 9, 25
Dai, Ning-Yi 25
Ding, Li 9
Dong, Cheng 65
Dong, Yinan 70
Fei, Yuan 1
Han, Ying-Duo 21
Han, Yu 94
He, Jin 94
Hong, Yibin 74
Hu, Guangming 41
Hua, Lin 78
Hussain, Arshad 82
Ieong, Chio In 65
Inoue, Yasuaki 41
Jia, Song 135
Kanjanop, Arnon 86
Kasemsuwan, Varakorn 86, 90
Kuo, Ron-Chi 37
Lai, Zongsheng 78
Lam, Chi-Seng 21, 25
Lande, Tor Sverre 115
Lee, Kin Keung 115
Li, Dongmei 5, 13, 17
Li, Ning 123
Li, Ting 13
Li, Wei 123
Lian, Yong 74
Liang, Wei 127
Lin, Xinnan 94
Liu, Chi 94
Liu, Fei 49
Liu, Jen-Wei 37
Liu, Liyuan 5, 13, 17
Liu, Miao 131
Liu, Shengfu 78
Liu, Tao 61
Liu, Ting 53
Liu, Wenjiang 61
Lok, U-Wai 119
Mak, Peng Un 65
Mak, Pui In 65, 119, 131
Mao, Zhigang 98
Martins, R. P. 1, 9, 25, 82, 119, 131
Muthukrishnan, Moorthy 106, 110
Næss, Øivind 115
Niu, Dan 41
Parlapalli, Sai Phaneendra 110
Rajendran, Iniyal 74

Ren, Junyan 123, 127
Rong, Mengtian 61
Ruan, Ying 78
Shan, Junming 57
Sin, Sai-Weng 1, 9, 25, 82
Srinivas, M. B. 102, 106, 110
Su, Jie 78
Sun, Bo 25
Tang, Kea-Tiong 70
Tseng, Hsin-Yuan 37
U, Seng-Pan 1, 9, 25, 82
Vai, Mang I. 65
Veeramachaneni, Sreehari 102, 106, 110
Vudadha, Chetan Kumar 102
Wan, Feng 65
Wang, Bin 98
Wang, Chua-Chin 37
Wang, Duyao 57
Wang, Guoxing 70
Wang, Qin 98
Wang, Rui 9
Wang, Ruolin 61
Wang, Yuan 135
Wang, Zhihua 9, 13
Wong, Chi-Kong 25
Wong, Man-Chung 21, 25
Wu, Fengfeng 135
Wu, Huan-Ming 45
Wu, Jo-Yu 70
Wu, Qi-Song 49
Xia, Jing 94
Xie, Jing 98
Yan, Qiong 78
Yang, Eryan 33, 53, 57
Yang, Hai-Gang 45, 49
Yao, Yuan 127
Ye, Fan 123, 127
Ye, Wenbin 29
Ye, Yafei 13
Ye, Zhen-Hua 49
Yin, Tao 45, 49
Yu, Wei-Han 119
Yu, Ya Jun 29
Zhang, Aixi 94
Zhang, Hao 61
Zhang, Hui 45
Zhang, Kaichen 123
Zhang, Shulin 78
Zhang, Wei 78
Zhang, Weifeng 70
Zhang, Yiwei 17
Zhao, Wei 94
Zhao, Yao Hua 131
Zhou, Libing 5

CURRAN ASSOCIATES INC.
proceedings
.com

9781457716089